Ius Gentium: Comparative Perspectives on Law and Justice

Volume 105

Ius Gentium is a book series which discusses the central questions of law and justice from a comparative perspective. The books in this series collect the contrasting and overlapping perspectives of lawyers, judges, philosophers and scholars of law from the world's many different jurisdictions for the purposes of comparison, harmonisation, and the progressive development of law and legal institutions. Each volume makes a new comparative study of an important area of law. This book series continues the work of the well-known journal of the same name and provides the basis for a better understanding of all areas of legal science.

The *Ius Gentium* series provides a valuable resource for lawyers, judges, legislators, scholars, and both graduate students and researchers in globalisation, comparative law, legal theory and legal practice. The series has a special focus on the development of international legal standards and transnational legal cooperation.

Matteo Nicolini

Legal Geography

Comparative Law and the Production of Space

Springer

Matteo Nicolini
Department of Law
University of Verona
Verona, Italy

ISSN 1534-6781 ISSN 2214-9902 (electronic)
Ius Gentium: Comparative Perspectives on Law and Justice
ISBN 978-3-031-19409-2 ISBN 978-3-031-19410-8 (eBook)
https://doi.org/10.1007/978-3-031-19410-8

This Springer imprint is published by the registered company Springer Nature Switzerland AG
The registered company address is: Gewerbestrasse 11, 6330 Cham, Switzerland

They also that dwell in the uttermost parts
are afraid at thy tokens :

thou makest the outgoings of the morning
and evening to rejoice.

Psalm *65:8 (King James Version)*

They also that dwell in the uttermost parts
are afraid at thy tokens :

thou makest the outgoings of the morning
and evening to rejoice.

Psalm *65:8 (King James Version)*

Dedicated to my beloved children,
Tobia Libero and Valeria Assunta,
whose very acts of geo-graphing
have completely reshuffled the world

Preface[1]

I had been pondering the idea of examining the 'special relationship' between law and geography for a long time before the Pandemic broke into our lives in 2020.

By then, I had already made a few presentations on the topic at the Universities of Nicosia (2013), Verona (2014 and 2019), Tübingen (2017), Lille (2018), and Preston (2019). In 2018, Gabriele Abels invited me as a guest lecture scholar at Eberhard Karls University Tübingen to deliver a course of lectures on 'The Legal Geography of Multi-Level Contexts: The Case of the EU Territorial Constitution'. Together with my friends and colleagues Chiara Battisti, Sidia Fiorato, and Thomas Perrin, I also hosted an International Conference on 'Islands and remoteness in Geography, Law and Fiction' in Verona (2019).

It goes without saying that, by then, I had also published some essays on the relation between comparative law and geography, addressing it through the angle of federal systems, such as in *Accommodating Divided Societies Through Constituent Units? A Comparative Survey of Regional Demarcation and Territorial Alteration* (2013); *Regional Demarcation, Territorial Alteration, and Accommodation of Divided Societies* (2015); *Boundaries and Identity: The Legal Geography of the European Union and the United States of America* (2015); and *Reforming the Territorial Constitution in Italy: Some Reflections on Durability and Change* (2019). I had also examined the impact of global ideo-logical spaces in '*Inequality of Goods and Lands' in Mortgaged Democracies: Paradigms and Effects of Global Comparative Law* (2020); whereas the spatiality of law had been explored in *Monstrosity 'Overseas'? Civilisation, Trade and Colonial Policy in Conrad's African Tales* (2019). Finally, I perused Cyprus as a case for legal geography in *Territorial and Ethnic Divide: A New Legal Geography for Cyprus* (2016) and *A*

[1] *This book pertains to the activities of the IEL (Innovative Education Laboratory) GEOLawB, which is part of the research excellence project 'Law, Changes and Technology', sponsored by the Ministry of Education and carried out by the Law School of the University of Verona.*

'New' Legal Geography for the 'Old' Cyprus Question: Regional Demarcation, Power-Sharing, and Divided Societies (2016).

Yet, the analysis of legal geography required a broader theoretical framework, which could allow me to reframe the 'special relationship' between law and geography in terms of legal geography. This would have meant engaging in a true cross-disciplinary project, following in the footsteps of, say, Irus Braverman, Nicholas Blomley, Sandy Kedar, and David Delaney.

Among lawyers, in particular, Irus Braverman showed us all how to achieve the goal of cross-disciplinarity. In a fruitful exchange of emails some years ago, she pointed me to her *Coral Whisperers. Scientists on the Brink* (2018), which I consider an impressive scholarly achievement in climate-change related research. There, she worked out how global warming had been affecting the 'political' communities of the seas which are coral reefs, also undertaking 'massive research ... that stretched across continents and disciplines' with coral experts, whom she labels 'the vanguard of conservation in the Anthropocene'. To this book and to the collection *Blue Legalities. The Life & Laws of the Seas* that she edited with Elizabeth R Johnson in 2020, I owe the approach I subsequently adopted in *Law, the Humanities and Political Incertitude in a Time of Climate Change* (2021).

While cross-disciplinary inspirations did not lack, I still had to make up my mind about the methodology applicable to my research. The occasion for doing so was prompted by the publication of the collection *Legal Geography: Perspectives and Methods* edited by Tayanah O'Donnell, Daniel F Robinson, and Josephine Gillespie in 2020. The book also comprised Liesel Spencer's chapter on *Comparative legal geography: context and place in 'legal transplants'*. Liesel Spencer quoted my essay *Territorial and Ethnic Divide: A New Legal Geography for Cyprus*, also pointing to its alleged weakness. Although 'illustrative of what legal geography has to offer to comparative law', she considered my chapter 'not in conversation with either legal geography or comparative law scholarship about developing new methodological approaches' (at p. 151). In particular, she highlighted that I did not quote Sandy Kedar's essay *Expanding Legal Geographies: A Call for a Critical Comparative Approach* published in Braverman et al (eds) *The Expanding Spaces of Law. A Timely Legal Geography* (Stanford UP, Stanford 2014) on transplants and the comparative legal approach to geography.

Liesel Spencer was quite right. And Sandy's essay is now quoted in this book. Yet, I remain convinced that cross-disciplinary research requires (or, at least, falls under) a good dose of 'disciplined subjectivity'.

As Ian Ward puts it, certainly 'every discipline has its own high-road, charted by generations of scholarly pioneers. Which occasionally intersect', thus generating fertile soil for cross-disciplinary conversation (*Law, Literature and History*, in *Law & Literature*, 2021). 'It is', Ian continues, 'at these intersections that you tend to find the greatest mess. All the litter lying about. And all the curiosity, the variously brilliant things glinting in the ditch, the fragments of this and that tossed aside by those otherwise distracted by the greater urge to get where they are going as fast as they can'. Cultural background (in my case, evidently, that of a comparative lawyer engaged with methodological issues) becomes our strength, allowing us to cross disciplinary borders without appearing amateurish. I subscribe to Ian's predicament in full: 'In the building trade we might be called skip-rats. Re-cycling sounds better; re-purposing perhaps. Re-shaping from fragments. Re-imagining'.

Whilst writing this book, I also had the possibility to present some chapters during conferences and seminars. Chapter 2 was discussed at the Society of Legal Scholars (SLS) Annual Conference, Durham University, under the banner 'Comparative Law and Geography: A Special Relationship' (September 1st, 2021). Chapter 3 was presented at the University of Turin (Italy) at the Conference 'Geographies of the law. Inquiries into the space-law tangle' as 'Law, power, and the production of natural Geography' (December 14th, 2021). On that occasion, Sandy Kedar and I acted as discussants of the keynote speech by Nicholas Blomley. Thanks to Nicholas for reading Chap. 5. Chapters 6 and 7 were delivered during the seminar 'Legal Geographies and The Production of Federal Spaces' at the University of Nicosia (March 16th, 2022). Chap. 4 was presented at the third Legal Histories of Empires Conference 'Beyond the Pale: Legal Histories on the Edges of Empires' (July 1st, 2022) held at Maynooth University (Ireland) with the title 'Manufacturing Imperial Spatialities. The British Integrated Legal Geography of Land and Sea'. Chap. 8 was finally presented at the SLS Conference (September 8th, 2022) held at King's College London with the title 'The Legal, the Digital, and the Global Production of Space'.

Writing a book is an individual journey. Yet, it is also a collective task encroaching upon (not to say interfering with) the lives of our families. A special thanks, then, to my wife Lisa for her endurance, patience, and resilience.

Thanks to Giuseppe Martinico for reading the whole manuscript.

I owe a big debt of gratitude to Francis Gordon, whose patience (again!) and good example of British humour assisted me in the exhausting activity of having the book proofread.

Verona, Italy

Matteo Nicolini

Contents

Part II Spatial Normativies

About the Author

Matteo Nicolini PhD, is Associate Professor of Public Comparative Law at the Department of Law of the University of Verona (Italy), where he teaches Comparative Constitutional Traditions and Global Comparative Law. He is Visiting Lecturer at the Newcastle University Law School (UK) and External Partner of the Centre for the Study of Law in Theory and Practice (a), Liverpool John Moores University (UK). He has held teaching and visiting appointments at the Universities of Stellenbosch, Barcelona, Valladolid, Tübingen, Lille, UNED (Madrid), Alberto Hurtado (Chile), and Innsbruck. He is an author and editor of several works written in Italian, Spanish, and English, which include: 'Law, Territory and Conflict Resolution. Law as a Problem and Law as a Solution' (Brill, 2016) (editor with Francesco Palermo and Enrico Milano); 'Law and Imagination in Troubled Times: A Legal and Literary Discourse' (Routledge, 2020) (editor with Richard Mullender, Thomas D.C. Bennett, Emilia Mickiewicz); 'Comparative Constitutional Justice' (Eleven, 2021) (author with Silvia Bagni); 'Islands in Geography, Law, and Literature. A Cross-Disciplinary Journey' (de Gruyter 2022) (co-editor with Chiara Battisti, Sidia Fiorato, Thomas Perrin).

Chapter 1
Law and Geography: A Special Relationship

1.1 Pluralising the Debate in Legal Geographical Studies

As a set of rules governing human activities, the law has always been in a special relationship with geography. In the course of human history, this relationship has taken different forms and served several functions, mainly revolving around the nation state as the politico-legal paradigm of modernity and the 'organizing principle of political and social life'.[1]

As the major output of the legal-geographical nexus, the nation state conveys 'the idea of exclusive sovereignty'.[2] With a twofold spatial articulation. Its inner dimension coincides with the 'well bounded order of politics'.[3] This cultivated image resonates with the state as a *hortus conclusus*, i.e. an enclosed garden developing its own politico-legal order as a separated process as regards the mayhem of the international community. Enclosure usually follows a process of demarcation or, as in the case of islands, depends on their being surrounded by water (see Sect. 3.5.1).[4] Whatever the process may be, the image delivers a spatial visualisation of the law. Carl Schmitt did the same; he explains that νόμος ('the law') derives from νέμειν ('to divide'), which points to 'the immediate form in which the political and social order of a people becomes spatially [and geographically] visible'.[5]

In its outer dimension, state power is projected over the world by geopolitics. This supplies further insights on the relationship between geography and the law. Since

---

[1] Anderson and O'Dowd (1999), p. 594. On the theory of state as the issue in political geography see Johnston (1981, 1984).

[2] Delaney (2005), p. 36. See also Moore (2015), pp. 26–28.

[3] Ashley (1988), p. 238.

[4] Britain is, for example, a 'sea-wallèd garden' (Shakespeare *Richard II*, 2.42).

[5] Schmitt (2006), p. 70.

M. Nicolini, *Legal Geography*, Ius Gentium: Comparative Perspectives on Law and Justice 105, https://doi.org/10.1007/978-3-031-19410-8_1

the Age of Discovery, the topography of the earth's surface has indeed been considered in 'bordering, ordering and othering' processes.[6]

Law and geography engage in a broader conversation with the spatiality of the political. Indeed, the legal-geographical nexus is an open texture. If it is 'pointless' to 'locate the boundaries of the discipline of geography',[7] it is even more difficult to draw a divide between the law and the non-legal variables (geographical and political ones) outlining the state as an enclosed order of politics.

This book aims to explore the open texture which is the 'special relationship' between law and geography. As its title suggests, *Legal Geography: Comparative Law and the Production of Space* invites readers to critically rethink legal-geographical research in its entirety. It is not just a dialogue among disciplines; the book poses methodological questions and advocates 'pluralising the debate' through cross-disciplinary research.[8] Pluralisation already populates post-modern geographies, which explore the 'ambiguities, contradictions or ruptures' percolating 'the "messiness" of [current] geographical research'.[9] Likewise, when we open the law to interdisciplinarity, it ceases to be dry, and the legal text becomes shaped also by geographical contexts and socio-political variables.

The legal-geographical nexus has attracted attention from scholars with sundry provenances. In the nineteenth and twentieth century, it was mostly populated by scholars with a background in geopolitics, owing to the utility of geography in supporting colonial enterprises. Historians, economists, and physical geographers soon ensued.[10] These cross-disciplinary conversations mainly dealt with matters of *representation*, as the study of *areal differentiation* in regional geography still upholds.[11] Only in the 1980s and 1990s geographers started examining the intersections between communities, law, power, and space. Among them, Blomley and Delaney became the forerunners of what is now legal geography.[12] We shall see in due course that the steady growth in cross-disciplinary research has made legal geography increasingly topical even among legal scholars (See Sect. 2.5). There is now a variety of approaches to legal-geographical research.

No wonder, therefore, that legal geographers have supplemented pertinent pieces of literature. Blomley's and Delaney's publications are considered books of authority by virtue of their theoretical and empirical dimensions. Besides them, legal geographers navigate discrete, albeit interrelated, topics, such as land use and law, indigenous legal geographies, anthropology and legal spatiality, post- and

[6]Tripathi and Chaturvedi (2020), p. 173.

[7]Nayak and Jeffrey (2011), p. 5.

[8]Baldwin (2014), p. 516.

[9]Nayak and Jeffrey (2011), p. 222. See also Delaney (2015), p. 673; Driver (2001).

[10]See Fleure (1996), Hartshorne (1996) and Massey (1996).

[11]See Castree et al. (2013), pp. 22–23 (areal differentiation) and 418–419 (regional geography). See also Wood (1992), Fleure (1996), Hartshorne (1996) and Massey (1996).

[12]See Blomley (1994, 2003). For more on geographers' engagement in legal geographic studies see Braverman et al. (2014), pp. 4–7.

non-human geographies;[13] they then devote themselves to the diachronic relations between law and geography.[14] The list is enriched by several publications, thus providing several disciplinary and methodological insights on how to practise legal geography.

1.2 'Understanding the Spatial Dimensions' in Law and Geography

In touring the fields of comparative law and geography, this book has cross-disciplinary ambitions. It goes beyond the limits marked by mono-disciplinary investigations. It shifts from law and geography to legal geography, bringing together different perspectives and methodologies whereby the legal-geographical nexus may be properly understood.

The operation is not without risk. Cross-disciplinary investigations peruse concepts and lexical items borrowed from different academic disciplines. The pendulum swings back and forth between two opposing poles. At one extreme, concepts and terms are often used without probing the specialised meaning they have within their epistemic community. At the other extreme, legal scholars may become 'amateurish' geographers.[15] Geographers, in turn, might think they 'are playing on another discipline's turf by another discipline's rules'.[16]

Certainly, scholars have already pluralised the debate. Yet, I assume that it is necessary to go further and expand the inter-disciplinary engagement with the legal-geographical nexus. Firstly, this critical engagement requires a wide-ranging examination of the 'special relationship' of law and geography. Secondly, we must shake off our methodological biases—and linguistic inaccuracies. In legal geography, concepts like space and place, for example, are used interchangeably (see Sect. 2.5). The academic distinction between physical and human geography, and between the *earth* and the *world*,[17] seems to lose sight of how geography should reintegrate societies and the environment. The resurgent formalism in legal studies has added fuel to the fire, drawing a divide between the law and the non-legal variables that affect the former. The 'closure of law' as a set of 'formalistic ... institutions' is particularly evident. Mainstream legal discourses are more interested in the legal representation of geographical features rather than in the relationship between law and geography.

Nor is the interdisciplinary research immune from methodological biases. Although they cover a host of different issues, cross-disciplinary publications on law and geography are predominantly collections written by authors with different

---

[13] Bartel and Carter (2021).

[14] See, respectively Platt (2014), von Benda-Beckmann et al. (2009) and O'Donnell et al. (2020).

[15] Husa (2014), p. 29; Husa (2022).

[16] Abler (1987), p. 517.

[17] Cosgrove (1994).

academic backgrounds. Cross-disciplinarity is restricted to the research team; each author is then free to explore the relations between law and geography from their academic angle. Although these perspectives are necessary to establish a cross-disciplinary enterprise, none of them is sufficient to define what legal geography is. This unavoidably loosens its methodological and practical dimensions and often dissuade from cross-disciplinary research.

This book contributes to this conversation by adopting a posture that is quite different from that usually deployed in cross-disciplinary research, addressing the 'special relationship' between law and geography in terms of *legal geography*.

The book thus aims to overcome the divide among both disciplines and the related methodological biases. It seeks to bridge different perspectives (geographical, legal, political, and so on) on this ambit. At the same time, it encourages geographers to reconsider their contribution to legal geography. Indeed, their pristine cultural ambition had been the integration of both human and physical geography within the production of space. This ambition was soon eclipsed by 'academic specialism'.[18] As David Delaney argues, the inter-disciplinary project which is legal geography 'is [now] aligned with the human side of the human/physical divide'.[19] The same holds true in legal research. And the book also challenges legal academic specialism and the idea of the separateness of the law.

References to the contexts where the production of space (also legal space) takes place disclose, *si parva licet*, the cultural ambitions of this book. These navigate the connectedness between the geo-ecosphere, human societies, and the law; but they also go above and beyond the 'spatial turn' in legal studies. Space and law are mutually constitutive;[20] but the book aims to engage in an even closer examination of how space is *also* produced through legal practices.

In sum, the book reappraises the mainstream assumptions we usually make about the law when contributing to the production of geographical space. Thus, the 'special relationship' between law and geography crosses the boundaries between disciplines and addresses their interrelations in terms of 'legal geography'.

1.3 Comparative Law and the Production of Space

Interested, as it is, in examining how *normative spatialities* are manufactured, legal geography is indebted to the cross-disciplinary potential of the 'spatial turn' that took place in human geography in the late twentieth century. Quite unsurprisingly, the stance was a cross-disciplinary one. The US geographer Edward Soja advocated for this turn after realising that non-geographers were also interested in 'space'.

[18] Bennett (2008), p. 36. On these ambitions see Mackinder (1887), p. 145; Livingstone (1992), p. 190.

[19] Delaney (2017), p. 672.

[20] Braverman et al. (2014), p. 1; Delaney (2015), p. 98.

Spatiality describes both the processes of spatial production and the role played by variables and practices (mainly social and geographical) within them. 'Space' thus labels the outcome of such processes; at the same time, as Lefebvre highlighted, it also requires its 'production' be taken into account: '*(social) space is a (social) product*'.[21] In sum, our concept of legal space extensively draws on Soja's and Lefebvre's cross-disciplinary approach to geography; for our purposes, it can be understood in terms of space *also* produced through legal practices.[22]

The book examines how comparative law may contribute to the production of space. In so doing, it explores the possibilities that comparative law offers when exploring the legal-geographical nexus. As we shall see, comparative law shares several features with geography (see Sect. 2.3). Perhaps most striking of all is that both disciplines have an empirical approach and show attentiveness to how humans process the knowledge of the world which is also shaped by the context where it is produced. Like comparative law, indeed, geography also focuses on context acting as a process whereby the law takes part in the production of space. A composite of variables and practices, space becomes a tangible scenario which requires attentiveness in the effort to identify the contingencies that govern the processes of its production also through legal operational rules.

In politico-legal terms, context supplies a 'cognitive mode of gaining control over the world', which makes it also possible to synthesise 'cultural and geographical information'.[23] Again, comparative law shares this way of 'ordering the world and rendering it meaningful for the particular audience'.[24] This was evident in the nineteenth century, when both geography and comparative law furnished the 'knowledge production' useful for manufacturing and establishing Western legal geopolitics around the world (see Sects. 2.2 and 4.1).

This way of ordering the world and rendering it meaningful is a real *paradigm*, as well as a 'disciplinary matrix', that assists us in identifying the rules of spatial production. Following in the footsteps of the English legal historian Frederick W Maitland, we shall also consider how this matrix turned the relation between human agencies and their territory into a concrete legal geography (see Sect. 2.6). The paradigm is not limited to the Western world. As a cultural competence, the pattern is iterative through time and place. Each civilisation has replicated it in an attempt to project its 'world view' onto the earth's surface, which means that the 'operation is never made from a neutral cultural standpoint'.[25] Comparative law helps to unveil these cultural standpoints and discloses the biases that align themselves in the use

---

[21] See Soja (1989). See also Warf and Arias (2008); Massey (2005), p. 91; Bennett (2008), pp. 111–112; Lefebvre and Enders (1976).

[22] See e.g. Lefebvre [1974] (1991), p. 26.

[23] Black (1997), p. 173.

[24] Nayak and Jeffrey (2011), pp. 241 and 249.

[25] Brotton (2013), p. 6.

(and misuse) of legal geographical devices by policy makers and academics, shedding new light on the relationship between geography, power, and the law.

The book thus considers legal geography a true interdisciplinary assessment, where geographers and legal scholars may jointly examine how political power codes the historical, cultural, political, and social variables, i.e. the geographical contexts, through the law.[26]

The book is organised into five parts. Part I (*Methodologies*) comprises Chap. 2 (*From Law and Geography to Legal Geography*), which critically rethinks the 'special relationship' between law and geography, reformulating it in terms of legal geography. It then points to the commonalities between geographical studies and comparative law, both desiring 'to find order in the world'.[27] It considers how power, space, place, and law are mutually constitutive in the production of space. After exploring the manifold interactions between the law and geography, it addresses the emergence of legal geography as an interdisciplinary intellectual project.

The interactions between law and geography disclose the presence of a *paradigm*, which reflects how the law is engaged in the production of space and turns spaces into *legal geography*. As said, the paradigm was devised by Maitland in his lectures *Township and Borough* (1898), where he defined 'legal geography' as a process of spatial production taking place in the legally relevant relationship between community and *its* territory. Maitland termed it 'belongs of public law' (see Sect. 2.6).

Part II (*Spatial Normativities*) addresses how the production of space takes place in legal geography. Political power usually codes geographical space (terrestrial, racial, or imaginative) through the law. Indeed, 'geography is rarely a simple matter of bare facts'; in manufacturing its own spatiality, political power exercises an 'imaginative leap';[28] with a good dose of imagination, it governs spatiality by taking into account physical and human geographical variables. To this end, geography and comparative law supply political power with the practical knowledge which policymakers need in order to manufacture their own legal spatialities and 'naturalise [their] particular representations of the world'.[29]

Chapter 3 (*Law and 'Geographies'*) focuses on the process of spatial production that fall under the umbrella of 'law of spatiality'. Through them, power processes physical geographical data in order to project its own imaginative geography onto the earth's surface. Not only does legal geography legally code physical geographical features, but it also adapts them (as well as the territorialised paradigm) to non-territorial spaces, thus generating new spatial normativities, such as outer space, and maritime, and polar, spaces. Whilst processing spaces, the law also regulates them and creates its own geopolitics of law. As Chap. 4 (*Colonial Underpinnings: Spatiality of Law*) discloses, legal geography also considers the mapping of legal

[26] Martin et al. (2010).

[27] Bennett (2008), p. 9.

[28] Bennett (2008), p. 7.

[29] Nayak and Jeffrey (2011), p. 240.

systems. Legal cartography has been servient to the dissemination and establishment of Western legal geopolitics, which means ranking legal systems after a hierarchy. With its critical perspective, the 'spatiality of law' shares features with critical legal geographies. Chapter 5 (*Critical Legal Geographies*) illuminates the politico-legal underpinnings which are present in colonial, imagined, and counter-geographies, as well as in those normative spatialities which, like the spaces of counter-terrorism law, still apply Maitland's paradigm.

Part III (*Territorial Nomenclatures*) examines the features of the territorialised paradigm in manufacturing legal spatialities. Chapter 6 (*Legal Geography, Linguistics, and Borders*) examines the lexicon of legal geography; in so doing, it considers some relevant legal-geographical concepts (among the others, placenames and borders). It also deals with bordering processes, which lie at the crossroads of several disciplines; human geographical studies equip lawyers with the material scenario for bordering processes. Linguistics then discloses the dynamics of boundary-making and bounded cultural spaces. Chapter 7 (*'Federal Legal' Geographies*) focuses on the nomenclature of the territorial constitution by placing emphasis on how federal and regional arrangements are demarcated on a geographical basis.

Part IV (*De-Territorialised Legal Geographies*) comprises a chapter that engages with de-territorialised processes of spatial production. Spaces like the Anthropocene, globalisation, as well as digital and climate-change spaces seem to depart from the territorial matrix. Chapter 8 (*Un-Bounded of Legal Spaces*) maintains Maitland's paradigm is still able to explain the production of these spaces.

In Part V (*Conclusions*) I shall draw some reflection on how the book has contributed to the debate on legal geography and the production of space also through legal practices. It indeed proposes a renewed methodology for exploring this interdisciplinary project. Certainly, we will have to revise our existing methodologies to face the challenges the global production of space stages before us. By then, I assume, legal geography will probably have acquired the status of an autonomous field of scientific research.

References

Abler RF (1987) Presidential address: what shall we say? To whom shall we speak? Ann Assoc Am Geogr 77:511–524

Anderson J, O'Dowd L (1999) Borders, border regions and territoriality: contradictory meanings, changing significance. Reg Stud 33:593–604

Ashley R (1988) Untying the sovereign state: a double reading of the anarchy problematique. Millennium 17:227–262

Baldwin A (2014) Pluralising climate change and migration: an argument in favour of open futures. Geogr Compass 8:516–528

Bartel R, Carter J (eds) (2021) Handbook on space, place and law. Edward Elgar, Cheltenham and Northampton, MA

Bennett A (2008) What is geography? SAGE, London

Black J (1997) Maps and politics. The University of Chicago Press, Chicago

Blomley N (1994) Law, space, and the geographies of power. The Guildford Press, New York

Blomley N (2003) From 'what?' to 'so what?': law and geography in retrospect. In: Holder J, Harrison C (eds) Law and geography. OUP, Oxford, pp 17–33
Braverman I et al (2014) Expanding the spaces of law. In: Braverman I et al (eds) The expanding spaces of law. A timely legal geography. Stanford University Press, Stanford, CA, pp 1–29
Brotton J (2013) A history of the world in twelve maps. Penguin, London
Castree N et al (2013) Oxford dictionary of human geography. OUP, Oxford
Cosgrove D (1994) Contested global visions: one-world, whole-earth, and the apollo space photographs. Ann Assoc Am Geogr 84:270–294
Delaney D (2005) Territory. A short introduction. Blackwell, London
Delaney D (2015) Legal geography I: constitutivities, complexities, and contingencies. Prog Hum Geogr 39:96–102
Delaney D (2017) Legal geography III: new worlds, new convergencies. Prog Hum Geogr 41:667–675
Driver F (2001) Geography militant: cultures of exploration and empire. Blackwell, Oxford
Fleure HJ (1996) Human regions. In: Agnew J et al (eds) Human geography: an anthology. Blackwell, Oxford, pp 385–387
Hartshorne R (1996) The character of regional geography. In: Agnew J et al (eds) Human geography: an anthology. Blackwell, Oxford, pp 388–397
Husa J (2014) Interdisciplinary comparative law between Scylla and Charybdis. J Comp Law 9:28–42
Husa J (2022) Interdisciplinary comparative law. Rubbing shoulders with the neighbours or standing alone in a crowd. Edward Elgar, Cheltenham and Northampton, MA
Johnston RJ (1981) Political geography. In: Wrigley N, Bennett RJ (eds) Quantitative geography: retrospect and prospect. Routledge & Kegan Paul, London
Johnston RJ (1984) Marxist political economy, the state and political economy. Prog Hum Geogr 8: 473–492
Lefebvre H (1974) La production de l'espace. Éditions Anthropos, Paris. English edition: Lefebvre H (1991) The production of space (trans: Nicholson-Smith D). Blackwell, Oxford
Lefebvre H, Enders MJ (1976) Reflections on the politics of space. Antipode 8:30–37
Livingstone DN (1992) The geographical tradition. Blackwell, Oxford
Mackinder HJ (1887) On the scope and methods of geography. Proc R Geogr Soc 9:141–160
Martin DG et al (2010) Making law, making place: lawyers and the production of space. Prog Hum Geogr 34:175–192
Massey D (1996) In what sense a regional problem? In: Agnew J et al (eds) Human geography: an anthology. Blackwell, Oxford, pp 398–413
Massey D (2005) For space. SAGE, London
Moore M (2015) A political theory of territory. OUP, Oxford
Nayak A, Jeffrey A (2011) Geographical thought. An introduction to ideas in human geography. Routledge, Abingdon and New York
O'Donnell T et al (eds) (2020) Legal geography. Perspectives and methods. Routledge, Abingdon and New York
Platt RH (2014) Land use and society, third edition: geography, law, and public policy, 3rd edn. Island Press, Washington
Schmitt C (1950) Der Nomos Der Erde: Im Volkerrecht Des Jus Publicum Europaeum. Duncker & Humblot: Berlin. English edition: Schmitt C (2006) The nomos of the earth in the international law of the Jus Public Europaeum (trans: Ulmen GL). Telos, New York
Soja EW (1989) Postmodern geographies: the reassertion of space in critical social theory. Verso, New York

Tripathi D, Chaturvedi S (2020) South Asia: boundaries, borders and beyond. J Borderl Stud 35: 173–181
von Benda-Beckmann F et al (eds) (2009) Spatializing law: an anthropological geography of law in society (law, justice and power). Ashgate, Farnham and Burlington, VT
Warf B, Arias S (eds) (2008) The spatial turn. Interdisciplinary perspectives. Routledge, London
Wood D (1992) The power of maps. The Guildford Press, New York

Part I
Methodologies

Chapter 2
From Law and Geography to Legal Geography

2.1 From Regulatory to Performative Practices

2.1.1 The Performative Practice of Comparative Law

Touring the fields of comparative law and geography has evident cross-disciplinary ambitions. This means going beyond mono-disciplinary investigations, shifting from law and geography to legal geography, and bringing together different perspectives and methodologies to better understand the legal-geographical nexus.

Thinking in terms of legal geography entails reappraising an assumption that is common in geographical studies, that is, that the law has a pure 'regulatory' role. Instead of seeing it as co-constitutive of spatial production, scholars usually consider it a force that is external to that process. Law 'inscribes spatial practices and orders the material landscape', codifies 'what can be done where and by whom'; finally, 'enables human territoriality through discourse and practice'.[1]

As a regulatory practice, the law triggers two cognitive biases. The first of these assumes that the law is an unfettered, not to say extrinsic, force as regards contexts. This practice, also termed *universalism*, emancipates the law from contexts, generalises it, and makes it applicable irrespective of social, cultural, and political backgrounds.[2] As it possesses universal substance, the 'unfettered' law denies local variations, ignores the environments where it is applied, and superimposes abstract legal derivatives on human and physical geographies of the world.

The 'objectivisation' of the law causes a second bias. The universal law is considered a Western derivative, and the geographical Westernisation of the world a 'cultural habit' related to it.[3] It is an ethnocentric attitude in the study of the law that reached its heights in the nineteenth and twentieth centuries (see Sect. 4.5.1).

[1] See Jepson (2012), p. 616. See also Delaney (2003), pp. 78–79; Feihle (2019).

[2] Husa (2015), p. 66. See also Siems (2022), pp. 36–39; Mousourakis (2019); Nicolini (2020).

[3] Husa (2015), p. 100. For more on such habit see Mousourakis (2007), p. 149.

M. Nicolini, *Legal Geography*, Ius Gentium: Comparative Perspectives on Law and Justice 105, https://doi.org/10.1007/978-3-031-19410-8_2

Legal geography, by contrast, shows attentiveness to contextual variables. This is, on its own, a process of co-production of law and space. Far from being merely regulatory, the law is first and foremost a performative practice creating spaces whose meaning is pragmatically enriched by socio-geographical contexts. Taking the 'spatialities of law seriously' means ceasing 'to look upon' it 'as a closed, formal, and acontextual system and see it instead as an assemblage of heterogeneous elements'.[4]

Comparative law is well-equipped when it comes to assessing the law as a performative practice. The assessment runs through multiple dimensions. The first of these disengages the law from the mere written text. Comparative law goes beyond the black-letter law and privileges a contextualised approach. Basic comparative legal concepts such as 'law in books', 'law in action', 'law in context', and 'legal formants' denote attentiveness to operational rules beyond the written law: 'rules alone, mere forms of words, are worthless.'[5] Processes of spatial production make the black-letter law meaningful (see Sect. 3.1). This resonates with another feature of comparative methodology, whose *empirical*, 'problem-based approach' is 'the starting point of a comparative analysis'.[6]

In its second dimension, this empirical approach allows scholars to grasp the 'deep structures of the legal systems', without ignoring its social, economic, and historical aspects. Contrariwise, they ascertain how, and to what extent, these aspects are 'meaningful' in law.[7] Comparative law ceases to be a set of academic speculations; in dialogue with non-legal disciplines, it gains control of the 'deep connection' with its cultural and environmental contexts.[8]

Its third dimension transcends the legal contexts of the West, revealing that each civilisation has its own conception of the law. All legal systems reflect culture-specific 'world views'. Not only do these cover the world map, but they often mirror several, and often conflicting, processes of legal (and spatial) production.[9] In sum, they embed a variety of relationships between power and the legal geographical dimension. Take, for example, the so-called 'chthonic', 'aboriginal', or 'autochthonous' legal traditions with their conception of the law and close engagement with the environment. They are said to be 'ecological', inasmuch as they develop 'in close harmony with the earth' and adopt a holistic approach to both the geographical and the legal.[10] They are geographical, since the interactions between communities and

---

[4]Hogg (2002), p. 34. See also Delaney (2010), p. 7.

[5]See Pound (1910), p. 12; Llewellyn (1930), p. 3; Sacco (1991). On contextualised approaches in comparative law and legal geography see Kedar (2014) and Spencer (2020).

[6]Siems (2022), p. 32. On geography as an empirical discipline see Johnston (1983).

[7]See, respectively, Legrand (1997), p. 56; Legrand (1996).

[8]Watt (2012), p. 84. Frankenberg (1985), p. 414, terms the process of deep appreciation as 'differencing'.

[9]On the varieties of world views in mapmaking traditions See Wood (2010), pp. 22–25.

[10]Glenn (2014), p. 62. The term derives from the ancient Greek *kthonos* ('earth'); hence mod. English 'autochthonous'. See Goldsmith (1992), p. xvii.

nature gain a real understanding of the earth in its diversity; they are legal, because these traditions distil their rules from the 'natural' patterns of the earth.

2.1.2 *'Integrated' Geographies*

Like comparative law, geography is also well-equipped in legal geographical terms. It aims to establish a bridge between the 'complex web of relations' that populate spatial contexts; it unfolds the processes of politico-legal Westernisation; it challenges the claims of objectivity and autonomy of the law, thus recognising 'the subsistence of other legal orders and other legal possibilities'.[11] Like in legal studies, a shift from regulatory to performative practices is required. Thinking 'in terms of' legal geography entails being cross-disciplinary also within geography.

It is not my intention to restore Mackinder's 'geographical experiment' as a 'bridge over the abyss . . . between the natural sciences and the study of humanity'.[12] I assume that the divide between the *earth* and the *world* (see Sect. 1.2), as well as the academic distinction between physical and human geography, are both untenable in legal geographic terms. This fissure loses sight of how both geographies should be 'integrative' of societies and the environment (see Sect. 3.2).

The shift towards 'integrative' geographies has two consequences. Like in comparative law, it helps to address some '*concrete* questions' related to how 'a wide range of phenomena connected to the Earth' occur.[13] Take, for example, geophysical phenomena such as 'eluviation', 'watershed', 'catchment areas', or 'anthropogenic global warming'. We shall see how they acquire new meanings in spatial production (see Sect. 3.5.1). Consequently, physical geography expands the scope of the cross-disciplinary research, assisting us in gaining knowledge of the variables implicated in spatial production.

The second consequence is related to core geographic lexical items. The 'spatial turn' favoured a reappraisal of the concept of space and the shift from regulatory to performative practices. Originally, space represented an 'absolute' and 'mathematical' concept with 'a strictly geometrical meaning'; 'the idea it evoked', as Lefebvre wrote, 'was simply that of an empty idea'.[14] In cartography, it pointed to a 'location . . . on the Earth's surface in relation to geographical coordinates', which was understood and measured 'through Euclidean geometry and mapped'.[15] I am not denying its importance. My point is this: in situating it within spatial production, legal geography contextualises and renders it more empirical.

[11] Respectively, Massey (1992), p. 66; O'Donnell et al. (2020), p. 4. See also Blomley (1994), p. 7; Bonnett (2008), p. 17.

[12] See Mackinder (1887), p. 145. See also Livingstone (1992), pp. 189–196.

[13] Agnew et al. (1996), p. 5.

[14] Lefebvre [1974] (1991), p. 1.

[15] Matthews and Herbert (2008), p. 11.

Last but not least, there is the concept of place, which 'includes the search for boundaries, edges, and limits that contain a definable and recognised territory'.[16] As noted, the concept of space is also related to the idea of territory, which in turn 'is a product of social translation, transformation and experience', as well as a repository of political connotations.[17] Place is, in other words, a space made meaningful by human agencies. As a bounded space, it also discloses the legal geographical territorial matrix I shall examine in due course (see Sects. 2.6 and 3.1).

2.2 Commonalities

Thinking in terms of legal geography has meant shifting from regulatory to performative practices in both comparative law and geography. Favoured by their empirical approach, the shift also accounts for their ability to address discrete issues, such as those triggered by climate change. Although global problem, I shall explain that global warming as the outcome of situated legal practices.[18] Responses to it must be concrete to reflect such a troubled reality (see Sect. 8.7).[19]

My approach has clear cross-disciplinary ambitions. Indeed, 'geography's defining act' lies in its 'ability to grasp the world intellectually'; and the approach also characterises comparative law.[20] For both disciplines, ordering the world knowledge gives sense to the complex 'space-place-environment nexus' present in legal contexts.[21] Geography has a strong empirical pedigree stretching back to fieldwork made up of explorations, travels, discoveries, land mapping, and surveying. The law has resort to sociology, ethnology, and anthropology to explain the legal interactions between ethnic groups, social contexts, and cultural spaces. When coining the term 'geography', Eratosthenes of Cyrene (288–205 BC) had already realised how complex 'Writing the Earth' would be. Geography 'writes' the earth in its complexity—even in its legal complexity, as in the case of the legal systems of the world (see Sect. 4.5).[22]

It is striking that their first commonality refers to their disciplinary boundaries, which are difficult to draw. In comparative law, these are usually demarcated on methodological grounds. As an academic discipline, it does not coincide with a set of legal rules applicable to a specific ambit. It is 'a field of legal research', whose method is 'an intellectual activity', as well as 'a form of study' which helps to

---

[16] Matthews and Herbert (2008), p. 83.

[17] Soja (1989), pp. 79–80. See also Delaney (2005), p. 13; Brighenti (2006); Gottmann (1975); Moore (2015), p. 66.

[18] Carlarne and Farber (2012).

[19] Etty et al. (2018), p. 200.

[20] Respectively: Bonnett (2008), p. 32; Husa (2015), p. 70.

[21] Matthews and Herbert (2008), p. 17.

[22] Eratosthenes (2010). See Brotton (2013), p. 35.

compare different legal systems.[23] Its boundaries are outlined by the very *process* of comparison. Not only does this make its boundaries fuzzy, but it also allows scholars to constantly redraw them. And this accounts for the 'several variations in which comparative law is presented in the academic literature'.[24]

Likewise, scholars consider it 'arbitrary' to draw firm boundaries in geography. There is nothing whimsical in this predicament. Firstly, the 'set of issues that constitutes scholarly geography' is still a matter of disagreement;[25] and its methodological approaches have changed over time. Furthermore, the disjuncture between physical and human geography and the variety of subdisciplines explain the difficulties with such demarcation. Finally, 'Disciplinary definitions' reinforce geographers' attempts to superimpose their own representation of the world onto 'ideas that extend beyond those who self-identify as geographers'.[26]

The porosity of disciplinary boundaries accounts for their relatively recent institutionalisation, which took place in the second half of the nineteenth century. Certainly, geographical and comparative practices had been applied well before. By 'institutionalisation' I refer to their emergence as academic disciplines. What counted as 'geography' or 'comparative law' was decided within the academia to give them consistency and scientific rigour.[27] At the same time, the birth of 'academic' geography and comparative law supplied the production of knowledge useful for establishing Western legal geopolitics around the world. In comparative law, this was favoured by the universalist approach and the proposed unification of national laws through generalisation. Distilled from the law of the European, civilised nations, these unified systems of laws would stimulate trade and facilitate commercial transactions. During the 1900 Paris Congress, Raymond Saleilles and Édouard Lambert paved the way for an ambitious project, i.e. entrusting comparative law with the discovery of universal principles common to 'civilised' systems.[28] Likewise, courses of geography were created in several Western universities; Mackinder's appointment as reader in Geography at Oxford (1887) reflected these universal ambitions, satisfying 'the practical requirements of the stateman and the merchant, the theoretical requirements of the historian and the scientist, and the intellectual requirements of the teacher'.[29]

The connection of both disciplines outside academic circles is also well-documented. The worlds of trade and politics relied on the practical implications of comparative law in drafting international trade agreements boosting legal unification and harmonisation. In promoting knowledge-based economic relations and overseas trade, this empirical pedigree is attested by the proliferation of

---

[23]Respectively, Kamba (1974); Husa (2018), p. 37; Mousourakis (2019), p. 1; Gordley et al. (2021).

[24]Siems (2022), p. 6.

[25]Nayak and Jeffrey (2013), p. xiii.

[26]Respectively, Livingstone (1992), p. 5; Nayak and Jeffrey (2013), p. xiii. See also Aay (1981).

[27]See Robic (2003).

[28]Peters (1999).

[29]Mackinder (1887), p. 159. See Brotton (2013), p. 350.

Geographical Societies around Europe; and the creation of the aforementioned Readership in Geography at Oxford was funded by the Royal Geographic Society.[30]

The open texture of comparative law and geography, as well as their attentiveness to real-world implications, explain the variety of methodological approaches comparative law and geography have displayed over the last 150 years. In making up the discipline, these have been influenced by cultural and political contexts. For example, their ethnocentric and universalist, approach was forged in the nineteenth century but still prevails in global comparative law and geography (see Sect. 4.6).

In the second half of the twentieth century, a well-defined *scientific state of mind* emerged beneath the surface of both disciplines. In the 1960s, geography veered towards quantitative approaches, whose 'rigorous scientific method' has now become so 'powerful that it can determine even how we allow ourselves to imagine the world'.[31] Comparative law did the same at the end of the previous century. The last few decades have experienced several 'turns' in both disciplines: geography underwent spatial, humanistic, and cultural turns. Together with several other subjective strands (such as geographies of race and sexuality, for example), these turns go hand in hand with critical approaches in legal studies.[32] The law is also experiencing a 'mathematical turn'.[33] Recent studies have joined hard sciences in employing quantitative approaches, and 'a growing field of empirical legal studies' applies 'statistical methods to legal questions.'[34] Through quantitative methods, scholars assess the 'quality' of the legal systems of the world and rank them, say, according to several benchmarks (see Sect. 4.6).

However 'messy and imperfect',[35] the methodological approaches now available in comparative law and geography are often mixed and jointly applied, fitting into a wide range of legal and geographical projects. They also represent a toolkit capable of enhancing a critical thinking in both disciplines and challenging the 'isolationist' attitude of lawyers and geographers. In so doing, they open up both disciplines to external connections, pluralise the methodological debate, and trigger 'internal critiques of the core principles' of both disciplines.[36] Thinking outside the box thus entails a good dose of subversion, which describes comparative law, geography, and their approaches as 'critical'.

Such critical potential unveils the ideological presuppositions of spatial production, which assisted the Europeans in drawing hierarchies among geographical areas and legal systems (see Sects. 4.3 and 5.1). Suffice it here to remember the racialised hierarchies lying beneath the legal-geographical representations of the Orient. Said's

[30]For more on the creation of geographical societies see Driver (2001), pp. 27–37; Bonnett (2008), pp. 102–103; Brotton (2013), p. 337; Nayak and Jeffrey (2013), pp. 11 et seq.

[31]Hammond (1968), p. 113; Levine (1987), p. 8.

[32]Orzeck and Hae (2020), p. 838.

[33]Restrepo Amariles (2015), p. 14.

[34]Siems (2022), p. 207.

[35]Adams (2012), p. 5.

[36]Blomley (2003), p. 22.

Orientalism and Ruskola's *Legal Orientalism*, for example, disclose how Asia has been studied by adopting 'an implicit colonialist attitude'.[37] The same holds true for 'tropical geography'. Far from being part of regional geography, the field had colonial underpinnings, later handed over to less overtly 'colonial' experiments. Geographies of modernisation and development, indeed, reflect neo-colonial relations of power and domination.[38]

2.3 Imbrications

2.3.1 Power, Comparative Law, and Geography

Comparative law and geography share a further commonality. Both engage in conversation with power, i.e. the ultimate source of the authority of the law. Power has been widely studied in geography. Thinking in terms of legal geography also means critically reappraising how power pursues its agenda through the 'special relationship' between law and geography.[39]

The topic is not a fresh one in legal studies, but legal formalism has drawn a divide between the law and the non-legal variables that influence it. Legal scholars teach law students that such 'closure' to societal contexts secures the value-neutral character of the law.[40] Following in the footsteps of Callon, Nicholas Blomley terms the closing the law 'bracketing', since it constructs 'a boundary within which interactions take place more or less independently of their surrounding context'. the meaning of such interactions is 'self-evident to the protagonists', i.e. the lawyers.[41] Bracketing is usually ideologically motivated (see Sect. 5.2.3); once enshrined in legal texts, ideology is carefully crafted so as to secure its allegedly objectivity.

My interdisciplinary research runs somewhat counter to the idea of the autonomy of the law, which cannot be reduced to a neutral set of rules to be 'understood and applied without reference to immediate economic, social, or political needs.'[42] Legal devices should be assessed, taking into consideration that their full understanding entails a full understanding of all the variables that affect them. The law is not an empty box incapsulating 'anonymous authority and power'.[43] Quite the opposite: external factors pragmatically enrich and influence its meaning.

[37] See Husa (2015), p. 138. See Said [1978] (2003) and Ruskola (2013).

[38] Power and Sidaway (2004).

[39] See Blomley and Bakan (1992) and Blomley (1994). For recent assessments see e.g., Coleman and Agnew (2018).

[40] See also Husa and Van Hoecke (2013).

[41] Blomley (2015), p. 169; Callon (1988), p. 249.

[42] See Frier (1989). See also Moore (1973) and Lempert (1988).

[43] Rissanen (2000), p. 121.

With its regulatory character, the mainstream relation between law and power may be termed 'law as ideology'. Ideology is broader than power and encompasses all agencies which try to influence the content of the law. The concept at stake is closer to *implicatures*, i.e. the interpretation of meaning that linguists work out in pragmatics.[44]

Likewise, legal methods are seldom value neutral. When ordering the world knowledge, they do not act as neutral forces; by contrast, their outputs are 'value-laden', reflecting the interests pursued by ideology. Take, for example, the aforementioned mathematical turn in law. Quantification is never value neutral; 'calculative practices' and indicators provide 'an alternative source of legitimisation', that alters the geographical representation of the law (the so-called spatiality of law: see Sect. 4.6.2) through a deliberate 'plan of . . . change based upon an imitation of laws,' transplants, and borrowings.[45] The legal frame thus encapsulates the relations between society, power, law, and geography.

The quantitative turn in geography triggered similar consequences. The incorporation of scientific techniques gained ground in the 1950s and 1960s. Space became an abstract and absolute 'container' of social-environmental interactions allowing scholars to present 'their work as dispassionate and neutral representations of social life'.[46] The abstract, measurable, and geometrical conception of space also had a strongly regulatory essence, enabling geographers to generalise patterns of spatial relations; spatial laws explained and predicted space-related phenomena. Quantitative practices did not represent a neutral-laden approach to geography; generalised spatial processes were applicable everywhere triggering changes in human-environmental relations. In sum, giving sense to the world means delivering particular representations of it, which are often 'a direct result of the craft competences of the investigators rather than a portrait of reality'.[47] The legal and geographical co-production of space is a discursive practice through which scholars 'naturalise', i.e. generalise, their own representation of the world (see also Sect. 3.3.3).[48]

Thinking in terms of legal geography unveils the legal implicatures behind the law. When assessing its authoritative meaning, it discloses its relationships with power and its performative role in shaping meaningful geographical places. And, again, we shift from regulatory to performative practices, from 'law as ideology' to 'law as performance'. Like law and ideology, the latter is also related to the content and meaning of the law. Unlike law and ideology, however, within 'law as performance' they do not stem from a textual (i.e. abstract) interpretation of legal provisions; rather, they are inferred from contexts.

---

[44] On ideological representation of geographical spatial knowledge see Lefebvre [1974] (1991), p. 45.

[45] Ajani (2017), p. 1. See also Nelken (2018).

[46] Nayak and Jeffrey (2013), p. 45.

[47] Livingstone (1992), p. 16.

[48] Respectively Nayak and Jeffrey (2013), pp. 240–241; Ó Tuathail and Agnew (1992), p. 190.

2.3.2 *Desired Geographies and the Cartesian Mind*

As performative practices, comparative law and geography reveal how power is constitutive of spatial production. Through geography, power makes its own representation of the world, gains control of it, and entrenches its own spatial hierarchies.

Spatial production has traditionally taken place through legal cartography. Mapping the world may be highly ideological since it fixes spatial hierarchies in an allegedly scientific way. In so doing, the geographical 'reality' is coded through the law, which naturalises its own representation of geography. In this by-product of Western political thought, the influx of Cartesianism is patent. Descartes had separated the human and the environmental by interrupting the *continuum* between the subject and the object and turning the world into a separate entity. By *acts of simplification*, cartography and the law reduce the inherent complexity of the world to a picture by stripping away whatever does not match the Cartesian mind, or power considers ideologically useless or dangerous.

The relation between power, law, and geography is therefore a matter of representation. Through legal-spatial production, the world (and the Earth's surface) is reshaped, rewritten, and altered to match the desired ends of political power. I label them 'desired' geographies. In the Bible, for example, God promises His Chosen People Israel the land of Canaan, which He will assign them once liberated from the enslavement in Egypt. God has His own desired geography: He will give Israel this land if it leaves Egypt thus complying with His 'political' agenda. Geographically, the land is a 'good land and ... large', also 'flowing with milk and honey'; politically, the territory will be demarcated at the expense of 'the Canaanites, and the Hittites, and the Amorites, and the Perizzites, and the Hivites, and the Jebusites'.[49] Law and geography produce a new space. Firstly, this is created for Israel through a highly discursive practice. Secondly, a clear politico-legal title is established in favour of Israel: 'I have said unto you, Ye shall inherit their land, and I will give it unto you to possess it'. Finally, God physically separates Israel 'from other people'.[50]

Desired geographies reflect the concept of 'imaginative geographies'. Coined by Edward Said in his seminal book *Orientalism*, these geographies of power are produced to match political desired ends. Again, God's discourse represents the land of Canaan and allows the Israelites to set up their own politico-legal space and imagine it in their minds.[51] This is sufficient for turning the desire into action—the conquest of Canaan by Israel is narrated in the Book of Joshua.[52] Imaginative geographies also uphold a variety of practices that create spatial relations based on hierarchical social structures and cultural imbalance. In *Orientalism*, for example,

[49] *Ex* 3.8; see also *Ex* 3:17, 13:5; *Num* 13:27, 14:8, *Deut* 6:3, 11:9. On boundaries of the Land of Canaan see also Kallai (1997), pp. 70–72.

[50] *Lev* 10:24.

[51] Said [1978] (2003), p. 54.

[52] Paton (1913).

they are the outcomes of spatial production. Said explains that 'Orientalism' is a Western political construct, i.e. a 'particular, power-laden view of the East'.[53] In order to contain and subjugate the Orient politically, the West has clearly put it in an 'enclosed space', whose 'representation is a theatrical one'. Within these spaces, 'Europe is powerful and articulate; Asia is defeated and distant'. Europe acts as a 'genuine creator', 'whose life-giving power represents, animates, and constitutes the . . . dangerous space beyond [its] familiar boundaries'.[54]

During colonialism, the interactions between law and imaginative geographies secured the superiority of the West regards the subordinate rest of the world. The law played a pivotal role in maintaining the hierarchies underpinning these colonial legal spaces. On the one hand, European powers shaped African and Asian legal cartography and superimposed their own spatiality of law onto these continents; peoples, communities, territories, and their collective legal wisdom still bear the consequences of colonial domination (see Sect. 4.4). On the other hand, modern legal practices create further racialised spatial relations reinforcing hierarchical social structures and inequalities. Within Western countries, for example, urban spaces reproduce the racial hierarchies between the metropolis and its colonies. As a result, these spaces often legally embed social, as well as cultural, segregation of homeless people and immigrants, relegating them into an 'enclosed' and marginal space (see Sect. 5.3.6).

The projection of imagined geographies onto the Earth's surface has created a variety of legal spaces. Take, for example, the 'Age of Discovery', i.e. the interweaving of voyages and explorations that led to the expansion of European geographies of power through the world in the fifteenth and sixteenth centuries. To grasp the world intellectually and politically, European geography had to be rewritten, and the earth remapped.[55] Being the major European powers, the Spaniards, the Portuguese, the English, and the Dutch had to accommodate newly discovered territories into a new political order made visible through maps; newly created colonial empires partitioned the world among themselves by representing it along global lines (see Sect. 4.2.3).

In the eighteenth century, new spaces were processed. Navigating towards the Southern seas in search of the *Terra Australis Incognita*, i.e. the fabulous southern continent, Captain John Cook provided Great Britain with new geographical imaginaries. Political power added a new variable to (and a new justification for) such geographical expeditions. This produced new scientific knowledge, which coalesced towards the commercial opportunities favoured by these discoveries. Both Cook's first (1768–1771) and second expeditions (1772–1775) had, among their objectives, the discovery of new lands and the making of observations 'as may be useful either to Navigation or Commerce'.[56]

---

[53] Castree et al. (2013), p. 279.

[54] Said [1978] (2003), p. 63.

[55] Brotton (2013), pp. 186 *et seq.*

[56] Williams (1996).

In this journey around the globe, power, geography, and the law have constantly assumed 'a *territorial* conception of space'. Evidently, territory is 'an implicit rather than an explicit paradigm' in legal geographical investigations.[57] I shall constantly peruse this paradigm throughout this book: for example, Maitland's paradigm of legal geography is territorialised (see Sect. 2.6.2); concepts such as territorial constitution and identity are patently indebted to it (see Sect. 7.2.4). Power is the most significant driver in legal and spatial production; in pointing to the 'geographical domain of a political entity', the concept of territory vividly represents the relationship between power and geography. No wonder, therefore, that territory still haunts legal geographical studies as a 'hidden geography'.[58]

2.4 Interactions

As a 'hidden geography', territory is still our dominant paradigm. It possesses strongly political connotations, such as in the case of territoriality, one of its most widely used derivatives. Whereas territory refers to a tract of land under the domain of a political authority, territoriality points to a territory made politically meaningful through processes of spatial production.[59] Located at the crossroads of politics, law, and geography, both territory and territoriality have cross-disciplinary potentials. This explains why legal geography considers their manifold 'territorial' interrelations.

These interactions are usually studied from monodisciplinary angles, such as international law, international relations, or geopolitics. Monodisciplinary research is not able to grasp the manifold interactions between law and geography at the territorial scale. By solely focusing on bounded political spaces, they have fallen prey to what Agnew terms the 'territorial trap', whose cardinal tenet is 'territorial integrity'. This has three corollaries, which revolve around 'sovereignty as territorial, the domestic-foreign opposition, and the state-society match', all reflecting the Western 'obsession with territory as the exclusive spatial modus operandi of world politics'.[60]

The following sections 'pluralise the debate' in cross-disciplinary terms.[61] Therefore, I shall consider some interactions, namely constitutional (Sect. 2.4.1), historical (Sect. 2.4.2), and ideological (Sect. 2.4.3). Deploying cross-disciplinary methodologies sheds new light on the political, legal, and cultural repercussions, which might prelude the true interdisciplinary intellectual project which is legal geography (Sect. 2.5).

---

[57] Agnew et al. (1996), p. 7.

[58] Agnew (1994).

[59] See Gottmann (1973). See also Diener and Hagen (2012), p. 6; Storey (2020).

[60] See also Agnew (2015).

[61] Baldwin (2014), p. 516.

2.4.1 Constitutional Interactions

Territory, Andrea Brighenti highlights, 'is an *imagined* entity'.[62] Carving it out of the earth's surface, power 'territorialises' it and makes this bounded space meaningful: 'What makes an enclosed space a territory is, first, that it signifies, and, second, that the meanings it carries or conveys refer to or implicate social power'.[63] This spatial combination is usually coded through the law and enshrined in constitutional arrangements, where legal-geographical interactions take different forms.

Legal scholars mainly adopt a formalistic approach to this topic focusing on whether (and how many) constitutions enshrine territorial arrangements, constitutionalise secession, or embed some form of annexation.[64] I term the entrenchment of state territory *constitutional regime*. This ranges from territorial borders to boundaries,[65] from demarcation processes[66] to the list of geographical features,[67] and references to, say, the territorial sea, airspace, and seabed.[68] It may acquire relevance where integrity of states is concerned.[69] The constitutional regime may also coincide with the *scope* of a legal system, such as in the EU, where national jurisdictions delineate the ambit where the EU Treaties have to be applied.[70] In federations, constituent units are demarcated on a geographical basis thus paving the way to the concept of 'territorial constitution' (see Sect. 7.2.4).

The constitutional regime also extends over waters, which are territorialised for demarcation purposes. States exert their sovereignty on vast stretches of open water. The national sea may broaden territory of insular states and turn them into 'gigantic miniatures'.[71] As Art. I s 1 Const. Micronesia affirms, for instance, state jurisdiction 'extends to a marine space of 200 miles measured outward from appropriate baselines, the seabed, subsoil, water column, insular or continental shelves, airspace

[62] Brighenti (2006), p. 68.

[63] Delaney (2003), p. 17. See also Delaney (2010), p. 4 discussing Lefebvre [1974] (1991), p. 142.

[64] See Mancini (2012) and Palermo (2019); Doyle (2018), p. 888 (87 percent of constitutions refer to territory; 14 percent delineate its features).

[65] See e.g. Art. 42, Const. Mexico; Art. 2 Const. Nepal; Art. 1(2) Const. Eritrea; Art. 2 Const. Cambodia; Sch. 1 to the Const. Belize.

[66] See e.g. Art. 75 Const. Argentina.

[67] See e.g. Art. 142 Const. El Salvador (which refers to several international treaties on territorial limits, as well as to islands, islets and cays enumerated by the Judgment of the Central American Court of Justice, pronounced on March 9, 1917); Art. 1 Const. Namibia (references include 'the enclave, harbour and port of Walvis Bay', the 'the off-shore islands of Namibia'; the 'southern boundary [extends] to the middle of the Orange River'. See Hangula (2010).

[68] See e.g. Art. 142 Const. Guatemala.

[69] See e.g. Art. 4(3) Const. Russian Federation. See also the Preamble, Const. German Empire (1871); Art. 79(1) Const. Austria; Art. 31(3) Const. Mexico; Arts. 5(2), 16, and 89(4) Const. France; Art. 115a(4) German Basic Law.

[70] See Art. 52 TEU and Art. 355 TFEU. See Jimena Quesada (2013).

[71] See e.g. Art. 6 Const. Comoros; Art. I s 1 Const. Palau. See Hanlon and Eperiam (1988), p. 98.

over land and water'. This provision results in an array of watery statutes and reflects the processes of oceanic territorialisation (see Sect. 3.6.2).

Not all constitutional texts explicitly outline the constitutional regime. But, the 'silent constitution of territory' is not 'constitutional absence'; each legal system always has a geographical referent 'reflecting the geographic range of the officials whose actions constitute the ultimate rule'. And the referent determines the territorial scope of the constitutional system.[72]

This predicament, however, assumes an undisputed congruence between territory, its imagined geography, and the scope of constitutional regimes. In many cases, congruence is absent. Within aggregative federations, for example, the territorial regime constituted a 'work in progress'. In the USA, Australia, and Canada, not only did pre-existing units come together to create the federation, but the federal territory was accrued by admitting new states usually carved out of federal territories. Within the British colonial Empire, this also occurred in Malaysia and India.[73] Secondly, Anglo-Saxon constitutional arrangements in the U.S.A, Canada, Australia, India, Pakistan, and Malaysia list a variety of units (states, territories, and so on) which have different constitutional statuses (see Sect. 7.3.3). The U.S. Constitution uses, for instance, the term 'territory' in relation to those tracts of land 'over which the Government of United States exercises jurisdiction' under the 'Territorial Clause'.[74] As for federal and state governments, 'territory' is replaced by 'land'. No wonder, therefore, that the 'Law of the Land' is synonymous with 'due process of law', which points to the well-bounded order of constitutional liberties.[75]

Finally, territorial constitutional silences (and regimes) may conceal highly controversial situations lacking territorial congruence. Different agencies may territorialise the same tract of land, and therefore interpret it in conflicting ways. 'Territorial contestation'[76] causes flaws between a given constitutional regime and its operational rules. Two or more states, even groups of insurgents, may consider a territory as *the* constitutive feature of their constitutional legal systems. Secession threats, territorial disputes, and geopolitical changes affect state territorial integrity.

Formalistic readings of constitutional regimes fall prey to Agnew's territorial trap, according to which each tract of land is meaningful (and therefore assigned) to only one political authority to the exclusion of all the others. Congruence is preserved by legally ignoring the flaws between constitutional regimes and operational rules. A state may lose control over both the disputed territory and its population; yet, it is still part of the constitutional regime. In terms of legal geography, these flaws lead to the *creation of new boundaries* and new *operational*

[72] See Doyle (2018).

[73] See Art. IV s 3 cl 1 US Const.; s 146 Constitution Act of Canada, 1867 and Newfoundland Act 1949; s 121 Commonwealth of Australia Constitution Act 1900; s 2(a) Const. Malaysia.

[74] Art. IV, s. 3, cl. 2. See Bacon (1901), p. 101.

[75] See Mullen (1953), pp. 736–737.

[76] Doyle (2018), p. 901.

regimes, generating new relationships between space, law, and territory. Processes of spatial production like these are '*constitutive* of [new legal] orders'.[77]

Separated politically and territorially, several organised communities claim *dominium* over their respective *imagined* territorial spaces. The most vivid example is Cyprus. The 1974 Turkish invasion split the island into two parts; the north fell under a new operational rule that superseded the regime enshrined in the 1960 Constitution; (see Sect. 6.1). The same occurs in Taiwan, whose 1947 Constitution still refers to the pre-1949 territorial arrangements (Art. 4). The establishment of the People's Republic of China (1949) made these claims 'illusory' and 'fictional.' Adapting the constitutional borders to the current situation, however, would entail 'renouncing the 'One China' legacy by claiming independence from continental China.[78]

Likewise, in Argentina, the territorial constitutional regime still projects its imaginative geography onto the Falkland Islands (*Islas Malvinas*), South Georgia (*Georgias del Sur*), the South Sandwich Islands (*Sandwich del Sur*), and their dependencies (First Additional Provision of the Constitution). The approach towards these territories is patent. Maintaining the constitutional reference to them entails reaffirming Argentina's *imagined* territory and production of its spatial normativity over them.

The cross-disciplinary reading of the constitutional interactions between law and geography eludes the territorial trap and acknowledges the presence of multiple spatialities. And different borders: besides the *de jure* one, there is also a *de facto* border running along the 'Green Line' (in Cyprus), the Taiwan Strait (in Taiwan), or encompassing the whole of Argentina's imaginative geography. In formalistic terms, the outcomes are 'illegal' spaces and borders (the north occupied by Turkey; mainland China; British Overseas Territories in Antarctica). However, they draw an illegal-albeit-legally-relevant *imagined* space; interferences and influences cross illegal borders, which are not in themselves obstacles to communication. The Israeli-Palestinian situation, which is a 'legal' geography of 'illegal' spaces, and where the relations between community and territory are extremely controversial (see Sect. 6.2).

2.4.2 Historical Interactions

Law and geography also interact at the historical and ideological levels. Political power has always sought to make territory meaningful through spatial production. It has also coded its own desired social orders through the law and suffused these manufactured places with a set of 'hierarchical relations [...] objectified by the

[77] Delaney (2003), p. 10.
[78] Liu (2014), p. 52.

territorial fixation'.[79] Formalistic law has been servient to process these spaces as geographies of domination.

Take, for example, the origins of modern legal geography. In the 1980s, Nicholas Blomley explored the enforcement of retail legislation in England and Wales.[80] He soon realised that 'the law was implicated in all sorts of dynamic social relations'. This piece of legislation was 'supposed to be uniform', yet 'spatial variation' featured its implementation. Bearing the responsibility to enforce it, local districts had adopted differentiated approaches which ranged from aggressive to more relaxed enforcement policies, 'even turning a blind eye to [its] violations'. Blomley assumed that the 'law ... was presented as an extra-local obligation' to be implemented locally. Far from being a pure regulatory force, it conformed to 'local dynamics' contributing to the production of local spaces.

A critical examination of how law, power, and the use of geography intersect destabilises 'the normativity and objectivity of law',[81] unveiling how space is produced through 'the twin imperialisms of law and geography'.[82] Quite unsurprisingly, legal-geographical representations of the earth are ideologically saturated; spatial legal practices draw hierarchies and 'geopolitics of social life'; and the law enforces several spatialities through colonial domination, racial segregation, and urban environments that are hostile to immigrants (see Sect. 5.3).

Geographical features are also deployed to critically examine legal cultures. David N Livingstone tells us that French lawyer Jean Bodin (1530–1596) attributed 'the global patterns of ethnological', as well as legal, 'diversity to the influence of environment, particularly climate'.[83] 'The laws ought to be relative . . . to the variety of those tempers', warned Montesquieu in *The Spirit of the Laws* (1748). His 'relativist', and highly contextual, approach to legal-climatic interactions conceal several drawbacks, which forestalls nineteenth-century environmental determinism, according to which 'landscape and climate [are] wholly productive of certain forms of social and cultural organization'.[84] At the same time, Montesquieu is the forebear of relativism in comparative law: the production of cultural- and environmental-specific legal spaces makes each system unique, thus acting as a hindrance for legal transplants and borrowing.[85] There are then hints of ethnocentrism, such as in his analysis of, say, the 'Immutability of ... Laws ... in Eastern Countries' or the unfitness of Muhammadan law for Europe. Orientalist biases are also present, inasmuch as Oriental despotism is opposed to (we assume European) monarchies,

---

[79] Brighenti (2006), pp. 75–76.

[80] See the Shops Act 1950 (Eliz II Ch 28). Quotations and further references are from Blomley (2003), pp. 17–19.

[81] Jones (2003), p. 186.

[82] Delaney (2003), p. 69.

[83] Livingstone (1992), p. 97.

[84] Nayak and Jeffrey (2013), p. 21; Coleman and Agnew (2018), p. 11.

[85] Book XIV, Chapter I. Pierre Legrand is the champion of contemporary relativism: see Legrand (1996, 1997). See also Kedar (2014), pp. 99–101.

aristocracies, and democracies as far as slavery is concerned.[86] This ideologically motivated use of 'comparative law and climate', indeed, is servient to rank political governments after an ethnocentric approach that resonates with that applied when classifying the sundry systems of the world (see Sect. 4.6).[87]

That geography influences legal cultures is also at the heart of John H. Wigmore's *A Panorama of the World's Legal System* (1930); in considering geographical features as 'part of the backdrop against which legal systems evolved', he did not pay adequate attention to how law and geography are spatially co-constitutive.[88] Like Montesquieu, he assumed that natural geography had immediate influence on legal systems.[89] Although he acknowledged that 'it was virtually impossible to disentangle [natural geographical influence] from other factors contributing to legal development', he did not grasp what Soja would explain five decades after, i.e. that such an entanglement is the pattern leading to the construction of space.[90]

Likewise, Marc Desserteaux's *Droit comparé et géographie humaine* (1947) and Bernhard Grossfeld's *Geography and Law* (1988) assume that geography influences legal cultures.[91] Unlike Wigmore, though, they consider them mutually co-constitutive. Grossfeld lists several examples of laws acquiring 'a characteristic coloring from the change of environment',[92] such as water laws. There were obstacles to transplanting the common-law riparian rights theory from a country rich in water (England) to the arid, and dry, State of Colorado. The U.S. common law had to develop a new rule, that departed from that devised in England. This new rule is the 'law of appropriation': *prior in tempore, potior in iure*. With the words of the Supreme Court of Colorado (1879): 'We conclude, then, that the common law doctrine giving the riparian owner a right to the flow of water in its natural channel upon and over his lands, even though he makes no beneficial use thereof, is inapplicable to Colorado. Imperative necessity, unknown to the countries which gave it birth, compels the recognition of another doctrine in conflict therewith.'[93]

Grossfeld also borrows an intriguing, but lengthy, quotation from Pascal, where the geographical colour of the law is addressed 'sarcastically'. 'We see neither justice nor injustice which does not change its nature with change in climate. Three degrees of latitude reverse all jurisprudence; a meridian decides the truth.

---

[86] Respectively: Book XIV Chapters IV and X; Book XV Chapters 1–2.

[87] See also Siems (2022), p. 370; Husa (2015), p. 174.

[88] Blacksell et al. (1986), p. 377.

[89] Wigmore (1928), I.5.

[90] Blacksell et al. (1986), p. 377.

[91] Desserteaux (1947) and Grossfeld (1984).

[92] *Javellana v. Mirasol*, 40 Phil. Rpts. 761, 775 (Phil. Sup. Ct. 1920) also quoted in Grossfeld (1984), p. 1513.

[93] *Coffin v Left Hand Ditch Co.*, 6 Colo. 443, 447 (1882) also quoted in Grossfeld (1984), p. 1514, where are recorded several changes in the common-law rules once transplanted into the United States.

[. . .] A strange justice that is bounded by a river! Truth on this side of the Pyrenees, error on the other side.'[94]

The reference to the Pyrenees is particularly interesting; owing to their being the 'natural' border between the Kingdoms of Spain and France, they are an example of modern bordering processes (1659–1660). During the Middle Ages, the politico-legal organising theme which is feudalism was more familiar with 'areas' than with lines. Mediaeval 'areal' order was superseded by the Westphalian order and its equation between sovereignty and territory (see Sect. 3.5.1).

Here lie the origins of the conceptual mistakes present in the territorial trap. The rise of sovereignty led to the territorial fixation of borders, which outlined the formalistic domestic-foreign opposition, as well as the relativistic idea of universal justice ridiculed by Pascal. Well-demarcated states were spatial 'containers' to be filled. The territorial trap added new sense to territory, which became the *imagined* entity of an *imagined community*. This resonates with Benedict Anderson's theory of *Imagined Communities*; the decline of Mediaeval empires and their porous borders were replaced by new normative spatialities; and political power made territory meaningful as an imagined national entity. Inaugurated by the 1648 Treaty of Westphalia, the spatial container which is the state was saturated with the ideology of nationalism, i.e. a new belief capable of securing the desired futures of communities within 'each square centimetre of a legally demarcated territory'.[95]

I shall revisit how European powers projected linear borders onto America, Africa, and Asia (see Sect. 3.6.1). In the aftermath of WWII, decolonialisation confirmed them, as well as the 'hidden geography' and 'trap' into which European powers had led the world during the colonial era. Former colonial empires split into several nation states. When meeting in Addis Ababa (1963) to create the OAU, the representatives of the newly independent African states conformed to colonial borders. The 2000 Constitutive Act of the African Union does the same. The 'hidden geography' of territory gained new ground after the Cold War, when the dissolution of the Communist block and the collapse of the Iron Curtain led to the creation of new independent states carved out of former multinational entities (Yugoslavia, Czechoslovakia, and the USSR). Its revival is a constant in bordering processes. Demarcated upon ethnic, religious, and cultural lines, the trap still presides over the creation of several states, such as Kosovo, East Timor, and South Sudan, to mention just a few.

2.4.3 Ideological Interactions

The historical interrelations between law and geography point to a further legacy of the hidden geography of territory, i.e. their ideological interactions.

[94] Pascal [1670] (1958), p. 83 (fragment 294).

[95] Anderson (1991), p. 19. See also Brotton (2013), p. 335.

My cross-disciplinary approach discloses how these ideologies are intertwined with the territorial trap. The point is this: ideological interactions between law and geography are servient to paradigms of domination, which assert inequality between different groups, and in which one group endeavours 'to impose its cultural order to the subordinate' ones.[96] Said's *Orientalism* clearly explains how it is possible to draw lines separating 'superior' and 'inferior' legal cultural spaces. Again, orientalism and colonialism—now, globalisation—are species of the broader paradigm of domination.

Geographies of power have provided several 'ideological tools':[97] the extensive dominion of science in the acquisition of world knowledge; a Westernised scientific approach to geography; the advent of cartography. The territorial map fits the idea that geography is 'the art of the mappable', which sanctions the superiority of the Western, i.e. rational, conception of geography. On maps, for example, 'the size of the temperate zones' is expanded, thus placing Europe at a 'centre stage, at the top and centre of the world'.[98]

Radiating from Europe, colonial powers reimagined and appropriated geography and space for radically new purposes. Among the others, the establishment of a European-centred world geopolitics: 'as Europe moved itself outwards, its sense of cultural', as well as political, 'strength was fortified'.[99] The law supplied further ideological tools functional to secure its domination. As a nascent discipline, comparative law conveyed a cartographic visualisation of the several legal systems of the world. Although keenly interested in legal mapping, it exhibited a colonial attitude towards non-Western conceptions of the law. Its ethnocentric approach advocated the superiority of European legal paradigms, favouring the transplant of the hidden geography of the territorial trap outside Europe (see Sect. 4.1).

The territorial trap receives a further ideological application within the spaces of globalisation. Global markets require new processes of spatial manufacturing; commercial transactions surpass state borders, which means refashioning the world after politico-legal uniformity. Legal geographers must critically explore the flattened and smooth transnational, global, and digital spaces (see Sects. 8.3 and 8.4). Behind them, multiple spatialities and legal discontinuities still disturb the uniform patterns of globalisation. These new spaces have indeed reinforced the permanent dialectic between geophysical discontinuity (namely, the presence of borders) and the legal connections caused by globalisation (see Sect. 8.2.2). Certainly, globalisation disregards the role territory has traditionally played in organising communities, because communities and territories are striations upon which globalisation produces ostensible effects. Here lies a further flaw in global (transnational) legal scholarship. The narrative of globalisation is a narrative of superiority, which covers an uneven relationship between the actors of globalisation (see Sect. 4.6).

---

[96] Merry (2003), p. 890.

[97] Bonnett (2008), p. 93.

[98] Bonnett (2008), p. 93. See also Sect. 4.1.

[99] Said [1978] (2003), p. 117.

Nothing new under the sun, then. Globalisation and colonialism share common features: the epistemologies and hierarchies underpinning the processes of domination. Like in the nineteenth century, traders and agents are still entrusted with a kind of civilising mission, which was (and still is) the main justification used to exploit labour force and natural resources that are territorially located in areas with—in Eurocentric terms—limited economic development (see Sect. 4.2).

2.5 A 'True Interdisciplinary Project'. The Emergence of Legal Geography

In the previous sections, I have engaged in a critical assessment of the 'special relationship' between law and geography. My primary purpose has been to revise their interactions in terms of legal-spatial production. I have discarded the regulatory approach to law and geography and adopted a more contextual, and performative, posture. This has meant challenging the 'approaches to law which idealize' its 'separateness, rationality, and reflexivity, and which portray' it 'as deaf to material, physical, spatial, and cultural influences'.[100] I have also run counter to the 'internalist' and 'isolationist' attitude of both lawyers and geographers, circumventing their 'deep suspicion of that deemed to lie outside [their academic] boundaries'.[101] I have focused both on spatial production and the role of power in legal-geographical interactions. Consequently, I have read the legal in terms of the spatial, and vice versa. Finally, I have disclosed how spatial engagements are often ideologically saturated, as the concepts of 'geographies of power' and 'territorial trap' still demonstrate.

I noted that comparative law and geography are well-equipped for cross-disciplinary investigations. Their empirical potentials are particularly apt for assessing their reciprocal production. At the same time, their interplay engages in a dialogue with other disciplines. To mention just a few, Said's *Orientalism* and the spatial turn are the outcomes of such fruitful encounters, also accounting for legal-geographical interactions as open textures. Anthropology has joined the interdisciplinary project of law and geography,[102] which has also been examined through the lenses of history and literature. There is the conversation within the scientific ambit that falls under the umbrella of 'Island studies', which focuses on islands on their own terms (so-called *nissology*) across the disciplines.[103] Island Studies are at the crossroads of several academic disciplines and examine many facets of islands: their shape, size, flora, and fauna, and islanders' life, i.e. their economic development, politics, communications, and services. There has been a sort of 'relational turn' in

[100] Holder and Harrison (2003), p. 3. See also Friedman (2019), p. 326; Monateri (2021), p. 49.

[101] Agnew et al. (1996), p. 7.

[102] See e.g., Benda-Beckman and Benda-Beckman (2014).

[103] See Baldacchino (2004, 2007, 2019); Depraetere (2008), pp. 3–16; Ginoza (2020).

island studies, which has shifted from the figure of insularity and remote, and often isolated, islands to archipelagic connections: islands, oceans, and ships cannot 'be reductively conceptualised in isolation, because they are often inextricably interwoven into complex, multifaceted and shifting arrays of relations and assemblages'. Such a 'relational turn' in island studies 'blurs the boundaries between land and sea.'[104]

Likewise, this relational turn is present in three recent cross-disciplinary collections, *Islands and insularity: between law, geography, and fiction* (2020); *Geographical Connections: Law, Islands, and Remoteness* (2021); and *Islands in Geography, Law, and Literature. A Cross-Disciplinary Journey* (2022). There, islands act as metaphors of the intersections between law, geography, and literature. Lying at the crossroads of several areas of research, they offer an insight into how societies have manufactured and organised their spatial normativities throughout history. The entanglements of different scientific perspectives place their investigations 'at the crossroads of an emerging ... field concerning the cross-disciplinary analysis of islands and insularity'.[105]

A further imaginative metaphor is supplied by Manderson and van Rijswijk's *Introduction to Littoral Readings: Representations of Land and Sea in Law, Literature, and Geography* (2015). The collection compares cross-disciplinary interactions to 'the inter-tidal waters that ebb and flow between law, literature, and geography'. Not only does the collection contain references to Schmitt's *The* Nomos *of the Earth*; but the image of the littoral points to processes of spatial production, which political power codes through the law and makes legally, spatially, and territorially meaningful. Tidal waters reflect the porosity of the disciplinary boundaries of both law and geography. The ebbs and flows also portray the bordering and re-bordering processes that both disciplines have undergone since their academic institutionalisation. This accounts for the dynamics of legal spaces; these are manufactured, reshaped, and processed on 'terrestrial, riparian, maritime, and finally oceanic basis'.[106]

Likewise, in a recent edited collection on the *Laws of the sea*, Irus Braverman peruses the 'interdisciplinary currents' of the oceans by gathering scholars from sundry disciplines such as law, geography, anthropology, and environmental studies. Her ambition is to go beyond the binary divide between law and sea so as to detect processes of spatial production that recreate 'amphibious legal geographies' as a 'call to recognize the shared material and symbolic qualities of both land and sea and our continued existence as an "edge species" that thrives not only *by* the sea but also *with* the sea'.[107] This is a process of maritime spatial production that I shall investigate in terms of legal hydrology and legal geopolitics (see Sects. 3.5.2 and 4.3).

[104]Pugh (2016).

[105]Nicolini and Perrin (2020), p. 7; Nicolini and Perrin (2021); Battisti et al. (2022).

[106]Manderson and van Rijswijk (2015), pp. 167–168.

[107]Manderson and van Rijswijk (2015), pp. 167–168. 'Edge species' is in Gills (2012), p. 198.

Legal-geographical studies like these allow us to represent the authoritative meaning hinted at by the law within the processes of spatial production, describing legal geography as 'a truly interdisciplinary intellectual project'.[108]

Yet, this project lacks a clear academic status. Firstly, it has been argued that our 'transdisciplinary project' is 'trapped' by academic fissure that opposes law and space. Legal and geographical scholars examine how law and geography intersect but then turn to monodisciplinary investigations, triggering what Delaney labels as the 'archipelagic feel' to legal geography.[109] Secondly, legal geography has 'never been the monopoly of professional geographers or legal scholars'.[110] Academics with discrete backgrounds have joined it. Thirdly, the 'ebbs and flows' caused by the interweaving of different disciplines have widened its scope, which has a 'remarkable thematic breadth'.[111] In expanding its scope, though, legal geography now extends over an extremely differentiated set of topics, such as non-human geographies, posthumanism, law and nature, and animal geographies.[112] Fourthly, the proliferation of legal geographical studies has overshadowed the methodological issues. Of these, several are left unanswered; whether legal geography is a 'coherent enough body of scholarship'; whether 'itinerant participants whose primary intellectual concerns are elsewhere' might contribute to the methodological debate in law and geography; whether the 'self-declaration of identity' as legal geographical scholars is a pre-requisite for joining the cross-disciplinary project.[113]

Three major points deserve to be considered.

The first of these refers to *jurisdictions*. Within legal geography, the mainstream demarcation 'roughly corresponds to where' legal geographers 'live and work'.[114] Mainly operating in Anglophone jurisdictions, they are familiar to legal-spatial production taking place, say, in Canada, the USA, and Australia. Undoubtedly, legal geography should explore new avenues above and beyond the common law jurisdictions and therefore engage in closer dialogues with legal geographers operating in discrete legal regimes and contexts, both Western and non-Western. The process of 'de-provincialising Western Law' is currently on the right track; legal geographers have also started considering contexts where the production of spatial normativies is plural and environments are governed by non-Western conceptions of the law. These analyses, though, have often been conducted by either merely documenting diverse (and often incommensurable) legal realities or applying our distinctive Western 'cultural habit' (i.e. Ethnocentrism). When, for example,

---

[108] Braverman (2022), p. 2.

[109] Delaney (2010), p. 12.

[110] Braverman et al. (2014), p. 10.

[111] Orzeck and Hae (2020), p. 832. See also Delaney (2017).

[112] For more on these topics see Braverman et al. (2014), Ojalammi and Blomley (2015), Orzeck and Hae (2020), and Bartel and Carter (2021).

[113] Respectively: Delaney (2010); Hatcher (2012); Braverman et al. (2014), p. 2; Orzeck and Hae (2020), p. 832.

[114] Delaney (2017), p. 668. On Australia see Bartel et al. (2013).

Josephine Gillespie examines land property rights in Cambodia, she seems to unconsciously share Montesquieu's hierarchies and epistemologies. It is as if the production of cultural- and environmental-specific (i.e. non-Western) legal spaces acts as a hindrance for transplanting the common-law 'dominant property narratives' to Cambodia. These narratives do not necessarily fit into 'other social and physical contexts, particularly in developing countries'.[115] The way comparative law contributes to spatial production is particularly productive here. On the one hand, it points to the 'sustainable diversity' of the legal systems of the world, thus critically examining their ideological underpinnings. On the other hand, in examining how spatial differentiation takes place, it considers whether (and to what extent) Western legal derivatives pragmatically enrich or alter non-Western processes of spatial production.[116]

The second point that is worthy of consideration is *disciplinary construction*. In their introductory Chapter to *The Expanding Spaces of Law*, Irus Braverman, Nicholas Blomley, David Delaney, and Alexandre Kedar list the stages of development whereby the legal-geographical project has evolved over time.[117] The first phase is 'import and export'. The presence of disciplinary divides allowed legal scholars to find 'space without having found geography' and geographers to pay limited attention to the field of the law.[118] At the same time, geographers rediscovered the nation state as the core problem in political geography but assumed that the 'legal' was a constant in several areas of human geography, closely intertwined with social contexts, such as redistricting and urbanism, which meant neglecting its performative role in spatial production.[119] The second stage of development saw legal geography undergo a process of 'cross-disciplinarisation'. Deserting the formalistic approach to both law and space, critical legal scholars and critical geographers started mapping the implications of law and geography, as well as considering the ideological implications of the legal geographies power.[120] The third stage refers to the establishment of a truly cross-disciplinary project, with its own specialised lexicon, as David Delaney's *The Spatial, the Legal and the Pragmatics of World-Making* (2010) demonstrates. There, he coins the term 'nomosphere'; again, a loan from Schmitt's *The* Nomos *of the Earth* describes the processes of legal and spatial co-production. A further lexeme is represented by 'nomic settings', i.e. social constructs like territory, place, and space that become legally meaningful 'by way of inscription or assignment of legal meanings'.[121]

[115] Gillespie (2016), p. 3. See also Hatcher (2015).

[116] On the sustainable diversity of comparative law see Glenn (2014).

[117] A historical account on how legal geography has evolved over time see also Delaney (2010), pp. 9–14. For more on the development of the special relationship between law and geography see Blacksell et al. (1986).

[118] Blomley (1989).

[119] Blomley (1989). For more on such mutual See also Clark (1989), pp. 315–318.

[120] Blomley (1994).

[121] Delaney (2010), p. 59. See Nayak and Jeffrey (2013), pp. 285–266. See Knight (1982), p. 517.

Processes assigning legal meaning to nomic settings cannot be reduced to pure regulatory practices. In our cross-disciplinary project, the law is first and foremost a performative practice constitutive of normative spatialities. If I understand Delaney's predicament correctly, nomic settings become meaningful through spatial production and are subsequently coded through legal devices deeply connected with these contexts.

The regulatory practice of the law is probably represented by the third lexeme used by Delaney, the *nomoscape*. Nomic settings are constellations assembled by 'pervasive cultural processes and forces associated with ideological projects'. These are precisely what we have termed ideological interactions. No wonder, therefore, that several ideologies may act at the same time within the same space; several nomoscapes may encroach upon the others, thereby giving a clear ideological consistency to the constellations of nomic settings.[122] This overlapping of nomoscapes takes place in multinational federations, where different identities and interest encroach upon the same territory; this leads to what I will term as the demarcation and delimitation processes, which aim to govern a national society which is both ethnically and economically divided, thus paving the way to multi-level/territorial identities (see Sect. 7.4.4).

2.6 Creating Spatial Normativities

2.6.1 Thomas Kuhn's Paradigm and Legal Geography

The third point that deserves our attention operates at the epistemological level. We have already noted that comparative law and geography aim to gain control, and therefore order, the knowledge of the world. Their ultimate ambition is the production and the growth of knowledge that might be useful for scientific advancements. This predicament has always been part of the geographical discourse, which requires a 'social and conceptual framework' capable of setting adequate 'standards for relevant research'.[123] Several geographers—in particular, historical geographers—have found these standards in Thomas Kuhn's idea of *paradigm*. His *The Structure of Scientific Revolutions* sets, in broader terms, 'the framework within which a science proceeds'.[124]

Explaining the basic features of geographical research through the history of science had several discontents and gave rise to manifold difficulties. Kuhn's paradigm appears too simplistic to fit into geographical patterns; the spatial and cultural 'turns' of the 1990s made it even more difficult to mechanically transplant

[122] Delaney (2010), p. 100; Braverman et al. (2014), p. 11; Orzeck and Hae (2020), p. 839.

[123] Livingstone (1992), p. 13.

[124] Kuhn [1962] (1970). The quotation is from Johnston (1978). After a warm reception in geographical research (see Livingstone 1984, 272n 11 and 1990; Johnston 1978), its use progressively declined in the 1980s and 1990s.

Kuhn's paradigm into geographical studies. The idea of paradigm has also been deployed to trigger transformative changes in the field of geography, making it a scientifically respectable academic discipline. This 'reform' of geography resonates with the mathematical turn in comparative law (see Sect. 2.3); it is also in line with the assumption that hard-science paradigms are more scientific and objective than those practised in social sciences and the humanities. Yet, the extent to which scope and purposes of scientific and geographical research coincide are still a matter of scholarly disagreement. A prescriptive application of Kuhn's model may lead to considering geography a type of scientific 'revolutionary progress'—yet geography cannot be understood 'in the same terms of scientific [i.e. hard-science] knowledge'.[125]

What favoured the warm reception of Kuhn's paradigm was its ability to link the different stages around which scientific research may be arranged. For our purposes, the paradigm gives consistency to the cross-disciplinary intellectual project which is legal geography. Johnston's essay *Paradigms and Revolution or Evolution?* lists these stages as follows: (1) 'the accepted facts', (2) the 'puzzles which remain to be solved', and (3) 'the procedures by which solutions to the puzzles are sought'.[126]

Let us begin with 'the accepted facts'. Kuhn's paradigm points to the description and explanation of the growth of scientific knowledge. As said, this objective is shared by both comparative law and geography. The same applies to legal geography, whose goal is 'the production of holistic knowledge about the place and function of law' in present and past societies.[127] The second stage points to the 'puzzles which remain to be solved', i.e. the aims of scientific endeavours. In legal geography, these focus on what we have labelled *normative spatialities*, i.e. the processes of spatial production and differentiation, as well as 'the structural conditions' within which 'human agency [. . .] operates' and makes the space legally meaningful.[128] The third stage comprises of 'the procedures by which solutions to the puzzles are sought'. Spatial production takes place within contexts that are an admixture of discrete variables; consequently, the examination of their operational rules requires a wide range of approaches which fit into the rich subject of legal geography and the diversity of its findings.

2.6.2 *The Territorialised Paradigm of Legal Geography*

A further feature of Kuhn's paradigm appealed to geographers, i.e. its ability to match cross-disciplinary research. As Livingstone puts it, the idea of paradigm

[125] For such critical assessments see e.g. Young (1979), Wheeler (1982) and Mair (1986).

[126] Johnston (1978).

[127] Orzeck and Hae (2020), p. 833.

[128] Orzeck and Hae (2020), p. 833.

'introduced a thoroughly relativist note into the philosophy of science'.[129] This reminds us of the idea of 'paradigm shift', which points to the changeover to new scientific programmes and models, as well as to the criteria and procedures (the methodology) applied within a given epistemic community.

We have already examined how spatial normativities are made meaningful (and therefore 'are *lived*' by its agents, as Delaney puts it) through the processes of spatial production. Through these, political power encodes its own hierarchies within an enclosed space thus making it politically meaningful. A tract of land becomes a bounded territory; in Said's *Orientalism*, the East is turned into a legal space that reflects Western geographies of domination; Delaney's nomic settings are made meaningful through processes of legal and spatial production. And, quite evidently, the same goes valid for Agnew's territorial trap. Bounded, as it is, the notion of territory denotes 'a calculative matrix'; which serves 'specific social functions associated with modern norm and institutions' that are replicable on the Earth's surface irrespective of time and space.[130]

There are hints of 'territoriality' in all these processes. Indeed, the idea of making spaces meaningful for human agencies resonates with patterns of social and political appropriation. At least at an imaginative level, the space is supposed to belong to the human agency that has made it meaningful. It might come as a surprise, but the German verb *erreichen* (which means 'to reach', 'to be able to touch', and 'to get hold of' something) shares with *Reich* (state) and *Bereich* (realm, domain) the same lexeme and the related idea of physical apprehension—and political appropriation. Although the effective control over these spaces entails demarcating them to the exclusion of other human agencies, these political and social orders cannot be reduced to the 'territorial trap'. There is nothing ideological in such processes of apprehension and control; rather, there is only the human need to make the context in which we live meaningful and useful to us. Again, these hints of territoriality represent, with Kuhn's parlance, the basic paradigm whereby we, as humans, attempt to grasp and take control of the knowledge of the world.

In Kuhn's terms, there is consensus around this 'disciplinary matrix'; the territorialised paradigm is a cultural competence, a faculty of mind, a capacity for comprehension and synthesising the knowledge of the world. It is not, in Lyotard's words, a *grand narrative*, or one of the 'metadiscourses [. . .] which have provided ideologies with a legitimating philosophy of history'. If it were the case, this would undermine the relativist tone of legal geographical methodology.[131] The matrix gives consistency to our scientific efforts to identify the contingencies, variables, and operational rules of spatial production. And this matrix undoubtedly accounts for the complexity of the relations between law, geography, power, and territory.

Nulle terre sans seigneur, nul seigneur sans titre: this seems the historical matrix of the processes of spatial production. This paradigm was originally a

[129] Livingstone (1992), p. 14.

[130] Steinberg (2018), p. 218. See Elden (2013).

[131] Chandler and Munday (2011). See Lyotard (1979).

'territorialised' paradigm. Whether imaginative or physical, the process of apprehension νέμει (i.e. divides) and creates differentiated spaces, as well as their νόμος (the law). When someone gets hold of the spaces thus produced, *that* someone attaches a politico-legal meaningful order to them; and it was generally assumed that *that* someone 'owned them'. With these words the English legal historian Maitland explains how social orders become spatially and geographically visible, and how the territorialised paradigm turns law and geography into *legal geography*.

Maitland was probably the first scholar to use the expression legal geography in his *Township and Borough: The Ford Lectures 1897*. But this went unnoticed. His course of lectures examined a legal dispute arising after the West Fields, that laid 'immediately outside the town of Cambridge', had been enclosed under an Act of Parliament in 1802.[132] In directing the enclosure, the Act assigned several 'ring-fenced plots' to the owners of the strips that had lain 'intermixed' in the common arable fields (University, Colleges, Municipal Corporations, and Ecclesiastical Bodies). Allotments were also made to those who had been exercising rights of pasture or had been owners of the tithe thereon.[133] Yet, Maitland reminds us, 'dispersed among the arable strips there were many small pieces of waste land [...] Who owned them? Who could claim an allotment in their stead?' Litigation before the Commissioners of the Enclosure soon ensued; and the municipal corporation of Cambridge was declared 'lord of this field'. Together with the Enclosure Act of 1807 related to the Barnwell (or East) Fields,[134] Maitland tells us that the litigation under scrutiny 'decided a question, or one part of a question, that had been simmering for centuries'.

What has been simmering for centuries was—probably—the effort to restore the thorough application of the operational rules governing legal spatial production in Cambridgeshire. Nothing is more human than making the context meaningful.[135] In an attempt to take control of the knowledge of Cambridgeshire, the matrix helped to disentangle the complex relations between law, geography, power, and territory, which Maitland termed 'legal geography'.

Like the feudal adage *Nulle terre sans seigneur, nul seigneur sans titre*, Maitland's usage of legal geography is twofold. On the one hand, it reflects the operational rule of the territorialised paradigm in Mediaeval England: 'Time was when we in England had a respectably neat system of legal geography [...] the whole country (this was the theory, if not precisely the fact) was cut up into vills or towns. The law assumed that every acre of land lay in some town'. 'Their boundaries', Maitland continues, 'have provided wide enough room for fields and meadows and pastures'. Clearly, *nulle terre sans seigneur*. On the other hand,

---

[132] An Act for dividing, allotting, laying in severalty and inclosing the open and common fields, common meadows and other open and commonable lands and waste grounds, within the Parish of St Giles, in the Town of Cambridge, in the County of Cambridge, 42 GEO III, 1802.

[133] Maitland [1897] (1964), p. 2. See also Guillebaud (2005).

[134] Act of Parliament 47 George III, session 2, cap 60, 1807 –Act for enclosing the Parish of St Andrew the Less or Barnwell with parts of St Andrew the Great, St Mary the Great and St Mary the Less, Cambridge.

[135] Gottmann (1975), p. 29.

Maitland states that 'this was the theory'. In other words, he acknowledges that the scope of the territorialised paradigm is broader and can therefore be generalised: 'the relationship of the town community to the town lands, the relationship of the oldest *burgenses* to arable strips and green commons' defined 'legal geography' as the relationship between community and *its* territory.[136]

The process of imaginative, political, and factual apprehension soon ensued: *nul seigneur sans titre*. Maitland extends the paradigm to various actors: tribes, clans, families, villages, and ethnicities are claimants asserting an exclusive close relationship with a specific territory. In a court of law, Maitland reminds us, they were 'bound to name the *villa* in which [the acre of land] lay'; which means, for our purposes, naming the politico-legal (or imaginative) order that made the land meaningful to the claimant. No wonder, therefore, that Maitland labels this legal-spatial relationship as 'belongs of public law'.[137] In these 'belongs', concepts like ownership, property, lordship, rulership, and sovereignty merge in a matrix that transcends time and space: within the territorialised paradigm, ancestral physical apprehension, colonial domination, modern political appropriation, and globalisation may be 'regarded as being very like any other belongs'.

This explains how the territorialised matrix has incrementally evolved over time, and why it has been applied to the most diverse processes of spatial production. As humans, we have rearranged knowledge supplied by tradition to new contexts, variables, and contingencies. Like in Kuhn's 'normal' phases of science, the varieties of actors, methods, variables coded through the law generate a 'cumulative picture of scientific progress'. This picture accounts for the unceasing regeneration of the territorialised paradigm. And this demonstrates that, at least as a nascent field of research, legal geography has a 'reasonable chance of solving the puzzle', which also depends on the cumulative additions of self-appointed legal geographers and itinerant contributors.[138]

Throughout this book, I shall interpret the process of spatial production through the lenses of the territorialised paradigm. My point is this: the processes of spatial production, even those of globalisation, are still explicable in terms of 'normal' (and cumulative) phases of science. This cumulative and cross-disciplinary paradigm is still able to provide an explanation for the processes of spatial production. As legal geographers we have to revise our existing methodology to allow our discipline to acquire the status of an autonomous ambit in scientific research.

---

[136] Maitland [1897] (1964), pp. 6–9.

[137] Maitland [1897] (1964), pp. 11 and 29.

[138] Kuhn [1962] (1970), p. 92.

References

Aay H (1981) Textbook chronicles: disciplinary history and the growth of geographic knowledge. In: Blouet BW (ed) The origins of academic geography in the United States. Archon Books, Hamden, CN, pp 291–301

Adams A (2012) Comparing law: practice and theory. In: Adams M, Bomhoff J (eds) Practice and theory in comparative law. CUP, Cambridge, pp 1–21

Agnew J (1994) The territorial trap: the geographical assumptions of international relations theory. Rev Int Polit Econ 1:53–80

Agnew J (2015) Revisiting the territorial trap. Nord Geogr Publ 44:43–48

Agnew J et al (1996) General introduction. In: Human geography. An essential anthology. Blackwell, Malden, MA, pp 1–16

Ajani G (2017) Ruling by indicators: how the use of vague notions and quantitative indicators facilitates legal change. Cardozo Electron Bull:1–24

Anderson B (1991) Imagined communities. Reflections on the origin and spread of nationalism. Verso, London and New York

Bacon S (1901) Territory and the constitution. Yale Law J 10:99–117

Baldacchino G (2004) The coming of age of island studies. Tijdschrift voor economische en sociale geografie 95:272–283

Baldacchino G (ed) (2007) A world of islands: an island studies reader. Institute of Island Studies–University of Prince Edward Island and Agenda Academic, Carlottetown, Canada, and Luqa, Malta

Baldacchino G (ed) (2019) The Routledge international handbook of island studies: a world of islands. Routledge, London and New York

Baldwin A (2014) Pluralising climate change and migration: an argument in favour of open futures. Geogr Compass 8:516–528

Bartel R, Carter J (eds) (2021) Handbook on space, place and law. Edward Elgar, Cheltenham and Northampton, MA

Bartel R et al (2013) Legal geography: an Australian perspective. Geogr Res 51:339–353

Battisti C et al (eds) (2022) Islands in geography, law, and literature. A cross-disciplinary journey. de Gruyter, Berlin and Boston

Benda-Beckman F, Benda-Beckman B (2014) Places that come and go: a legal anthropological perspective on the temporalities of space in plural legal orders. In: Braverman I et al (eds) The expanding spaces of law: a timely legal geography. Stanford University Press, Stanford, pp 30–52

Blacksell M et al (1986) Human geography and law: a case of separate development in social science. Prog Hum Geogr 10:371–396

Blomley (1989) Text and context: rethinking the law-space nexus. Prog Hum Geogr 13:512–534

Blomley N (1994) Law, space, and the geographies of powers. The Guilford Press, New York

Blomley N (2003) From 'what?' to 'so what?': law and geography in retrospect. In: Holder J, Harrison C (eds) Law and geography. OUP, Oxford, pp 17–33

Blomley N (2015) The ties that blind: making fee simple in the British Columbia treaty process. Trans Inst Br Geogr 40:168–179

Blomley N, Bakan JC (1992) Spacing out: towards a critical geography of law. Osgoode Hall Law J 30:661–690

Bonnett A (2008) What is geography? SAGE, London

Braverman I (2022) Amphibious legal geographies: toward Land–Sea Regimes. In: Braverman I (ed) Laws of the sea: interdisciplinary currents. Routledge, Abingdon/New York, pp 1–24

Braverman I et al (2014) Expanding the spaces of law. In: Braverman I et al (eds) The expanding spaces of law: a timely legal geography. Stanford UP, Stanford, pp 1–29

Brighenti A (2006) On territory as relationship and law as territory. Can J Law Soc 21:65–86

Brotton J (2013) A history of the world in twelve maps. Penguin, London

Callon M (1988) An essay on framing and overflowing: economic externalities revisited by sociology. In: Callon M (ed) The laws of the markets. Blackwell, Oxford, pp 244–269
Carlarne C, Farber D (2012) Law beyond borders: transnational responses to global environmental issues. Transnatl Environ Law 1:13–21
Castree N et al (2013) The Oxford dictionary of human geography. OUP, Oxford
Chandler D, Munday R (2011) A dictionary of media and communication, 1st edn. OUP, Oxford
Clark G (1989) The geography of law. In: Peet R, Thrift N (eds) New models of geography, vol 1. Routledge, Abingdon and New York, pp 310–337
Coleman M, Agnew J (eds) (2018) Handbook on the geographies of power. Edward Elgar, Cheltenham and Northampton, MA
Delaney D (2003) Beyond the word: law as a thing of this world. In: Holder J, Harrison C (eds) Law and geography: current legal issues. OUP, Oxford, pp 67–83
Delaney D (2005) Territory. A short introduction. Blackwell, Oxford
Delaney D (2010) The spatial, the legal and the pragmatics of world-making. Blackwell, Oxford
Delaney D (2017) Legal geography III: new worlds, new convergencies. Prog Hum Geogr 41:667–675
Depraetere C (2008) The challenge of nissology: a global outlook on the world Archipelago. Part I: scene setting the world Archipelago. Island Stud J 3:3–16
Desserteaux (1947) Droit comparé et géographie humaine. Annales de Géographie 56(302):81–93
Diener AC, Hagen J (2012) Borders. A very short introduction. OUP, Oxford
Doyle O (2018) The silent constitution of territory. Int J Const Law 16:887–903
Driver F (2001) Geography militant. Cultures of exploration and empires. Blackwell, Oxford
Elden S (2013) The birth of territory. University of Chicago Press, Chicago
Eratosthenes' Geography. English edition: Eratosthenes' Geography (2010) (trans: Roller DW). Princeton UP, Oxford and Princeton
Etty T et al (2018) Transnational climate law. Transnatl Environ Law 7:191–200
Feihle P (2019) Territoriality. In: Grote R, Lachenmann F, Wolfrum R (eds) Max Planck encyclopaedia of comparative constitutional law. OUP, Oxford
Frankenberg G (1985) Critical comparisons: rethinking comparative law. Harv Int Law J 26:411–456
Friedman LM (2019) A history of American law, 4th edn. OUP, New York
Frier BW (1989) Autonomy of law and the origins of the legal profession. Cardozo Law Rev 11: 259–286
Gillespie J (2016) A legal geography of property, tenure, exclusion, and rights in Cambodia: exposing an incongruous property narrative for non-western settings. Geogr Res 54:256–266
Gills JR (2012) The human shore. University of Chicago Press, Chicago
Ginoza A (2020) The challenges of island studies. Springer, Cham
Glenn HP (2014) Legal traditions of the world. Sustainable diversity in law, 5th edn. Oxford, OUP
Goldsmith E (1992) The way: an ecological world view. Rider, London
Gordley J et al (2021) An introduction to the comparative study of private law: readings, cases, materials, 2nd edn. CUP, Cambridge
Gottmann J (1973) The significance of territory. University of Virginia Press, Charlottesville, VA
Gottmann J (1975) The evolution of the concept of territory. Soc Sci Inf 14:29–47
Grossfeld B (1984) Geography and law. Mich Law Rev 82:1510–1519
Guillebaud P (2005) Cambridge St Giles: Cambridge University and the Parliamentary Act of 1802. Proc Camb Antiquarian Soc 94:185–198
Hammond BS (1968) Scriblerian self-fashioning. Yearb Engl Stud 18:108–124
Hangula L (2010) The constitutionality of Namibia's territorial integrity. In: Bösl A et al (eds) Constitutional democracy in Namibia: a critical analysis after two decades. KAS, Windhoek, pp 191–198
Hanlon D, Eperiam W (1988) The evolution and development of the Federated States of Micronesia. Micronesian Polit:85–106

Hatcher C (2012) Reviewed work: the spatial, the legal and the pragmatics of world-making: nomospheric investigations by David Delaney. Geografiska Annaler Ser B Hum Geogr 94: 199–201

Hatcher C (2015) Globalising homeownership: housing privatisation schemes and the private rental sector in post-socialist Bishkek, Kyrgyzstan. Int Dev Plan Rev 37:467–486

Hogg R (2002) Law's other spaces. Law/Text/Culture 6:29–38

Holder J, Harrison C (2003) Connecting law and geography. In: Holder J, Harrison C (eds) Law and geography. OUP, Oxford, pp 3–16

Husa J (2015) A new introduction to comparative law. Hart, Oxford

Husa J (2018) Advanced introduction to law and globalisation. Edward Elgar, Cheltenham-Northampton, MA

Husa J, Van Hoecke M (eds) (2013) Objectivity in law and legal reasoning. Hart, Oxford

Jepson W (2012) Claiming space, claiming water: contested legal geographies of water in South Texas. Ann Assoc Am Geogr 102:614–631

Jimena Quesada (2013) Article 52 TEU. In: Blanke HJ, Mangiameli S (eds) The Treaty on European Union (TEU): a commentary. Springer, Berlin, pp 1433–1449

Johnston RJ (1978) Paradigms and revolutions or evolution?. Observations on human geography since the second world war. Prog Hum Geogr 2:189–206

Johnston RJ (1983) Texts, actors and higher managers: judges, bureaucrats and the political organization of space. Polit Geogr Q 2:3–19

Jones GA (2003) Camels, Chameleons, and Coyotes: problematizing the 'histories' of land law reform. In: Holder J, Harrison C (eds) Law and geography: current legal issues. OUP, Oxford, pp 169–189

Kallai Z (1997) The Patriarchal Boundaries, Canaan and the Land of Israel: patterns and application in Biblica historiography. Isr Explor J 47:69–82

Kamba WJ (1974) Comparative law: a theoretical framework. Int Comp Law Q 23:486–489

Kedar A (2014) Expanding legal geographies: a call for a critical comparative approach. In: Braverman I et al (eds) The expanding spaces of law: a timely legal geography. Stanford UP, Stanford, CA, pp 95–119

Knight DB (1982) Identity and territory: geographical perspectives on nationalism and regionalism. Ann Assoc Am Geogr 72:514–531

Kuhn T [1962] (1970) The structure of scientific revolutions. Chicago UP, Chicago

Lefebvre H (1974) La production de l'espace. Éditions Anthropos, Paris. English edition: Lefebvre H (1991) The production of space (trans: Nicholson-Smith D). Blackwell, Oxford

Legrand P (1996) European legal systems are not converging. Int Comp Law Q 45:52–66

Legrand P (1997) The impossibility of 'legal transplants'. Maastricht J Eur Comp Law 4:111–124

Lempert RO (1988) The autonomy of law: two visions compared. In: Teubner G (ed) Autopoietic law: a new approach to law and society. Walter de Gruyter, Berlin, pp 152–190

Levine G (1987) One culture. In: Levine G (ed) One culture: science and literature. Essays in science and literature. The University of Wisconsin Press, Madison, pp 3–32

Liu JCH (2014) The Taiwan Question: border consciousness intervened, inverted and sideplaced. In: Liu JCH et al (eds) European-East Asian borders in translation. Routledge, Abingdon and New York, pp 38–62

Livingstone DN (1984) The history of science and the history of geography: interactions and implications. Hist Sci 22:271–302

Livingstone DN (1990) Geography, tradition and the scientific revolution: an interpretative essay. Trans Inst Br Geogr 15:359–373

Livingstone DN (1992) The geographical tradition. Blackwell, Oxford

Llewellyn K (1930) Brumble Bush: on our law and its study. Oceana, New York

Lyotard J-F (1979) La condition postmoderne. Les éditions de minuit, Paris

Mackinder HJ (1887) On the scope and methods of geography. Proc R Geogr Soc Monthly Record Geogr 9:141–174

Mair A (1986) Thomas Kuhn and understanding geography. Prog Hum Geogr 10:345–369

Maitland FW (1964) Township and Borough: The Ford Lectures 1897. CUP, Cambridge

Mancini S (2012) Secession and self-determination. In: Rosenfeld M, Sajó A (eds) The Oxford handbook of comparative constitutional law. OUP, Oxford, pp 481–500

Manderson D, van Rijswijk H (2015) Introduction to littoral readings: representations of land and sea in law, literature, and geography. Law Lit 27:167–177

Massey D (1992) Politics and space/time. New Left Rev 196:65–84

Matthews JA, Herbert DT (2008) Geography. A very short introduction. OUP, Oxford

Merry SE (2003) From law and colonialism to law and globalization. Law Soc Inq 28:569–590

Monateri PG (2021) Advanced introduction to comparative legal methods. Edward Elgar, Cheltenham and Northampton, MA

Moore SF (1973) Law and social change: the semi-autonomous social field as an appropriate subject of study. Law Soc Rev 7:719–746

Moore M (2015) A political theory of territory. OUP, Oxford

Mousourakis G (2007) Comparative law and legal universalism: an historical perspective. J Jurid Sci 32:144–162

Mousourakis G (2019) Comparative law and legal traditions. Historical and comparative perspectives. Springer, Cham

Mullen JM (1953) The supreme law of the land. Va Law Rev 39:729–751

Nayak A, Jeffrey A (2013) Geographical thought. An introduction to ideas in human geography. Routledge, Abingdon and New York

Nelken D (2018) The legitimacy of global social indicators: reconfiguring authority, accountability and accuracy. Les Cahiers de Droit 59:35–84

Nicolini M (2020) Methodologies of comparative constitutional law: universalist approach. In: Max Planck Encyclopaedia of Comparative Constitutional Law [MPECCOL]. OUP, Oxford

Nicolini M, Perrin T (eds) (2020) Focus: islands and insularity: between law, geography, and fiction. Pólemos 14:209–336

Nicolini M, Perrin T (eds) (2021) Geographical connections: law, islands, and remoteness. Liverpool Law Rev 42:1–14

Ó Tuathail G, Agnew J (1992) Geopolitics and discourse: practical geopolitical reasoning in American foreign policy. Polit Geogr 11:190–204

O'Donnell T et al (2020) An Australasian and Asia-Pacific approach to legal geography. In: O'Donnell T et al (eds) Legal geography. Perspectives and methods. Routledge, Abingdon and New York, pp 3–15

Ojalammi S, Blomley N (2015) Dancing with wolves: making legal territory in a more-than-human world. Geoforum 62:51–60

Orzeck R, Hae L (2020) Restructuring legal geography. Prog Hum Geogr 44:832–851

Palermo F (2019) Towards a comparative constitutional law of secession? In: Martinico G, Delledonne G (eds) The Canadian constitution to a comparative law of secession. Legacies of the Quebec secession reference. Palgrave MacMillan, London and New York, pp 265–282

Pascal B (1670) Pensées. English edition: Pascal B (1958) Pensées (trans and ed: Eliot TS D). EP Dutton & Co., Inc., New York

Paton LB (1913) Israel's Conquest of Canaan: Presidential Address at the Annual Meeting, Dec. 27, 1912. J Biblic Lit 32:1–53

Peters A (1999) Universalist assumptions and implications of comparative law: should they be deconstructed? Proc Annu Meet (Am Soc Int Law) 93:366–369

Pound R (1910) Law in books and law in action. Am Law Rev 44:12–36

Power M, Sidaway JD (2004) The degeneration of tropical geography. Ann Assoc Am Geogr 94: 585–601

Pugh J (2016) The relational turn in island geographies: bringing together island, sea and ship relations and the case of the landship. Soc Cult Geogr 17:1040–1059

Restrepo Amariles D (2015) Legal indicators, global law and legal pluralism: an introduction. J Leg Plur Unoff Law 47:9–21

Rissanen M (2000) Standardisation and the language of early statutes. In: Wright L (ed) The development of standard English 1300-1800: theories, descriptions, conflicts. CUP, Cambridge, pp 117–130
Robic M-C (2003) Geography. In: Porter TM, Ross (eds) The Cambridge history of science. Vol 7. The modern social sciences. CUP, Cambridge, pp 379–390
Ruskola T (2013) Legal orientalism: China, The United States, and modern law. Harvard Up, Cambridge, MA
Sacco R (1991) Legal formants: a dynamic approach to comparative law (Instalment I of II). Am J Comp Law 39:1–34
Said E [1978] (2003) Orientalism. Penguin, London
Siems M (2022) Comparative law, 3rd edn. OUP, Oxford
Soja EW (1989) Postmodern geographies: the reassertion of space in critical social theory. Verso, New York
Spencer L (2020) Comparative legal geography. Context and place in 'legal transplants'. In: O'Donnell T et al (eds) Legal geography. Perspectives and methods. Routledge, Abingdon and New York, pp 149–166
Steinberg P (2018) The power of water. In: Coleman M, Agnew J (eds) Handbook on the geographies of power. Edward Elgar, Cheltenham and Northampton, MA, pp 217–228
Storey D (ed) (2020) A research agenda for territory and territoriality. Edward Elgar, Cheltenham and Northampton, MA
Watt G (2012) Comparison as deep appreciation. In: Monateri PG (ed) Methods of comparative law. Edward Elgar, Cheltenham-Northampton, MA, pp 82–103
Wheeler PB (1982) Revolutions, research programmes and human geography. Area 14:1–6
Wigmore JH (1928) A panorama of the world's legal systems. 3 vols. West Publishing Company, St Paul, MN
Williams G (1996) 'To make discoveries of countries hitherto unknown'. The admiralty and pacific exploration in the eighteenth century. Mariner's Mirror 82:14–27
Wood D (2010) Rethinking the power of maps. With J. Fels and J. Krygier. The Guildford Press, New York and London
Young RW (1979) Paradigms in geography: implications of Kuhn's interpretation of scientific inquiry. Aust Geogr Stud 17:204–209

Part II
Spatial Normativies

Chapter 3
Law and 'Geographies'

3.1 Manufacturing Spatial Normativities

Part II of this book examines how spatial normativities are manufactured. We have already ascertained how the territorialised paradigm devised by Maitland still percolates spatial production (see Sect. 2.6). In embedding this paradigm, spaces become 'places'—or, as David Delaney puts it, 'nomic settings' (Sect. 2.5).[1] Both expressions reflect the need to make context meaningful to human agencies. When space becomes 'meaningful', it develops into places; these, in turn, act as 'repositories' of shared knowledge, whose 'basic essence . . . lies in the largely unconscious intentionality that defines places as profound centres of human experience'.[2]

'We manipulate space', Ronald J. Johnston warns us, 'in the creation of places', which are 'culturally defined milieux within which economic, social and political decisions are made'.[3]

The idea of place resonates with the territorialised paradigm—at least when this is related to the pristine (and radical) idea of physical and political apprehension. There is something ancestral in this. Place 'involves an appropriation and transformation of space and nature that is inseparable from the reproduction and transformation of society'.[4] Space is infused with experience and knowledge; and its demarcation is an attempt to take control of its societal repository. In asserting their relationship with a specific territory, groups and individuals arrange their own hierarchies within a physical setting, which thus turns out to be a politically meaningful context (see also Sect. 5.2.2).

---

[1] Respectively: Nayak and Jeffrey (2011), p. 59; Delaney (2010), p. 59.

[2] On places as repositories see Relph (1976), p. 43; Johnston (1986), p. 452; Johnston (1991), pp. 50–55; Black (1997), p. 26.

[3] Johnston (1986), p. 542.

[4] Pred (1984), p. 279.

M. Nicolini, *Legal Geography*, Ius Gentium: Comparative Perspectives on Law and Justice 105, https://doi.org/10.1007/978-3-031-19410-8_3

The territorialised paradigm does not necessarily convert all spaces into territorially 'bounded' places. Some of them are well-defined locales; others do not possess clearly demarcated boundaries; other areas may experience a variety of processes of appropriation. Edward Said has demonstrated that, albeit politico-legal, apprehension may also be imaginative. In claiming a territory, an institutionalised group projects their own territorial demarcation on it; and this imaginative order makes the land meaningful to the claimant (see Sect. 2.6).

The way the territorialised paradigm is applied depends on how communities arrange economic, geographical, and social variables to make it meaningful. Ultimately, it is contingent on how power spatially codes its own geographies through the law and adapts the territorialised paradigm to its own desired legal spaces.

Spatial production undergoes several processes of legal coding: the 'law of spatiality', the 'spatiality of law', and counter legal geographies.

The 'law of spatiality' refers to how physical geographical features acquire significance in legal geography. It considers how the physical environment is processed by political power and coded through the law. It examines how the environment (and its physical features) are implicated in the production of a given spatial normativity. Like in geopolitics, it is a matter of discourse; formally abstract and neutral, the law justifies the political process of spatial construction.[5] These canons of legal-geographical spatial production are craftily applied by state actors. The legal coding, which is used for processing geographical information, provides the universal grammar and lexicon used in cartography, allowing states to organise their practical affairs. For example, the components of the toposphere (e.g., the watersheds) and hydrosphere (e.g., the rivers) are turned into 'natural' borders; the seabed into EEZs; outer space into 'global commons'. For the sake of prestige, every state projects its own imagined legal geography onto them.

There is, then, the 'spatiality of law',[6] which acknowledges the 'importance of geographical factors in explaining differences between legal families, which gather legal systems sharing common features.[7] Natural, i.e. physical, and human geographical attributes have been used (and are still used) to assert the superiority of a given legal, or constitutional, system. Easterly terms this process of spatial production 'geojurisprudence'; although he grasps the point, he still confines them within the field of political geography.[8] In my research, by contrast, I bestow the spatiality of law a cross-disciplinary connotation, which reflects the join application of law and geography to be more precise, legal cartography) in disseminating and establishing Western legal geopolitics. Indeed, legal systems are hierarchically ranked to reflect colonial relations of domination. Like in classical geopolitics, the legal geopolitical struggle between systems still saturates the current state of world affairs (see Sect. 4.6).

---

[5]For more on geopolitics as discourse see Nayak and Jeffrey (2011), p. 235; Dodds (2019), p. 5.

[6]Blomley et al. (2001), p. xv.

[7]Economides et al. (1986), p. 164.

[8]Easterly III (1977), p. 209.

The ‘spatiality of law’ has traits in common with the process of spatial production I term ‘counter legal geographies’. In sharing the perspective of resistance (see Sect. 5.2.1), it reveals the role played by ideology in the production of space, whose enforcement makes it possible ‘to track the effect that a given law or ruling has on some specific structure’.[9] Critical legal geographies illuminate the politico-legal underpinning in colonial and urban geographies and examine normative spatialities of ‘resistance’ where the territorialised paradigm receives ideological application.

3.2 The Law of Spatiality

Interactions between human and physical geography lie at the heart of the law of spatiality. These extend over the processes of spatial production whereby geographical features become meaningful in legal geographical studies.

The law of spatiality is not the revival of Mackinder’s ‘geographical experiment’ (see Sect. 2.1.2). The idea of ‘integrated geography’—which is advocated for by geographers[10]—must be reframed in terms of legal geography. The human-physical geographical nexus is part of a broader and ideological politico-legal discourse revolving around geographical features and entities (e.g., mountains, forests, and rivers) that populate the natural environment. But *that* environment is also occupied by individuals who superimpose their representations onto the earth’s surface. The environment becomes meaningful because the ideological discourse about it provides the practical knowledge needed by policymakers to manufacture their own legal spatialities. Being part of the ‘interplay’ between the natural environment and human agencies, this spatial production is servient to the establishment of legal geographies of power.[11]

In legal geographical terms, the spatiality of law presupposes the interactions between human and physical geography. Ronald J Johnston, though, raised several objections as regards their usefulness in geographical research.[12] Reframing the idea of ‘integrated geography’ in terms of ideological discourse helps us to sidestep these objections—at least in legal geographical investigations.

Johnston terms this ‘quest for unity in geography’ an ‘environmental fix’. He assumes that the debate on the ‘separation/integration of physical and human geography’ is no more ‘than talking across rather than at each other’, without a real conversation between both subdisciplines. This is due, Johnston maintains, to the semantic dualism of the word ‘geography’, which has two usages: the academic and the vernacular. The former describes both ‘geographies’ in terms of scientific disciplines, whereas the latter refers to their practical application. Consequently, the

[9] Nayak and Jeffrey (2011), p. 240.

[10] Matthews and Herbert (2008), p. 83. See also Bonnett (2003).

[11] Kitchin and Dodge (2007), p. 334.

[12] Johnston (1986), p. 449.

geographical experiment is not a form of integration; academics (physical or human geographers alike) who oversee the process of integration also govern which vernacular usage of the 'other' geography is relevant for their research. In human geography, for example, human (academic) geographers use the physical (vernacular) geography in examining the transformation of spaces into meaningful places.

It is still possible, we assume, to integrate physical and human geography into a whole. When emphasising how human agencies impact on the environment, Johnston grasps 'the importance of studying the human/physical interface'.[13] The missing point in his arguments is the definition of what *acts as the interface*. Such a role is probably played by the third 'geography' that populates the law of spatiality, i.e. the imaginative geography. Through the law of spatiality, power processes physical geographical data and naturalises its own representation of the world, projecting its own imaginative geography onto the earth's surface—even onto outer space (see Sect. 3.4.2). Unlike its physical and human counterparts, imaginative geography does not have two usages. As noted above (see Sect. 3.1), the spatiality of law provides policymakers with the practical knowledge needed for manufacturing legal spatialities. This 'practical knowledge' is nothing more than the sum of physical and human (vernacular) usages that are integrated when manufacturing spatial normativies. Or, paraphrasing Lefebvre, the interface of my third geography blurs the line that distinguishes 'mental space ... and physical space'. The mental reframing of the human/physical interface is thus 'deployed in space'.[14]

What really matters is the practical application of physical and human geography alike. Both are processed to create, rather than simply reveal, a new 'practical' and ideologically motivated knowledge of the world. In the production of space, indeed, several decisions are taken about: (1) what to include in (and to exclude from) the manufacturing processes; (2) how to frame the politico-legal narrative which assigns a sense to the newly created space; (3) how to code and store geographical information; finally, (4) how to represent this new reality objectively and truthfully 'using scientific techniques that capture and display spatial information'.[15]

Political power frames a narrative through which 'the differences between [geographical] reality and the descriptions of [such] reality' fade.[16] Maps turn into 'propositions–that is, they are statements that affirm or deny the existence of something'.[17] Two mechanisms allow us to achieve this goal. On the one hand, there is the *normative approach*. The concept of 'spatial data' is crucial; in geographical terms, these represent 'real-word' phenomena in terms of position (e.g., through locational coordinate systems), attributes, and 'spatial interrelations with

---

[13]Johnston (1986), p. 449. On the importance of the human/physical interface see also Johnston (1983).

[14]Lefebvre [1974] (1991), pp. 27–28. See also Bachelard [1942] (1986), p. 6.

[15]Kitchin and Dodge (2007), p. 333.

[16]Westlake (1985), p. 139 also quoted in Wood (1992), p. 26.

[17]Wood (2010), p. 41.

each other' (so-called topology).[18] Abstracted from their geographical environment, spatial data are processed and coded: 'the light reflected from the earth [. . .] is transformed into an electric signal, it is recorded, it is played back, it is transmitted, it is received [. . .] it is used to drive an electron beam recorder to produce . . . a picture'.[19] The alleged technical and scientific objectivity of Earth Observation Data, GIS, and maps fill the gap between geographical facts and their representation. Acting as their interface, these devices do not merely represent, but also produce, them.[20]

On the other hand, there is the *legal approach*. The law of spatiality means making physical geographical features meaningful for political purposes. Their legal consideration complements the normative approach and makes a deliberate use of it, and the outcomes resemble a 'reshuffling of the spatial furniture as a result of some external intervention'.[21] These features are already present in nature; what power does is 'to shift and reshuffle [them] in space as [it] transforms the natural environment into a cultural environment'.[22]

Consequently, imaginative geographies become imaginative *legal* geographies. Physical features become part of an '*imagined* entity, a space that is carved out, excerpted' from the physical earth, and 'circumscribed' for political purposes.[23] The features of the geo-ecosphere thus generate a variety of legal geographies: for instance, mountains, watersheds, and rivers are assigned a legal meaning, which supplies political power with the legitimacy to regulate, i.e. intervene in, the world.

The law of spatiality is therefore a politico-legal 'way of seeing, a way of framing and imagining the environment'. It is a practice that orders the world and naturalises its representations: 'naturalizing the map has the effect of *universalizing* it'.[24] The law 'draws lines, constructs insides and outsides, assigns legal meanings to lines, and attaches legal consequences to crossing' and disregarding them.[25] At the same time, it 'acquires complexity and subtlety'; it is 'intrinsically geographical'. Space is 'clearly far more than a passive surface', and its 'specificities [. . .] may serve to shape the law'.[26]

Take, for example, geographical primitives, such as points, lines, and polygons. The polygon, in particular, is the 'homogenous representation of a two-dimensional space' demarcated by a boundary. The concept is useful in terms of spatiality of law,

---

[18] Borrough et al. (2015), p. 2 and Chap. 2.

[19] Wood (1992), p. 54.

[20] Respectively Kitchin and Dodge (2007), p. 334; Pickles (2004), p. 146. See also Edney (1996), p. 187.

[21] Blomley (1989), p. 517.

[22] Kristof (1959), p. 275.

[23] Brighenti (2006), p. 68.

[24] Respectively: Brighenti (2006), p. 69; Nayak and Jeffrey (2011), p. 240. See also Delaney (2005), p. 12. See Wood (2010), p. 19.

[25] Delaney (2015), p. 99.

[26] Blomley (1989), p. 517.

as it provides a geographical explanation of states as spatial 'containers' for politico-legal constructions.[27] The role of primitives does not end here. The idea of contiguous polygons is replicable everywhere and informs 'political boundary-drawing'.[28] Like state confines, they are demarcated by unchanging borders; like polygons, states-as-containers are 'unchanging, and do not contain any information about temporal of spatial variation'.[29] Geographical primitives supply a further explanation for the 'territorial trap': maps use states (i.e. polygons) as their 'building block' and reinforce the ideological discourse about geography, which is to be explained in terms of codes and symbols (which are both signifiers and signified), 'territorial control', 'political authority', and 'assertions of sovereignty'.[30] Polygons also distil real-word data: as on maps, these 'are never absolute but mark the center of transitional zones between categories'.[31] Despite this, the spatiality of law processes it as lines. With a good dose of imagination, and 'having insisted that a boundary is not a line', this 'kind of nominalism ... feels perfectly free to draw it as nothing else'.[32]

3.3 Processing Geographies

The law of spatiality is a process through which spatial data are coded and projected as particular representations of the world onto its surface. We have also noted that spatial creation consists of many steps (see Sect. 3.2). Firstly, it is necessary to decide how to arrange, and what to include in (and exclude from) the space. Secondly, the law of spatiality assigns a meaningful sense to these newly created spaces, also by reshuffling its components. Finally, data are stored in maps and GIS, whose 'scientific' character displays spatial information.[33] In so doing, they do not merely represent legal space; rather, they are constitutive of it.

In the following subsections I shall consider these phases in detail.

27 Taylor (1994).

28 Miller (2016), p. 34.

29 Respectively: Miller (2016), p. 34; Borrough et al. (2015), p. 33.

30 Wood (2010), p. 36; Black (1997), p. 12.

31 Espenshade and Morrison (1982), p. 73.

32 Wood (1992), p. 25.

33 On how mapping became scientific see Crampton (2010), p. 53.

3.3.1 Representing the World Geographically

Legal scholars and geographers examine the law-space nexus and how power affects the processes of spatial production.[34] Whereas lawyers mainly focus on the law as its internal regulatory force, geographers usually address how power, space, and the law interweave as patterns of social construction. To this extent, even the physical environment becomes a by-product of human agents; whether we like it or not, human geography transforms physical geography into 'nature'. Not only is the environment completely mediated by human activity, but this approach also blurs the line between 'culture' and 'nature'. As it transforms environmental spaces into meaningful places, it allows us to be 'everywhere'. Consequently, 'Where is this pristine nature to be found now?'[35]

These interactions are assessed in relation to the production of maps. The emphasis is placed on the rendering of spatial data, cartographic techniques, and models that describe geographical phenomena. Geographies are traditionally coded through maps; mapmaking—and, more recently, GIS[36]—makes geomorphological features relevant; geodetical surveying, cartography, and photogrammetry allow us to record and represent 'locations and characteristics of well-defined natural and anthropomorphic phenomena'.[37] Spatial data are then transfused into repositories, such as maps and GISs. Through them, geography acts as 'a substitute for the physical space it claims to show', as well as a 'powerful imaginative act' which organises the 'sensuous variety of the earth's surface'.[38]

No wonder, therefore, that our academic speculations revolve around the 'power of maps'; and that we consider the map-maker a 'creator rather than [a] reflector' of the world.[39] Such creativity is favoured by the constraints of geographical knowledge, such as maps and GISs. 'Deformations and contortions, distortions and misrepresentations ... are the unavoidable consequence of attempting to display simultaneously on a plane the entire surface of the sphere' which is the earth.[40] Maps supply 'unchanging' and 'static' spatial representations, where 'frozen scenes' are displayed without any reference to the processes of creation and people that participated in it, or to the changes that spatial data have undergone over time. Time is considered crucial in managing geographical data models and the examination of legal spaces; the latter are indeed socially bound, and therefore subjected to the

---

[34] See Sect. 2.5. See also Driver (1992, 1994), Braverman et al. (2014) and Bartel and Carter (2021).

[35] Whatmore (2014), p. 153. On nature as a social construct see e.g., Whatmore (2002) and Castree (2013).

[36] Tobler (2004).

[37] Borrough et al. (2015), p. 9.

[38] Brotton (2013), pp. 7 and 337.

[39] Black (1997), p. 11.

[40] Wood (1992), p. 56; Black (1997), p. 11. On the distortion related to map projection see e.g., Tobler (1961, 1966); Wood (1992), p. 47; Black (1997), p. 35; Crampton (2014); Wood (2010), p. 37; Tobler (2018).

temporalities of space. Temporally located in a particular position, these data are made static to conceal the idea of legal spaces as contingent practices; at the same time, this reinforces the 'polygonal' explanation of the 'territorial trap'.[41]

Map-making is, in sum, a matter of 'choice', which extends over 'scale, projection, orientation, symbolization, key, colour, title and caption', as well as the intentional suppression of knowledge, which Harley terms 'cartographic silence'.[42] Political power normally takes into account factors before deciding whether to display them on maps. In so doing, it usually points to permanent geographical features; and 'what we . . . sense is the presence of a code: permanent = important'.[43] Power makes an ideological, i.e. calculated, use of such cartographic limitations, turning challenges (i.e. the choice) into opportunities (i.e. the objectivity of the cartographic representation). Indeed, '*every* projection is like this, good for one thing, but not another'.[44] Its evolution reinforces the idea that cartography is somehow a technique *bonne à tout faire*; it serves an interest, which is usually related to the fact that 'others [power, groups, or civilisations] see the planet in different terms'.[45]

3.3.2 *The 'Logical Space' of Geographies*

Legal scholars and geographers pay less attention to power acting as the interface between physical and human geographies. Despite the aforesaid distortions and limitations, mapping is considered a 'universal' cultural competence crossing time and places.[46] Undoubtedly, cartography is of the utmost importance to the law of spatiality, as it helps us to code geographical patterns and arrange the world knowledge in ways that fit into our conceptual framework. This coding of the world also has its drawbacks. Let us consider how we represent spatial data. Its systematic use makes us feel as if we were constricted within a 'prison chamber' imposing a 'preordained hierarchical framework for ordering the world'.[47] Even the distortions we use to gain control of the world are arranged after such hierarchies; likewise, our cartographic 'choices' and 'mistakes' are bound by them.

---

[41] See Sect. 3.2. On the extrasignificance of time to maps see Wood (2010), p. 94. On the contingency of places see e.g., Pred (1984), p. 279. On the relevance of time in geographical data models see Borrough et al. (2015), p. 38. On how time reinforces the contingency of socially constructed legal places see von Benda-Beckmann and von Benda-Beckmann (2014).

[42] Harley (1988).

[43] Thompson (1979), p. 27; Wood (1992), p. 80; Wood (2010), p. 36.

[44] Wood (1992), p. 57. See also Crampton (2010), p. 9.

[45] Wood (1992), p. 68.

[46] Blaut et al. (2003).

[47] Respectively: Crang (2014), p. 130; Marston (2000), p. 220.

The way we frame our geographical conceptions of reality is neither preordained nor ontologically given. The use of geographical devices establishes its scales 'through the geographical structure'—not to say stricture—'of social interactions.'[48] In pointing to geographical features, they certainly point to *somewhere*. That *somewhere*, though, marks the transition to something else, which is usually hinted at by the cartographic representation. Again, maps represent—i.e. make present to us—a particular, i.e. ideological, representation of the world.

Some underpinnings lie beneath this preordained framework. Following in the footsteps of Crampton, Kitchin, and Dodge, I assume that cartography is a repository of geographical knowledge built on 'a core foundational knowledge' that 'unquestioningly underpins' the cartographic representation of the word.[49]

'What lies beneath' is, I argue, the decision on how to reshuffle the 'spatial furniture', whereby space is produced, and a given meaning is assigned to the physical world. The idea of a 'core foundational knowledge' refers not to the ontology of physical features, but to their representation; not to their 'essence', but to their 'existence'. It is, in sum, an ontology proposed in order to normalise a spatial reshuffling, which comes into existence once it eventually finds its place on the map.[50] This new 'ontological proposition' (i.e. what the real-world object *really* is according to our proposed reshuffling) depends on its 'locative proposition'. *This is* a mountain, a watershed, and a river, our map says, because we have ideologically probed, manipulated, or disregarded these physical elements,[51] making them part of our own politico-legal decision cartographically localised on the map.

Law and power redefine, reimagine, and appropriate geography and space for radically new purposes.

The concept of foundational knowledge echoes Ludwig Wittgenstein's 'logical space'. In his *Tractatus Logico-Philosophicus*, he used this expression to signify the 'place' where facts must be actually displayed as parts of the real world.[52] Once set there, 'The facts . . . are the world' (1.13). Take, for example, the earth's surface. The reshuffling of its spatial furniture occurs in logical space: such a surface is unbounded; 'the land isn't divided except in our heads and on our maps', Daniel Wood correctly argues; 'it's we', he continues, 'who break it up and bring the folk ontological world of things into being'.[53] In terms of spatial reshuffling, human agencies ideologically arrange their discourses about geography to make them fit into their 'head and maps' and achieve their own desired political ends and legal geographies. Logical space is thus the 'place' where physical facts become politically meaningful. Its 'existence ... is guaranteed by the mere existence of the constituents – by the existence of the proposition with a sense', which is formulated

[48] Smith (1992), p. 73. See also Taylor (1987).

[49] Crampton (2003), p. 7; Kitchin and Dodge (2007), p. 333.

[50] See Schuurman (1999), pp. 80–81; Couclelis (2010), p. 1787.

[51] Wood (2010), pp. 56–57.

[52] Wittgenstein (1922). For more on Wittgestein's logical space see Pilch (2017).

[53] Wood (2010), p. 49.

by political power (3.4). At the same time, logical place is a process of rationalisation and management of physical geographical facts. Spatiality is reshuffled, and every fact meaningfully ordered. Indeed, 'A proposition can determine only one place in logical space' (3.42).

'Nevertheless', Wittgenstein argues, 'the whole of logical space must already be given by it' (3.42). Once inscribed in this imagined spatiality, every fact is assigned its own meaning which is indispensable to gain control of the world. This imagined geography is the 'The logical scaffolding surrounding the picture [that] determines logical space' (3.42) Logical place is thus both relative and absolute. It is *relative*, inasmuch as it projects the world's view of a specific political interface between human and physical geographies; it is *absolute*, because 'The force of [power's] proposition reaches through the whole of logical space' and seeks to transform the way we see the world according to the 'reshuffling' proposed within the logical place.[54]

At the same time, logical space has an ontological nature; it is interested in determining what, in the real world, counts as 'geographical things', thus giving coherence to physical and human spatial features. In so doing, it creates its own understanding of the world 'to be shared with different users'.[55] In superimposing its own law of spatiality, power defines the classes of entities and identifies which entities correspond to them. The real world is aligned to logical space by both the *normative approach* and the law. As noted above, 'real-world' phenomena are defined in terms of position, attributes, and topology, which become part of the legal discourse (see Sect. 3.3.2).

Maps represent logical space, i.e. the ideologically reshuffled state of the world. Paraphrasing Wittgenstein, the map is a 'picture' which 'presents a situation [already arranged] in logical space', as well as 'the existence and non-existence of states of affairs' which have already been decided (2.11). To a broader extent, maps capture and portray both the normative and the legal approach, in a way that we can understand (see Sect. 3.3.3). Within the law of spatiality, geographical and legal considerations of geographical features (physical, anthropic, economic, and social) overlap, and power creates its own legal cartographic representation of the earth. This is apparent in the works of Manfred Langhans-Ratzeburg and of Walther Merk—the latter expressly used legal geography when referring to the cartographic representation of the law (*Rechtsgeographie*).[56]

Owing to its normative and legal representation, logical space counts as the real world.

Unsurprisingly, mapmaking is not a 'real' universal competence; what is really universal is logical space. Again, the 'spatial competence to make a map'[57] points to the physical apprehension of the earth embedded in the territorialised paradigm. As a

[54]See also Farinelli (2003), p. 3.

[55]Borrough et al. (2015), pp. 22 and 36.

[56]Langhans-Ratzeburg (1928) and Merk (1926).

[57]Wood (1993), pp. 54–55.

faculty of mind, logical space grasps and reshuffles world knowledge. As a cultural competence, it is present in different cultures, reflects sundry epistemological structures, and expresses differing ways of seeing and representing the world.

3.3.3 *Spatiality as a Politico-Legal Discourse*

The way power arranges ideological discourses about geography resonates with critical geopolitics, which assesses how world affairs might be regulated. Like critical geopolitics, the law of spatiality is interested in the ideological use of geographical knowledge; interactions between human and physical geographies are reframed so as to serve the interests of specific human agencies.

Taking place in logical space, spatial reshuffling shares several commonalities with Daniel Wood's legal-cartographic earth's representation as 'a social construction of reality'.[58] Imaginative, as it is, spatial pattern framed in the logical space must be turned into a real-world geography. Prompted by the 'desires to induce changes in others', power transposes its 'ideational plane in some kind of sign system'; it selects its desired ends; it assigns geographical features a meaningful role within the imagined space; it chooses which features have to be displayed. The purpose is to make its own imaginative geography be accepted as *the* common understanding and representation of the world. No wonder that 'maps will vary' so as to reach this end.[59]

When considering the sign systems suitable for geographical purposes, Wood refers to both language and map. As ways of expressing the logical space, though, their use depends on the 'social situation' we represent, and varies in relation to the discourse function they fulfil. Wood also assumes that language and mapping are 'different': whereas 'speech and writing are developed by a logic of concatenation,' maps reflect the 'logic of juxtaposition'. The former 'unfold over time'; the latter 'unfurl over space'. They develop in parallel 'and bifurcate into writing and mapmaking'.[60]

Wood does not explain, though, whether their parallel development intertwine; or whether social situations they represent calls forth their joint application. Owing to his being a geographer, he also sets a disciplinary hierarchy between language and mapmaking; maps supply the internalised representation of space as a privileged form of expression.

Our argument runs counter to Wood's predicament and assumes that, in transposing its ideational plane into the world, logical space uses both language and maps. In order to become a real-world representation, it needs to arrange it through both the normative and the legal. Whether coeval or not, maps (the normative

[58] Wood (1993), p. 52.

[59] Wood (1993), pp. 52 and 105. See also Wood (2010), pp. 58–61.

[60] Wood (1993), p. 57 n6.

representation) and language (the legal discourse) coalesce towards a legal cartographical representation of reality.

When creating and disseminating its own understanding of geographical data, logical space needs to develop a politico-legal discourse. Not only does it reproduce its ideas of how to regulate the world affairs, but it also has to convince us that the proposed spatial reshuffling serves the interests of all human agencies. Such 'reciprocal transfer of meanings between speakers and listeners' is 'the metabolism of language', which also deploys a 'communifying' function. Language, not maps, performs such function and creates 'bonds of social cohesion'; without it 'there can be no community . . . the community itself provides the sources of those words, makes them available to its members, in the language that it transmits'.[61]

The role played by both language and discourse is apparent in cultural geography. As pointed out above (Sect. 3.1.1), the geographical discourse about the environment is completely mediated by human activity, and 'naming and sense-making' processes are all 'attributes of human cultures'.[62] The same occurs in legal geography, where the territorialised paradigm displays territorial spaces that are imagined by communities. As Pred puts it, language is crucial in the social creation of places, as 'it provides a foundation for describing, grouping, and differentiating things'. Indeed, 'the limits of a population's language mean the limits of their place [. . .] at the same time that the limits of their place mean the limits of their language', which usually evolves or declines as its experience and knowledge of the world progresses.[63] Nothing is more human than governing the processes of legal spatial production by making them linguistically meaningful.

Several cultural geographers have tracked the relations between language and maps. Both are part of the discourse that power makes about the world; both complement its strategies to turn a specific vision of world affairs into common understanding.[64] In particular, John Pickles's treatment of maps as 'problematic texts' is particularly interesting. As problematic texts, indeed, maps cannot be 'read in simple ways'; like legal interpretation, their construction cannot be limited to either the original intent of lawmakers (and mapmakers) or the *ratio legis*; it asks 'what is hinted at by an utterance in its particular context, what the sender's "agenda" is'.[65] Not only does Pickles recognise 'the multiple, institutional and contextual nature of mapping', but he also implicitly upholds the application of linguistic (and legal) canons of construction to it.[66]

There is, then, a common legal-geographical lexicon applicable to both sign systems which also represents geography as a discourse. Consequently, Wood's assumption that language and mapping are 'different' has to be reframed. Legal

---

[61] Berman (2012), pp. 59, 38, 69.

[62] Whatmore (2014), p. 156.

[63] Pred (1984), p. 285.

[64] Pickles (2004), pp. 67, 146.

[65] Griffiths (2006), p. 7.

[66] Kitchin and Dodge (2007), p. 334. On maps as grammars see Edney (1996), p. 188.

geography follows logics of both 'concatenation' and 'juxtaposition'. The former is patent in the construction of geography as a discourse. The law of spatiality is a way of knowing the world 'geographically' conveyed through an agreed lexicon. From a normative perspective, concepts like geographical primitives, georeferencing systems, attributes and their properties, entities defining geographical things, and the way we structure such data already constitute a common lexical index. Through them, it is possible to represent meaningful geographical data and favour the transfer of their meanings between different agencies (and their logical spaces). 'Juxtaposition' presupposes common lexicon; and ideological struggles between agencies (and their logical spaces) will take place within a reproduction of the world that has been previously agreed.

Nayak and Jeffrey use the verb 'to naturalise' when pointing to these processes of political appropriation and naming of the geo-ecosphere. The verb is intriguing, as it refers to geo-politics as a discourse whereby the political appropriation of physical geography becomes 'common sense'. Take, for example, the cardinal points; in Mediaeval Europe, the east was set at the top of the map because it encapsulated 'both the hopes [Christ's resurrection, Jerusalem, and the terrestrial paradise] and the deep-seated fears [the Orient and the Tribes of Gog and Magog] of Christian thought and belief'.[67] With the advancement of cartography, setting the north at the top became common sense;[68] in our globalised world, 'cartographers point' to the north and 'take its presence for granted', as 'a fact of nature, a common truth'.[69]

In other words, 'to naturalise' means assigning a new meaning to physical features; in becoming their new connotation, this meaning still conveys the idea that physical geography reflects a 'neutral' reality. What renders the new meaning 'common sense' is its cartographic and legal coding; through it, human agencies process data collected from the physical environment and project them onto the earth's component spheres. Through the law of spatiality, physical features undergo a 'semantic change'. The shift modifies the lexical semantics of geographical language, i.e. the function of its lexical items. In so doing, semantic change matches the purposes of political power and affects the existing lexemes; these adopt an additional meaning, which complements the old (geographical) one.[70]

Such a shift is due to the intrinsic features of the legal language; its 'neutrality and generality' make it applicable to whichever geographical environments; they avoid 'subjective and personal attitudes' towards geographical features, as well as their 'regional marking'. Yet, these features 'represent anonymous authority and power';[71] the change in meaning occurs because interests and perspectives of geopolitical actors are able to impose a new sense on existing lexical items.[72]

---

[67] This was apparent as regards the T-O maps: see Brotton (2018), p. 188.

[68] See Brotton (2013), p. 188.

[69] Wood (1992), pp. 22, 70.

[70] See Kastovsky (2006), p. 215.

[71] Rissanen (2000), pp. 120–121.

[72] On maps as a vehicle for 'creating and conveying authority' see Wood (2010), p. 52.

The legal coding is servient to this process. It provides a 'restrained, apolitical mode of social regulation' and an 'ideology of abstraction' that allows us to 'ignore the spatiality and contextuality of meaning' and consider only geographical 'facts' crucial for processing their new meaning.[73] Political power creates a disjuncture between the law and the environmental space, between the process of legal coding and the context to be processed. If this is true from a formalistic angle, it is quite the opposite in terms of legal geography. Despite its alleged objectivity and neutrality, the law 'is *not* constructed in some ontologically privileged domain'; as an 'absolute . . . a "thing in itself" with an existence independent of matter', the law resonates with the idea of the 'absolute space', i.e. the basic unity in geographical studies.[74] The spatiality of law, by contrast, rearranges all the 'facts' of the physical world after the logical space of political power and assigns them with a new meaning that fits into its agenda.

3.3.4 Uniform Design and Common Scale

Spatial reshuffling and its legal coding are processed in the form of legal cartography. Maps—not to mention the most modern digital geographical information systems—are '*about* the world that' they reveal; they are 'not the world – or not *just* the world – but also (and sometimes especially) the agency of the mappers'.[75] With these words Denis Wood y epitomises the characteristics of cartography, which makes visible the 'whole world at one view', embodying ideologies and partialities of geopolitical mappers.[76] Ultimately, it is servient to geopolitics.

The way spatial data are processed through the grids of maps and GIS is more than representation. The need for adopting uniform designs and common scales still reflects the idea that physical geography reflects a 'neutral' reality through a process of 'naturalisation'. It is not purely a matter of choosing what to show and which entities, scales, and projections to use. In legal cartographic terms, these choices are part of the politico-legal discourse about place. If we take it for granted that maps embody the legal coding of a given logical space,[77] the problem is not of representation of a partisan spatial reshuffling, but of its generalisation and acceptance by the broader community.

In sharing its own idea on how to arrange world affairs, legal cartography draws a divide between cartographic 'communication' and 'visualisation'. According to Wood, the former points to maps designed for specific purposes and hinting at

[73] Blomley (1989), p. 513.

[74] Blomley (1989), p. 513. See also Blacksell et al. (1986), p. 376; Harvey (2009), p. 13.

[75] Wood (1992), p. 24.

[76] Blundeville (1589) in Hartley (1981), p. 227.

[77] For the idea that maps display what power is interested in seeing e.g., Wood (1992), p. 70; Pickles (2004), p. 124; Crampton (2010), p. 18.

specific messages, whereas in the latter the message is *apparently* unknown, and its purpose is concealed under the neutral justification conveyed by maps and legal discourse.[78]

In generalising its own spatiality of the law, power certainly makes a strategic use of mapmaking. To be a convincing callout to the public, legal cartography sidesteps the limitations of geographical representations. For example, it justifies the generalisations of, say, its preferred symbols and projections. Although these are servient to its own sectional world view, they become the sole *graphic discourse* compatible with the strictures of data visualisation. It also restructures the data displayed. If the physical area of interest 'stretches across two or more images or map sheets', it has to amalgamate them by converting the data to the same coordinate system, by joining them and creating a 'seamless database' (edge matching and mosaicking). Otherwise, the credibility and integrity of the process would be undermined.[79]

The creation of a common legal cartographic lexicon has been favoured by the aforementioned logic of concatenation and juxtaposition. That fact that 'maps have been thoroughly naturalised within our society' and considered 'natural objects'[80] is a legacy of our Western, and dominant, cartographic tradition. In a logic of juxtaposition and competition among different cartographic traditions, states have used surveys in order to naturalise, i.e. generalise, their own geographical representations of the state of world affairs.

And their jargon, too. Mapmaking has always been a monopoly of a limited elite of cartographers—and these in turn produced maps for the rich and those in authority. Cartographers elaborated their own grammar, lexicon, and techniques for the creation of maps. As technology progressed, cartography went through new stages of development, supplying the spatiality of law with increasing accuracy and comprehensiveness. Legal cartography was favoured by the narrative of progress related to the scientific and neutral character of mapmaking; this narrative complemented the politico-legal discourse aiming to naturalise what we term 'nature'. In so doing, cartographers generalised the idea of maps as 'realistic images', whose scientific approach to geographical information could be 'improved with repeated measurements over time'. Although the partiality of power may make 'each individual exercise in [legal] map making . . . potentially flawed,' the generalisation of scientific 'mapping exercises' was able to turn the history of cartography into a narrative of progress.[81]

The advent of GIS has boosted the 'scientific' character of cartography, which is now mostly dominated by technical issues, an even more accurate jargon, and academic specialisms. Needless to say, our society has also accepted the 'naturalisation' of such technical discourses. And the logical space has found new avenues whereby the spatial reshuffling of the environment may take place. This has

---

[78] See Wood (1994).

[79] Borrough et al. (2015), pp. 84–85.

[80] Edney (1996), p. 187.

[81] Edney (1996), p. 188.

had consequences on legal cartography, which has shifted from communication to 'mapped data quantifying a physical or abstract system in prescriptive terms'.[82] The computerised management of spatial data and the use of English as the global language of the scientific community have provided legal cartography with a universal (and univocal) linguistic and binary code. The spatiality of law has been experiencing an increase in the joint application of these sign systems in cartographic visualisation. Wittgenstein taught us that such linguistic paces triggered new geographical definitions: indeed, 'definitions are not mere abbreviations' nor a shorthand for non-spatial geographical and political features. By contrast, they mark the transition 'from one technique to another, projections from one technique into another'.[83] Or, in our case, from a physical to a digitalised logical space.

This is apparent as regards GIS. Created half a century ago, it has reshaped many aspects of legal cartography, as well as the special relationship between law and 'geographies'. In addition, GIS seems to be 'acentric', with a 'view from nowhere', dissimulating 'a perspective which the reader cannot evade'.[84] Like in traditional cartography, it reinforces the assumption that a 'god-like', 'neutral', and 'scientific' positionality may be provided to spatial-data observers.[85] The resort to computer systems facilitates the collection, examination, organisation, analysis, and visualisation of geographical information, whose digital representation affects logical space. The possibility of mathematically combining data and producing digital overlay analysis favours spatial 'simulation and modeling', which undoubtedly lie at the heart of contemporary spatiality of law (see Sect. 8.6).[86]

3.4 Cartographic Digitalisation, Democratisation, and Imagination

3.4.1 Lived Geographies and Digitalised Democracies

Cartography has benefitted from the increased use of computer programming in several ways. As 'consumers' of geographical data, we have swiftly internalised digitalised geography. New software and applications have equipped us with several open-access interfaces giving us the illusion of taking part in crowdsourced cartography and digital mapmaking through the collective generation of media.[87] *Digitalisation*, it goes without saying, has favoured the 'democratisation' of cartography, paving the way to what, not without irony, has been termed 'a new populist

[82] King (1991), p. 70; Berry (1986), p. 43.
[83] Wittgenstein [1939] in Diamond (1976), p. 43.
[84] Black (1997), p. 17; Edney (1996), p. 187.
[85] Kitchin and Dodge (2007), p. 332.
[86] King (1991), p. 69.
[87] Goodchild and Longley (2021).

cartography in which the public is gaining (some) access to the means of production of maps'.[88]

Since the advent of Google Maps and Google Earth,[89] there has been a proliferation of new types of GISs and online mapping platforms. Google platforms, OpenStreetMap, Wikimpia, Flickr, Housing Maps, MyMap, Platial, GeoCommons, ArcGIS, Miss Pronouncer, and MapQuest provide us with mapping 2.0. Mapmaking processes are no longer the precinct of academics and cartographers; nor are they the monopoly of politico-economic elites. They are now developed by computer programmers and enriched by individuals; acting as 'sensors' and 'local observers', they create 'their own map data . . . It's 6 billion pairs of eyes' participating in online mapping processes.[90] We take part in participatory mapping and 'volunteered' geographic information.[91] We are geo-referenced through our GPS mobile phones and devices, insert geotags to note our geographical location, and exchange data with people and providers.[92] Through this exchange, the broader public contributes to collecting spatial data subsequently stored in GIS; at the same time, individuals share what they do, feel, experience, and experiment with in everyday life.

In sum, we convert our lived geographies into the geographical knowledge of the world and share them as georeferenced spatial data through open-source mapping applications. Geographers term this process 'mashup', i.e. a web-based program that does 'map hacking', exploits mapping applications, and combines 'two or more sources of content into one tailor-made experience'. Through the easy access to free and open-source (FOSS) mapping tools, individuals experience the possibility of giving cartographic, and therefore scientific, consistency to their own logical spaces.[93]

We are all geographers now.

GIS democratisation does not come without a cost, though. Firstly, our logical spaces do not trigger any real spatial reshuffling. We accept to arrange our lived geographies after technical languages and binary codes that will be inscribed within the logical spaces of those that have created the mapping platforms. Secondly, as Crampton argues, 'free' does not mean *gratis*. Technical developments in digital computing have placed greater emphasis on the spatial data that we share with open-source applications in taking part in mashups. We make a bargain with the creators of digital logical spaces. When sharing our lived geographies, we also share our data with multinational media companies. Thirdly, and in terms of visualisation, their ideological discourse remains *apparently* unknown, and their commercial purposes are concealed under the technical justification provided with by the open-access

[88] Crampton (2010), p. 69.

[89] Ratliff (2007). See also Crampton (2010), p. 27; Brotton (2013), pp. 420–428.

[90] Goodchild (2007), pp. 217–218; Ratliff (2007). See also Hacıgüzeller (2017). On mapping 2.0 see Crampton (2009).

[91] See Chambers (2006); Goodchild (2007); Borrough et al. (2015), p. 77.

[92] Dodge and Kitchin (2013). See also Goodchild (2007), pp. 215–217.

[93] Crampton (2010), pp. 26 and 129; Goodchild (2007), p. 214; Butler (2006).

interface. Our data are meaningful to them; our geodemographic data are used to gain understanding of our behaviours as consumers. Commercial and consumerist interests of both providers and consumers elaborated a new ontology; 'what people choose to capture and how they categorise and describe phenomena' is nothing else but the reflection of a capitalist spatial reshuffling. What is displayed exists because it is present on the map. Yet, such a presence accounts for neither its ontological proposition (i.e. what the real-world object *really* is) nor its locative proposition (i.e. where the real-world object *really* is). By contrast, both propositions reflect that our lived geographies may have room on maps because of their commercial values for the providers. It is not relevant what the object is or where we experience it; it is *there*, in the map, since it is commercially laden.

The consequences are twofold. On the one hand, geodemographic information systems 'promise the capability to manipulate the market and consumer identity to enhance profitability'.[94] On the other, these practices resonate with the idea of the 'surveillance society'; albeit market-driven, geodemographics are acceptable to us because they are based on the same technologies applied for official 'census tabulation' planning.[95]

3.4.2 *Imagined Geographies by Proxy. Outer Space*

Processes of normalisation related to outer space have undergone democratisation and digitalisation. Together with its astronomical components, it has experienced several spatial reshufflings. The history of cartography teaches us that political and economic powers have integrated outer space into their logical spaces by applying the territorialised paradigm.

We may argue that 'with geography specifically meaning "earth writing" . . .' there is no room for the '*geographies* of outer space'. Quite the contrary. Firstly, as the 'newest of our frontiers', outer space has been a constant of our imaginative geographies.[96] Secondly, the same term 'space' extends over 'the cosmic void between planetary and other cosmic bodies'; it is a space in itself. Thirdly, the nothingness of cosmic space has also contributed to reshaping our conception of the earth. For example, the image AS17-148-22727 (labelled *The Whole Earth*): 'taken at some 21,750 nautical miles' from us, it 'captured . . . the full terracqueous disk without a solar shadow'.[97] Daniel Cosgrove explains that 'the floating Earth seems to merge into nothingness', thus reinforcing the human ambition of implementing our *One-World* reading of the Earth. The descriptor points to a Western geopolitical conception that 'signifies the expansion of a specific socio-economic order across

[94] Goos (1995), p. 131.

[95] Pickles (1992).

[96] Van Gent and Van Helden (2007). See also Dunnett et al. (2019), p. 316.

[97] Cosgrove (1994), p. 270.

space', erasing national borders through the technological and financial domination of transnational elites (see Sects. 4.4 and 8.5).[98] Fourthly, technical progress has made certain spatial elements extremely valuable real estate. The geostationary orbit, for example, provides satellites with a 'precise orbital position'; at 35,786 km over the Equator, 'the satellite moves at exactly the same speed as the Earth beneath it', which means that it 'stays in the sky above a single fixed point on the Earth'. In addition, its terrestrial footprint allows it to 'see 42 percent of the Earth's surface at once'. With only three satellites in this orbit, communications providers may 'cover almost the entire Earth'.[99] This has prompted the exploitation of the cosmos, making it (and outer space more generally) suitable for an extension of the territorialised paradigm of legal geography. No wonder that such reshuffling mimics the 'terrestrial' processes of spatial production.

Before satellites, rockets, and shuttles were launched into space, their apprehension had occurred by proxy and imagination. Accumulated geographical knowledge was displayed in charts and maps: from the seventeenth to the nineteenth century, geographical-spatial knowledge 'depended on the use of maps to structure and order spatial information collected through exploration, surveying, and censuses'.[100] The development of astronomy and telescopic observations were also favoured by the expansion of oceanic navigation for commercial purposes during the Renaissance. The detection of new stars and the forging of new constellations were progressively embedded into logical spaces.[101] Outer space was integrated into geography, which accounts for the application of the terrestrial politico-legal discourse, principles of legal cartography, and spatiality of law to it.[102] The firmament became meaningful to human agencies; and the imagined geographies of discovery and navigation adopted a holistic approach to the earth and sky, which made them part of a bigger logical space. In legal-cartographic terms, this prompted the extension of terrestrial taxonomies to the sky and the creation of a 'quasi-territorial jurisdiction' over spatial objects.[103] Potentially unbounded and infinite, outer space became a legal construct, which meant an imaginative process where law and geography were mutually co-productive.

In the second half of the twentieth century, the rush to outer space brought noticeable changes in spatial construction. On the one hand, geostationary satellites inaugurated a new era, which project our imagination onto the spatial firmament. Remote sensing also monitors the spatial reshuffling taking place on earth. Even when remote sensing is concerned, human-related activities are still conducted by proxy. It is now possible to grasp new geographical knowledge by virtue of Earth Observation Systems (EOS): through satellites and aircrafts we can measure 'our

[98]Cosgrove (1994), pp. 289–290.

[99]Collis (2009), p. 316; Collis (2017), p. 295.

[100]Edney (1994), p. 103.

[101]Edney (1994), p. 103.

[102]Friedman Herlihy (2007), p. 102.

[103]Von der Dunk (2020), p. 7.

influence on the range of Earth systems' such as mountains, glaciers, sea-level rise, and other geophysical phenomena. EOS has thus become pivotal for gaining control of the world. By proxy, remote sensing is carried out by 'a recording device' which has 'not physical contact with features under surveillance'.[104]

Applying the territorialised paradigm, the spatiality of the law generates two legal geographies. The first one is the imaginative geography of outer space, whereas the second is the legal geography of earth's spaces. By proxy we apprehend spatial data which are servient to framing our terrestrial logical spaces. Through remote sensing we reshuffle space; remotely sensed data are indeed geospatial, inasmuch as the observed areas and objects are referenced according to their geographical location in a geographical coordinate system. We revisit these areas 'on a regular cycle, facilitating the acquisition of data that reveals changing conditions . . . over time'.[105]

These imaginative activities and the legal geography of outer space fall under a common standard of rules and scales, which are universally known as 'space law', whose role is patent in the regulatory practices of remote sensing. Geostationary satellites now monitor areas—and their resolution can also reach 0.61 m. This has prompted the adoption of stringent rules, such as the 'Principles relating to remote sensing of the Earth from space'.[106] Principle XII, in particular, allows sensed states to have access to primary and processed data concerning their territory on a non-discriminatory basis and on reasonable cost terms.

Thus, outer space is accessed to collect remotely sensed data, explore stellar components, and exploit them. This requires the adoption of a legal cartography which is shared by all the potential politico-economic powers which may have an interest in outer space. Again, the process of its normalisation takes place through patterns of concatenation and juxtaposition (See Sect. 3.3.4). Concatenation percolates the construction of outer space as an embedded space regulated by tenets of international and space law. In so doing, the law of spatiality represents the world 'geographically' and has to rely on a lexical index common to political actors.

Juxtaposition presupposes this lexicon. No wonder that ideological struggles between agencies have been prevented by adopting a logical space that is broader, and therefore surpasses, each individual logical space. Outer space is designated as 'global commons' or 'free gifts of the cosmos'.[107] Outer space, the moon, and other celestial bodies cannot be 'subject to national appropriation by claim of sovereignty'. They must be used 'for the benefit and in the interests of all countries'. As a 'province of all mankind', they are 'free for exploration and use by all States without discrimination of any kind', which means 'free access to all areas of celestial bodies'.

---

[104] Khorram et al. (2012), p. 2.

[105] Khorram et al. (2012), p. 2.

[106] A/RES/41/65 95th plenary meeting 3 December 1986 41/65. See Harris (2003), pp. 539, 541–543; Khorram et al. (2012), p. 115.

[107] Respectively: Von der Dunk (2020), p. 16; Dunnett et al. (2019), p. 321.

Likewise, 'freedom of scientific investigation' and 'international co-operation in such investigation' are the basic tenets of space law.[108]

The outcome is a hyperregulated space, which is an admixture of five international treaties and several resolutions: the Outer Space Treaty (1967), the Rescue Agreement (1968), the Liability Convention (1972), the Registration Convention (1975), and the Moon Agreement (1979).[109] The 'territorial trap' is projected onto this imaginative space. As Art. VI of the Outer Space Treaty states, in outer space non-governmental entities may carry out their activities provided that these are authorised and supervised by the State Party to the Treaty which also assures and is liable if national activities are not carried out in conformity with the same Treaty. Thus, this broader logical space and its legal cartographic representation are the by-product of a limited set of states whose interests have been projected onto the firmament. This logical space avoids juxtaposition because the basic features of its legal geography have been previously agreed. The process of normalisation, then, allows new states to take part in the law-making process related to outer space. This occurs, however, within the parameters of the Treaties in force.

3.5 Applying Legal Cartography

In this section, I shall apply the tenets of legal spatial production (i.e. logical space, the politico-legal discourse, and legal cartography) to some of the six components of the geo-ecosphere. Physical geography subdivides the Earth's surface into the toposphere (earth surface), the biosphere (the 'totality of life on Earth'), the atmosphere (the 'lower layers of the gaseous envelope'), the pedosphere (the soil covering the Earth), the hydrosphere (the liquid waters), and the cryosphere ('the world of snow, ice, and "frozen" ground').[110] In particular, the following subsections will explore how the spatiality of law manufactures, thus legally coding, the toposphere, the hydrosphere, and the Antarctic cryosphere. The outcomes of the territorialised paradigm may be termed 'legal' geomorphology, hydrogeology, and cryogeology.

These processes do not occur in a legal vacuum. International law also examines them applying its own practices. Concepts like territorial acquisition, cession, and annexation have been vividly epitomised by the International Court of Justice in the seminal case *Burkina Faso v. Mali.*[111] Yet, international-law research is confined within the precincts of legal studies, which reduces its ability to grasp the cross-disciplinary implications of the spatiality of law. My primary purpose is therefore the pluralisation of the debate by boosting the cross-disciplinary ambitions of legal geography (see Sect. 2.4).

---

[108] See Arts I and II Outer Space Treaty (1967).

[109] See Khorram et al. (2012), pp. 113–129.

[110] Matthews and Herbert (2008), pp. 21–22.

[111] *Frontier Dispute, Burkina Faso v Mali*, Merits, Judgment, [1986] ICJ Rep 554.

I shall also examine how the spatial reshuffling of physical geographical features occurs within logical space. Applicable, as it is, to physical, human, and imaginative geographies alike, logical space is open to any form 'exploration', i.e. 'cultural practices' involving the 'mobilization of people and resources, especially equipment, publicity and authority'.[112] It is the interface between the materiality of the physical, i.e. real, world and our imaginative geographies. As Driver puts it, 'to lay eyes upon a lake, a peak, a species . . . was the dream of every aspiring explorer', as well as of political power.[113] As said, the way we reshuffle the world in logical space is usually represented through words and maps which normalise our geographical 'dream'. With a good dose of approximation, we may say that spatial reshuffling makes our imagined geographies present, i.e. it conveys the outcomes of our 'reproductive' practices. In so doing, we turn logical space into a lived geography, i.e. in a nomic setting politically meaningful to us.

3.5.1 *Legal Geomorphology and Natural Borders*

When considering the land surface, the first thing that comes to our mind is its demarcation through borders. This reflects the territorialised paradigm; carefully implemented through geographical primitives, it reinforces Agnew's 'territorial trap' and therefore the idea that states are bounded spatial entities. Boundaries, though, are not mere lines that turn state bounded territories into politically meaningful places. Quite the opposite. The possess a performative, relational function that shapes the territorial identities of those who are inside and outside them (see Sect. 6.5).[114] Located at the centre of geomorphological studies, dryland becomes territory, i.e. a highly politically and legally relevant concept.

International law, geopolitics, and international relations have traditionally focused on the concept of 'natural borders'; and bordering processes have followed physiographic elements such as rivers, mountains, and watersheds. The idea that borders are somewhat natural features is the outcome of spatial reshuffling taking place in Western logical space. The application of the territorialised paradigm has led to their association with the Westphalian system; borders have shifted from politico-legal artefacts to 'natural' elements. We take them for granted as if nature had already predetermined 'the limits for sovereignty'.[115] Like nature, borders are cultural artefacts that have been naturalised through the joint application of language and maps. In so doing, they provide human agencies with universal grammar and lexicon.

---

[112] Driver (2001), pp. 8–9.

[113] Driver (2001), p. 9.

[114] Taylor (1994).

[115] Diener and Hagen (2012), pp. 41–42. See also Johnson (2003), p. 195; Ochoa Espejo (2020), p. 96.

The spatial articulation of islands is somehow intriguing. 'Surrounded by wild waters, cliff-bound islands appear formidably awesome as well as starkly beautiful'; their location is paired with their geographical demarcation; insular 'boundaries are ... narrow and firmly defined' by open waters.[116] The 'island model' points to 'bounded territories as solid blocks of dry land apart from the variations in the water, air, or changing water'.[117] The 1982 United Nations Convention on the Law of the Sea (UNCLOS) holds the same assumption; according to its Art. 121(3), islands are tracts of land which can be the stage of human activity. This is a palpable application of the territorialised paradigm, which makes islands meaningful and useful for us. UNCLOS goes above and beyond the paradigm, inasmuch as it distinguishes islands from rocks; the latter are features that 'cannot sustain human habitation or economic life of their own.' There is something beneficial in turning a feature into an island; states expand their sovereign rights by increasing their exclusive economic zone (EEZ) and, at the same time, reduce the extent of the International Seabed Area. This triggers further maritime border disputes. Several coastal states have used their 'quasi-insular features' to broaden their own EEZ. From 1989 onwards, China 'turned seven reefs in the Spratlys into artificial islands'; it then 'occupied the islands it built, erected structures, paved runways, and installed military defenses'.[118] This 'veritable process' of spatial production is termed 'islandization'. Like nature and borders, the reshuffling of insular logical space generates new human artefacts; the idea that 'quasi-insular features' are islands on their own right has been naturalised to match UNCLOS.[119]

Legal geomorphological features of the earth's surface are constantly being altered under the combined influence of human and natural factors. Consequently, borders are not as 'stable' as geographical primitives traditionally assume. I have already referred to geophysical phenomena that acquire new meanings in legal geography (see Sect. 2.1.2). Together with atmospheric agents and climate-change related events, phenomena such as eluviation and accretion erode rocks causing the melting of glaciers and rising sea levels; rivers may change their course, water basins dry up;[120] areas become environmentally fragile, which calls for their protection. The 'removal of snow and ice by melting', which is ablation, impacts on glaciers, whose retreat changes borders affecting communities.[121] Demarcation becomes highly controversial. Even when the territorial trap superimposes a well-defined set of boundaries onto the land surface, such changes undoubtedly have some bearing on the spatial consideration of physical boundaries. For example, Pacific islands face the risk of becoming 'entirely underwater', losing their territory,

---

[116] Lovegrove (2012), p. 3.

[117] Ochoa Espejo (2020), p. 33.

[118] Gaynor (2020), p. 109.

[119] Faccio (2021), p. 17.

[120] Grainger and Conway (2014).

[121] Allaby (2020), p. 1 (ablation).

forfeiting their rights to the territorial sea (Art. 76 of UNCLOS).[122] When such geomorphological and hydrological changes occur, their legal consideration may trigger litigation. In Italy, for example, the Constitutional Court settled a dispute between the Veneto and the Trentino-Alto Adige/Südtirol regions concerning the border running through the Marmolada Glacier.[123]

Far from being permanent, borders are contingent human artefacts, whose meaning has evolved over time. They have served different functions among different civilisations; in our times, they unveil a 'particular cognitive mode of gaining control over the world, of synthesising cultural and geographical information.'[124] The change in the meaning of borders reflected (and still reflects) the evolution of how humans arrange their belongs of public law after reshuffling the spatial furniture in logical space.

In ancient civilisations, boundaries did not demarcate impenetrable jurisdictions. The Romans, for example, placed emphasis on natural features as defensive devices. In projecting their legal-geographical notion of the world, 'the concept of limit to power was foreign' to them.[125] Areas located beyond the *limes*, i.e. the borderland running at the margins of the Empire, were 'floating' zones without specifiable boundaries, which might potentially fall under Roman control and authority. In sum, the *limes* separated internal and outer spaces but could not formally be considered a true border, as it shaped the inner 'legal space' where Roman law applied.[126]

I understand that the unoccupied space beyond Hadrian's Wall was part of Roman logical space; made meaningful as an *erreichende* area (see Sect. 2.6.2), it was susceptible to physical apprehension—and political appropriation. This situation changed when the Germanic tribes became a threat to the *Res publica*. The conception of the *ager publicus* as an unlimited public domain was replaced by the idea of boundary separating 'legal Roman space' and Barbarian one; 'in between, stood the border, the new border.'[127]

During the Middle Ages, logical space still arranged the 'belongs of public law' around floating zones. Related to the exercise of political *dominium* over 'cities and territories', they did not have 'clear boundaries', but 'somewhat vague borderlands.'[128] Mediaeval 'areal' orders were superseded by the Westphalian system, based on the equation between sovereignty and territory. This shift was facilitated by

[122] Leal-Arcas (2013), p. 46.

[123] See judgement No 743/1988. See Rossi (1988).

[124] Black (1997), p. 18; Brotton (2013).

[125] Bowersock et al. (1999), p. 542; Kratochwil (1986), pp. 35–36.

[126] Kerneis (2016), pp. 3–4. See also Diener and Hagen (2012), pp. 22–35 (for the Aztec, Roman, Chinese, Assyrian, and Sassanid Empires).

[127] Kerneis (2016), pp. 3–4; Brunet-Jailly (2005), p. 634.

[128] Brunet-Jailly (2005), p. 635. See Sect. 6.5.

the power of mapping, which 'allowed rulers to have a spatial view of their possessions:' 'borderland or border regions ... became boundaries or frontiers.'[129]

The new state order prompted several international treaties whereby borders were demarcated. The first documented treaty is the 1323 Treaty of Pähkinälinna between Sweden and Novgorod. Several others ensued: the 1546 Anglo-French Treaty of Ardres, the 1555 Treaty of Augsburg, the 1659 Treaty of the Pyrenees, and so on.[130] This idea of natural borders proved a fortunate one because it allowed political power to manage spaces by spatially reshuffling physical features to meet state-related imagined legal geographies. Their normalisation triggered a variety of fictitious legal geographies: watersheds, basins, and glaciers were indeed manipulated. In Alpine space, for example, the border between Italy and its neighbouring countries (France, Switzerland, Austria, and Slovenia) does not always follow the Alpine watershed. Physical and human geography do not overlap, and this mismatch has legal consequences. For example, tracts of lands within the Italian Geographical Region are part of, say, France (as in the case of Nice, La Brigue, and Tende; the latter ones became French municipalities by virtue of the 1947 Paris Peace Treaties), Switzerland (e.g., Tessin and the Southern part of the Canton of Grisons), Slovenia and Croatia. By contrast, the Valley of Livigno and several other valleys in Piedmont and South Tyrol (e.g., the Toblacher Sattel and the Drau springs) spread well beyond the Alpine watershed.

The modern territorial order inaugurated by the 1649 Treaty of Westphalia reached its heights during the European colonial enterprises, when linear borders were projected onto America, Africa, and Asia. This is particularly apparent in the case of the Andes, which were constructed as natural borders to provide Chile with a discursive practice when it came to institutionalising its 'shared space on a national scale' in the nineteenth century.[131]

Areal geographical representations were replaced by a 'linear' system that disregarded borderless communities. Usually divided among different states with the aim of creating political entities based on jurisdictions, they fell prey to the territorial trap. The Berlin West Africa Conference (1884–85) led to the partition of Africa and the establishment of borders across the continent. In Asia, the India-Nepal border was consolidated between 1814 and 1857; and the Durand line between Pakistan and Afghanistan was servient to consolidate British control over India, thus artificially separating 'the Pathans of Afghanistan ... from their brethren in India'.[132] The 1869 Treaty of Peking established Russian and Chinese areas of influence in Central Asia. As Kratochwil states, the Treaty was based on the application of natural borders, fixing the boundary as 'following the mountains, great rivers and the present lines of Chinese permanent pickets', thus leaving 'a substantial part of the Sinkiang region on the Chinese side'. Even after the

[129] Brunet-Jailly (2005), p. 635.

[130] Black (1997), p. 123; Kratochwil (1986).

[131] Núñez et al. (2017).

[132] Tripathi and Chaturvedi (2020); Bhattacharyya (2021), p. 38.

annexation of this region into the Empire (1884), Chinese logical space was still areal; and the Sinkiang region was considered an 'outer region' inhabited by 'barbarians'. The border remained fuzzy until 1949, when China formally extended its sovereignty to the Sinkiang region, over which largely indirect control had already been exercised.[133]

Cartographic representation of borders encouraged abandoning bordering processes based on natural features. In providing states with new and powerful tools,[134] maps allowed states to select only natural features functional to their geopolitical goals. For example, African territories were mainly delimited after 'astronomic' and 'mathematical lines,' whereas 'geographical features' made up 'a mere twenty-six percent'.[135] Kratochwil terms these boundaries 'spheric boundaries', whose application in 'deserts or polar regions ... can be explained in terms of the costs of demarcation in an uncharted and hostile environment'.[136] Secondly, in the aftermath of WWII, decolonialisation confirmed the 'territorial trap' into which European powers had led the world during the colonial era. Former colonial empires, such as the British Raj, split into several nation states; and 'spheric demarcations ... became ... through state succession, international boundaries'; to name just a few, the Egypt-Sudan, the Tanganyika-Kenya, and Botswana-South West Africa borders. To this we may add the Tanzania-Malawi border; it was demarcated by the 1890 Anglo-German Treaty (also known as the Heligoland-Zanzibar Treaty), but it is still challenged by Tanzania.[137]

3.5.2 *Legal Hydrology*

Within logical space, waters resemble more 'a distinct space' capable of exerting 'an influence over ... societies' than 'a liquid medium separating' them. Indeed, the hydrosphere is in constant conversation with the toposphere.[138] Watersheds and coasts, for example, have been turned into 'natural' borders: as Ochoa Espejo reminds us, these features acted as defensive borders; in Roman times, they were labelled *arcifinious*, since they possess 'limits that form a natural defense'.[139] Indeed, 'The 'the sea is a natural barrier, by definition'.[140]

Yet, the toposphere is only a part—and not the widest one—of the geo-ecosphere. Not only have human agencies traditionally divided the earth's surface into nation

---

[133] Kratochwil (1986), p. 31.

[134] Thompson Ford (2012), p. 135.

[135] Abraham (2007), p. 163.

[136] Kratochwil (1986), p. 37.

[137] Maluwa (2016).

[138] Valková (2017), p. 44. See also Glassner (1991).

[139] Ochoa Espejo (2020), p. 33.

[140] Tombs (2022), p. 2.

states, but they have also projected their logical space onto the liquid waters that constitute the hydrosphere. On the one hand, they attempted to outline bounded spaces; and several disputes have arisen as regards boundaries drawn on lakes and river borders, as well as river basins.[141] The regime of internal waters is indeed crucial.[142] Within national contexts, for example, the U.S. Supreme Court usually settles disputes related to state borders, drawing them through lakes and rivers, demarcating their ripuarian jurisdictions, and addressing the equitable apportionment of basin waters.[143]

States ideologically rely on the idea of natural borders and try to normalise the meaning they assign to them. For example, France manipulated the River Rhine; through such a reshuffle it became country's 'natural eastern limit' and therefore 'a valid excuse for [its] territorial expansion' into Germany, i.e. well beyond the territorial limit along the Rhine.[144] But, there are also manifold disputes in Africa, where the legacy of Western colonisation, which is the spheric boundaries drawn by the Europeans, was extremely contentious, say, when partitioning India and West Pakistan (now Bangladesh) along the Ganges Delta in 1947.[145] And it is constantly challenged in the Great Lakes' area, as well as around the lakes Chad and Turkana, the Omo river, and the Nile basin. In a time of climate change, each country's logical space has also reshuffled the role of the hydrosphere; the water scarcity adds fuel to the fire of geopolitical disputes.

Unlike land, oceans seem elude the application of the territorialised paradigm. Marine water mass is not susceptible to being physically possessed. Oceans, indeed, cannot be demarcated. Even the wakes of ships disappear in the oceanic mass. To this extent, it is closer to outer space; human agencies may find it impossible to spatially reshuffle these unbounded areas in logical space.

There have also been several attempts to apply the territorialised paradigm to the oceanic part of the hydrosphere and to extend authority into the oceans.[146] The seventeenth-century debate on the legal regime of the seas reflected the juxtaposition between two contrasting 'laws' of spatialities and their logical spaces. The dominant conception in international law, i.e. the free use of the seas for the whole of humankind, was challenged, among the others, by the English lawyer John Selden:

---

[141] Thomas (2017), p. 34, reminds us that '263 river basins cross national boundaries and that waterbodies create unique challenges for the 145 countries whose territories sit partially or wholly within a transboundary watershed'.

[142] Caflish and Zacklin (1981); see also Brunet-Jailly (2015).

[143] See U.S. Supreme Court *Michigan v. Wisconsin*, 270 U.S. 295 (1926): the border was drawn on Lake Superior via the Montreal River, Lake of the Desert, and Menominee River to Green Bay, and thence through the centre of the most usual ship channel to the centre of Lake Michigan; *New Jersey v. Delaware*, 295 U.S. 694 (1934) and *New Jersey v. Delaware*, 552 U.S. 597 (2008) on states' ripuarian jurisdiction; *Florida v Georgia*, 584 U.S – (2018), as for the waters of the Apalachicola-Chattahoochee-Flint River Basin.

[144] Diener and Hagen (2012), p. 42; Ochoa Espejo (2020), p. 102.

[145] Thomas (2017), p. 45.

[146] Benton (2010), p. ix. See also Braverman (2022) and Steinberg (2018), p. 219.

'the sea, by the law of nature, or nations, is not common to all men, but capable of private dominion or proprietie, as well as the land'.[147] Such logical space was indeed able to give a topospheric title to the land, excluding the application of the institutes of maritime law, the *jus gentium*, as well as the jurisdiction of the High Court of Admiralty. The scope of its jurisdiction did not extend to topospheric 'maritime causes arising on Land in England or beyond the sea', but only to 'all maritime injuries, [for] causes arising on the sea, or in parts out of the reach of the common law . . . Its proceedings are according to the method of the civil law'.[148]

Selden's strategy aimed to prevent the application of non-English law, which, like maritime law, was deeply rooted in *jus gentium.* The free-seas doctrine led to rearranging England's logical space around a new *imaginative* legal geography of the hydrosphere, which aimed to bridge its overseas territories. No wonder, therefore, that the Empire, which Great Britain had fashioned as a 'well-earn'd empire of the deep',[149] displayed several forms of connectivity. No matter how geographically remote they were, overseas territories shared (and still share) manifold forms of engagement. For example, when navigating towards the South Atlantic Ocean, the 'path in the great waters' leads to 'a cascade of islands' and 'chains of islands and reefs'.[150] The insular constellation still stretches, say, from Saint Helena, Tristan da Cunha, and Ascension, which is the 'essential staging post for flighting to and from the Falkland Islands' and its former dependencies.[151] This testifies Britain's inherited gift of governing 'a sea of islands' rather than a pattern of 'islands in the sea', and of adapting to 'the sea ice zone of the Southern Ocean . . . [which] is a dynamic environment that can change rapidly' (see Sect. 4.3).[152]

Besides chains and cascades of islands, British territories were networked through a variety of geographical connections. The British Empire 'created several corridors'; 'persons, trade and power travelled in various directions to and from the United Kingdom, principally London, as the legal centre of the empire'. These corridors were aimed at linking colonies 'in a hub and spoke model, occasionally with clusters connecting discrete geographical areas together such as the British West Indies'.[153] In the late eighteenth century, 'British statesmen . . . were sometimes given to musing that a world-wide network of commerce was preferable to an Empire of rule over land and people'.[154]

The free-seas doctrine also created its own logical space, whose 'ocean law . . . ossified a regime of overexploitation and inefficient resource use'. Alison Reiser

[147] Selden [1635] (1652). See also Grotius (1609). See See also Lesaffer (2017).

[148] Baker (2019), p. 131; Blackstone [1770] (2016), p. 46.

[149] Thomson (1802): line 166.

[150] See respectively: Bottomley (2020), p. 18; Mentz (2015), p. 52.

[151] Lovegrove (2012), p. 68 (Ascension), 104 (Tristan da Cunha), and 136 (Georgia). See Fichter (2008).

[152] Nicol (2018), p. 25.

[153] Bromby (2021).

[154] Marshall (1999), p. 25.

contends that the battle for controlling the fisheries of *Clupea harengus* (Atlantic herring) in the North Sea basin 'sparked the ideational battle' between Selden's *Mare Clausum* and Grotius's *Mare Liberum.*[155] Migrating shoals of herring were therefore 'important to the oceanic imaginaries of Grotius, his clients in the merchant and political classes', whereas Selden's claims of British sovereignty over the *mare clausum* surrounding the British Isles meant that the 'the seas were capable of being owned'.[156] Being a migratory fish, ships had to be left free to chase herring wherever it migrated; in addition, the British had 'a patchwork of claims encompassing such things as rights to fishing, trade, or navigation on the Pacific Ocean'.[157]

Exploration and exploitation facilitated the creation of new logical spaces and oceanic imaginaries. As John Gillis puts it, 'the Western eye had to learn to see the oceans and their islands' thus acknowledging the existence of the 'ocean space'. The nineteenth-century scientific developments and the 'geopolitical ambitions of maritime nations' boosted 'a reconception of the ocean as a physical and intellectual space full of imperial and commercial significance'.[158]

The entry into force of UNCLOS in 1994 turned the sea into a hyper-regulated space. Besides the internal and archipelagic waters (Art. 2.1), there is the 'territorial sea' (Part II of the UNCLOS), the 'contiguous zone' (Art. 33), the *exclusive economic zone* (EEZ: Part V), and the *continental shelf* (Part VI). There is also the 'area' (or seabed: Art. 1) and the high seas, which are 'open to all States', thus revolving around the sanctity and integrity of borders and Grotius's predicament of *mare liberum*. On them, disputes are often addressed as if they were related to 'physiographic or geopolitical criteria' even when delimiting the 'Earth's water surface areas'.[159]

The territorialised paradigm is particularly apparent in Parts V and VI of UNCLOS; 'nature is represented as passive and instrumentalized;' 'the ocean is abstracted and flattened, and life-forms with no immediate commercial value are blindsided and discursively partitioned.'[160] And the delimitation of the 'Earth's water surface areas,' i.e. the EEZ and the *continental shelf*, follows 'physiographic or geopolitical criteria.'[161]

Commodification also refers to the exploitation of the seabed, which has opened new logical spaces: and the 'fixed Euclidean space'[162] dangerously transformed the ocean into a 'territorialised ocean space' suitable for mining.[163] This is patent in the Arctic Ocean, where the discovery of oil and gas resources has prompted two States

155 Reiser (2020), p. 202.

156 Reiser (2020), p. 212.

157 Korosy (2020), p. 221.

158 Gillis (2014), p. 109. See also Reidy and Rozwadowski (2014).

159 Hasan et al. (2019).

160 Sammler (2020), p. 64. See Jones (2016, 2022).

161 Hasan et al. (2019).

162 Sammler (2020), p. 76.

163 Reid (2020), p. 34.

of the Arctic-Eight group (namely, Russia and Denmark) to extend the territorialised paradigm to the seabed in order to claim some portions that are located well beyond the 200 nautical miles set forth in Art. 76 of UNCLOS. To be claimed, evidence must be supplied as regards the contiguity of the relevant portions of the seabed to the 'submarine ridge'. Denmark and Russia must demonstrate that 'the Lomonosov ridge and the Alpha-Mendeleev ridge constitute submarine continuation of their respective continental shelves'.[164] Perhaps most interesting of all is the process of normalisation that this claim discloses. The deep ocean is indeed 'nearly as rugged in its bathymetry as the terrain we see on land', with the lowest part of liquid logical space covered with 'high mountains, deep valleys and canyons . . . areas of flatter plains, with hills of varying heights'.[165] It goes without saying that geopolitical agencies treat this logical space as if it were located on a dryland; and yet, it is uninhabitable and hostile for us humans; nevertheless, this space is legally saturated to allow its exploitation and to match our desired commodifying ends.

Indeed, Deep-Sea Mining (DSM) is governed by Annex III of UNCLOS, whose legal framework is particularly difficult to enforce. As Susan Reid holds, UNCLOS places the burden of proof on corporations 'whose activities may pose a threat or cause irreversible harm'. Yet this proof is difficult to validate: as a self-regulated practice, DSM is in the hands of the miners, who 'establish the environmental baselines against which their enterprises are measured; monitor their own progress against these baselines; and report any changes'.[166] In addition, new GIS, such as Google Ocean, open up new possibilities (and new logical spaces) which should be used for pacific purposes (Art. 240) and for the benefit of all humankind (Art. 241) but not for covetous private exploitation.

These new logical spaces do not end here. Oceanic territorialisation has raised several issues and renovated bordering processes. Owing to rising sea levels, states face the risk of submersion and 'loss of territory,' as well as rights to their marine territory.[167] This will affect states' rights on their EEZ and trigger maritime border disputes under Arts. 55 and 56 of UNCLOS. Probably, submersion of islands and territories would expose their former seabed to pillaging by economic transnational actors, who may try to convince us of the necessity of unceasing economic growth.

3.5.3 *Antarctica*

Antarctica has undergone several spatial reshufflings in the course of human history. It has been described as a *res nullius*, i.e. one of the 'un-owned spaces which have

[164] Valková (2017), pp. 144 and 146–154 on the legal-geographical settings of Arctic claims.

[165] Holden (2017), p. 55.

[166] Reid (2020), p. 27.

[167] Leal-Arcas (2013), p. 46.

not yet been absorbed into the Westphalian' territorial trap.[168] Its discovery overlaps with the search for new logical spaces to be filled in the Southern Continent. Odd as it may seem, the quest for the imaginative logical space which was *Terra Australis Incognita* (see Sect. 2.3.2) prompted the discovery of a completely new physical space, where the toposphere, the hydrosphere, and the cryosphere merge, triggering an incredible environment. This testifies the uniqueness of the Antarctic sea-ice ecosystem, whose abundant wildlife and nutrients are now threatened by global warming: 'the ocean is getting stormier and becoming warmer and more acidic. Sea ice is becoming less extensive in summer, and, in some areas, is now absent in winter'.[169]

In this section, I shall focus on the eighteenth century, when new logical spaces began to be processed in the southernmost parts of the world.[170] The search for the 'mythical southern continent' was 'the single most potent receptacle for European geographical expectation about the southern hemisphere'.[171] As the shaping force of European modern history, the imagined geography of *Terra Australis* was hard to die. The progressive discoveries of the Solomon Islands (1567), Vanuatu (1606), Bouvet Island (1739), Tuamotu (1764), Tahiti (1767), and the Kerguelen Islands (1772) were enough to give materiality to the imagined logical space of such a continent.[172]

Navigating towards the Southern seas in search for the *Terra Australis Incognita*, Captain John Cook's discoveries and explorations provided Great Britain with new geographical imaginaries. Political power added a new variable to (and a new justification for) such geographical expeditions: the production of scientific knowledge coalesced towards new commercial opportunities.[173] Both Cook's first (1768–1771) and second expeditions (1772–1775) had, among their objectives, the discovery of new lands and the making of observations 'as may be useful either to Navigation or Commerce'.[174] According to Joseph Conrad, power gave a legal spatiality to these places through *Geography Militant*: exploration highlighted 'both the technical dimensions of geography', as well as 'its cultural dimensions': 'the arts of triangulation, navigation and cartography', on the one hand; 'the rhetoric and iconography of discovery', on the other.[175] After being detected in 1820, the 'last blank spot on the map', i.e. Antarctica, was eventually filled at the end of the nineteenth century. Scott's and Amundsen's achievements paved the way to the heroic age (1898–1916), which meant manufacturing the last earth's landmass left as

168 Collis (2017), p. 290.

169 Nicol (2018), p. 34.

170 On the twentieth-century explorations see Collis (2017), p. 291.

171 Stallard (2016), pp. xvi and 47–63.

172 Stallard (2016), pp. 14–18.

173 Stallard (2016), pp. 216–227; O'Reilly (2013), p. 386; Basberg (2017), p. 209.

174 Williams (1996).

175 Driver (1992), p. 24. See Conrad (1924).

a new spatial normativity. Conrad reminds us that this event marked the advent of *Geography Triumphant* and closed the 'epoch of open spaces'.[176]

All these lands and seas have traditionally lain at the crossroads of Southern geopolitical struggles. The most relevant one juxtaposes Argentina (and Chile) to the United Kingdom, whose title in these parts of the earth has constantly been challenged, as the dispute culminated in the 1982 Falklands War demonstrates. Conventional geopolitical understandings of territorial possession are not applicable to Antarctica. State claims implied discovery and effective occupation in accordance with the *uti possidetis* doctrine. In particular, Argentinian (and Chilean) claims over the Antarctic region were traced back to the 1493 Papal Bulls issued by Pope Alexander VI, and the 1492 Treaty of Tordesillas; besides partitioning the world between Portugal and Spain, the Pointiff entrusted them with the propagation of the Gospel (see Sect. 4.2.3). The Argentinian government has appealed to geographical titles, such as 'geographical proximity, and geological continuity' to support its 'natural right' to the so-called *Antártida Argentina*.[177] The British never considered the Bulls valid legal titles, because these 'could never have included [these territories] within its longitudinal demarcation of the then Spanish sphere of influence'.[178] In 1955, Great Britain submitted 'the question of title to the Falkland Dependencies (as they then stood) to adjudication by the International Court of Justice'. Argentina and Chile did not accept the Court's jurisdiction; from a British perspective, this 'in itself indicates a feel weakness on legal title by such South American claimants'.[179]

The International Geographical Year in 1957 inaugurated a new possibility, i.e. considering the Antarctica a *terra communis*, on its own not susceptible to state claims. Even during the Cold War, scientific cooperation among states took place for scientific purposes, thus changing the location of Antarctica within states' logical spaces. These were eventually enshrined into the 1959 Antarctic Treaty, which coded the new legal geography for Antarctica. This became a physical-but-non-militarised space where scientific information could 'circulate freely' for pacific purposes.[180]

Likewise, one of the most physical features of Antarctica, i.e. the 'oceanographic/climatic frontier' which is the Antarctic convergence, was used to create a politico-legal discourse about space. The convergence marks the place where the warmer waters of the Atlantic, Indian, and Pacific Oceans meet the colder waters of the Antarctic Ocean: 'it roughly coincides with the mean February isotherm (10 °C) and lies around 58° S, considerably north of the Antarctic Circle'. But it is also considerably south, inasmuch as this oceanographic frontier became the template for the 1959 Antarctic Treaty, whose Article IV embargoes sovereign claims on the

[176] See Conrad (1924). See also Bonnett (2003), pp. 83–94; Stoddart (1981).

[177] Howkins (2017), pp. 10–11.

[178] Symmons (1984), p. 731.

[179] Symmons (1984), p. 726.

[180] Collis (2010). See also Collis (2017), p. 292.

continent of Antarctica, its islands, ice-shelves, areas of continental shelf, all sea, and airspace 'south of 60° South Latitude'.[181]

References

Abraham G (2007) Lines upon maps: Africa and the sanctity of African boundaries. Afr J Int Comp Law 15:61–84

Allaby M (2020) Oxford dictionary of geology & earth science, 5th edn. OUP, Oxford

Bachelard G [1942] (1986) L'eau et le rêves. Essai sur l'imagination de la matière. José Corti, Paris

Baker JH (2019) An introduction to English legal history, 5th edn. OUP, Oxford

Bartel R, Carter J (eds) (2021) Handbook on space, place and law. Edward Elgar, Cheltenham and Northampton

Basberg BL (2017) Commercial and economic aspects of Antarctic exploration – from the earliest discoveries into the 'Heroic Age'. Polar J 7:205–226

Benton L (2010) A search for sovereignty. Law and geography in European empires, 1400–1900. CUP, Cambridge

Berman HJ (2012) In: Witte J Jr (ed) Law and language: effective symbols of community. CUP, Cambridge

Berry JK (1986) Learning computer-assisted map analysis. J For 84:39–43

Bhattacharyya H (2021) Federalism in Asia. India, Pakistan, Malaysia, Nepal and Myanmar, 2nd edn. Routledge, London and New York

Black J (1997) Maps and politics. The University of Chicago Press, Chicago

Blacksell M et al (1986) Human geography and law: a case of separate development in social science. Prog Hum Geogr 10:371–396

Blackstone W [1770] (2016) Commentaries on the Laws of England, III, of private wrongs. In: Prest W (ed) The Oxford edition of Blackstone. OUP, Oxford

Blaut JM et al (2003) Mapping as a cultural and cognitive universal. Ann Assoc Am Geogr 93:165–185

Blomley N (1989) Text and context: rethinking the law-space nexus. Prog Hum Geogr 13:512–534

Blomley N et al (2001) Preface: where is law. In: Blomley N et al (eds) The legal geographies reader: law, power, and space. Blackwell, Oxford, pp xiii–xxii

Bonnett A (2003) Geography as the world discipline: connecting popular and academic geographical imaginations. Area 35:55–63

Borrough PA et al (2015) Principles of GIS, 3rd edn. OUP, Oxford

Bottomley A (2020) Between islands: colonial legacies and cultural imaginaries. Pólemos 14:237–260

Bowersock GW et al (1999) Late antiquity: a guide to the postclassical world. Harvard UP, Cambridge, MA

Braverman I (2022) Amphibious legal geographies: toward land–sea regimes. In: Braverman I (ed) Laws of the sea: interdisciplinary currents. Routledge, Abingdon/New York, pp 1–24

Braverman I et al (2014) Expanding the spaces of law. In: Braverman I et al (eds) The expanding spaces of law: a timely legal geography. Stanford UP, Stanford, pp 1–29

Brighenti A (2006) On territory as relationship and law as territory. Can J Law Soc 21:65–86

Bromby M (2021) The Cayman Islands: paradoxes of insularity in the Caribbean and other British overseas territories. Liverpool Law Rev 42:35–49

Brotton J (2013) A history of the world in twelve maps. Penguin, London

Brotton J (2018) Trading territories: mapping the early modern world. Reaktion Books, London

[181] See Dodds (2012), pp. 4–5.

Brunet-Jailly E (2005) Theorizing borders: an interdisciplinary perspective. Geopolitics 10:633–649

Brunet-Jailly (ed) (2015) Border disputes. A global encyclopedia. 3 vols. ABC-Clio, Santa Barbara, CA

Butler D (2006) Mashups mix data into global service. Nature 439:6–7

Caflish L, Zacklin R (eds) (1981) The legal regime of international rivers and lakes/Le régime juridique des fleuves et des lacs internationaux. Brill-Nijhoff, Leiden and Boston

Castree N (2013) Making sense of nature. Routledge, Abingdon and New York

Chambers R (2006) Participatory mapping and geographic information systems: whose map? Who is empowered and who disempowered? Who gains and who loses? Electron J Inf Syst Dev Ctries 25:1–11

Collis C (2009) The geostationary orbit: a critical legal geography of space's most valuable real estate. Sociol Rev 57:47–65

Collis C (2010) Critical legal geographies of possession: Antarctica and the international geophysical year 1957–1958. GeoJournal 75:387–395

Collis C (2017) Territories beyond possession? Antarctica and outer space. Polar J 7:287–302

Conrad J (1924) Geography and some explorers. Natl Geogr Mag 45:241–356

Cosgrove D (1994) Contested global visions: one-world, whole-earth, and the apollo space photographs. Ann Assoc Am Geogr 84:270–294

Couclelis H (2010) Ontologies of geographic information. Int J Geogr Inf Sci 24:1785–1809

Crampton J (2003) The political mapping of cyberspace. Edinburgh UP, Edinburgh

Crampton (2009) Cartography: Maps 2.0. Prog Hum Geogr 33:91–100

Crampton JW (2010) Mapping: a critical introduction to cartography and GIS. Blackwell, Oxford

Crampton J (2014) The power of maps. In: Cloke P et al (eds) Introducing human geographies. Routledge, London and New York, pp 192–202

Crang M (2014) Representation-reality. In: Cloke P et al (eds) Introducing human geographies. Routledge, London and New York, pp 130–143

Delaney D (2005) Territory. A short introduction. Blackwell, London

Delaney D (2010) The spatial, the legal and the pragmatics of world-making: nomospheric investigations. Routledge, Abingdon and New York

Delaney D (2015) Legal geography I: Constitutivities, complexities, and contingencies. Prog Hum Geogr 39:96–102

Diamond C (ed) (1976) Wittgenstein's lectures on the foundations of mathematics, Cambridge, 1939. From the Notes of R.G. Bosanquet, Norman Malcolm, Rush Rhees, and Yorick Smythies. Cornell, Ithaca NY

Diener AC, Hagen J (2012) Borders. A very short introduction. OUP, Oxford

Dodds K (2012) The Antarctic. A very short introduction. OUP, Oxford

Dodds K (2019) Geopolitics. A very short introduction. OUP, Oxford

Dodge M, Kitchin R (2013) Crowdsourced cartography: mapping experience and knowledge. Environ Plan A: Econ Space 45:19–36

Driver F (1992) Geography and power: the work of Michel Foucault. In: Burke P (ed) Critical essays on Foucault. Scholar Press, Aldershot, pp 147–156

Driver F (1994) Making space for science. Ecumene 1994:389–393

Driver F (2001) Geography militant. Cultures of exploration and empires. Blackwell, Oxford

Dunnett O et al (2019) Geographies of outer space: progress and new opportunities. Prog Hum Geogr 43:314–336

Easterly ES III (1977) Global patterns of legal systems: notes toward a new geojurisprudence. Geogr Rev 67:209–220

Economides K et al (1986) The spatial analysis of legal systems towards a geography of law? J Law Soc 13:161–181

Edney M (1994) Mathematical cosmography and the social ideology of British cartography, 1780-1820. Imago Mundi 46:101–116

Edney MH (1996) Theory and the history of cartography. Imago Mundi 48:185–191

Espenshade EB, Morrison JL (eds) (1982) Rand McNally Goode's World Atlas. Rand McNally, Chicago

Faccio S (2021) "Human habitation or economic life of their own": the definition of features between history, technology and the law. Liverpool Law Rev 42:15–33

Farinelli F (2003) Geografia. Un'introduzione ai modelli del mondo. Einaudi, Turin

Fichter J (2008) The British Empire and the American Atlantic on Tristan da Cunha, 1811–16. J Imp Commonw Hist 36:567–589

Friedman Herlihy A (2007) Renaissance star charts. In: Woodward D (ed) The history of cartography, volume three (Cartography in the European Renaissance PART 1). Chicago UP, Chicago, pp 99–122

Gaynor JL (2020) Liquid territory, shifting lands. Property, sovereignty, and space in Southeast Asia's tristate maritime boundary zone. In: Braverman I et al (eds) Blue legalities. The life and laws of the sea. Duke UP, Durham and London, pp 108–127

Gillis JR (2014) Islands of the mind: how the human imagination created the Atlantic world. Palgrave Macmillan, New York

Glassner MI (1991) The frontiers of the Earth – and of political geography: the sea, Antarctica and outer space. Polit Geogr Q 10:422–443

Goodchild FM (2007) Citizens as sensors: the world of volunteered geography. GeoJournal 69: 211–221

Goodchild MF, Longley PA (2021) Geographic information science. In: Fischer MM, Nijkamp P (eds) Handbook of regional science. Springer, Berlin, Heidelberg, pp 1597–1614

Goos J (1995) Marketing the new marketing: the strategic discourse of geodemographic information systems. In: Pickles J (ed) Ground truth: the social implications of geographic information systems. The Guildford Press, New York, pp 130–170

Grainger S, Conway D (2014) Climate change and international river boundaries: fixed points in shifting sands. WIREs Climate Change 5:835–838

Griffiths P (2006) An introduction to English semantics and pragmatics. Edinburgh UP, Edinburgh

Grotius U (1609) Mare liberum. Lodewijk Elzevir, Leuven

Hacıgüzeller P (2017) Collaborative mapping in the age of ubiquitous internet: an archaeological perspective. Digit Classics Online 7:5–16

Harley JB (1988) Silences and secrecy: the hidden agenda of cartography in early modern Europe. Imago Mundi 40:57–76

Harris R (2003) Earth observation and principles on data. In: Holder J, Harrison C (eds) Law and geography. OUP, Oxford, pp 539–555

Hartley TE (ed) (1981) Proceedings in the Parliaments of Elizabeth I, volume 1: 1558-1581. Michael Glazier, Wilmington, DE

Harvey D (2009) Social justice and the city. University of Georgia Press, London and Athens

Hasan M et al (2019) Protracted maritime boundary disputes and maritime laws. J Int Marit Saf Environ Aff Shipp 2:89–96

Holden J (ed) (2017) An introduction to physical geography and the environment, 4th edn. Pearson, London

Howkins A (2017) Frozen empires. An environmental history of the Antarctic Peninsula. CUP, Cambridge

Johnson KR (2003) Open borders. UCLA Law Rev 51:193–265

Johnston RJ (1983) Resource analysis, resource management and the integration of physical and human geography. Prog Phys Geogr: Earth Environ 7:127–146

Johnston RJ (1986) Four fixations and the quest for unity in geography. Trans Inst Br Geogr 11: 449–453

Johnston RJ (1991) A question of place. Exploring the practice of human geography. Blackwell, Oxford

Jones H (2016) Lines in the ocean: thinking with the sea about territory and international law. Lond Rev Int Law 4:307–343

Jones H (2022) Commodifying the oceans. In: Braverman I (ed) Laws of the sea: interdisciplinary currents. Routledge, Abingdon/New York, pp 49–67
Kastovsky D (2006) Vocabulary. In: Hogg R, Denison D (eds) A history of the English language. CUP, Cambridge
Kerneis S (2016) State, law, border in ancient Rome. Kobe Univ Law Rev:3–8
Khorram S et al (2012) Remote sensing. Springer, New York
King GQ (1991) Geography and GIS technology. J Geogr 90:66–72
Kitchin R, Dodge M (2007) Rethinking maps. Prog Hum Geogr 31:331–344
Korosy Z (2020) Whales and the colonization of the Pacific Ocean. In: Braverman I, Johnson ER (eds) Blue legalities. The life & laws of the seas. Duke UP, Durham and London, pp 219–235
Kratochwil F (1986) Of systems, boundaries, and territoriality: an inquiry into the formation of the state system. World Polit 39:27–52
Kristof LKD (1959) The nature of frontiers and boundaries. Ann Assoc Am Geogr 49:269–282
Langhans-Ratzeburg M (1928) Begriff und Aufgaben der geographischen Rechtswissenschaft (Geojurisprudenz). K. Vowickel, Berlin-Grunewald
Leal-Arcas R (2013) Climate change and international trade. Edward Elgar, Cheltenham and Northampton, OR
Lefebvre H (1974) La production de l'espace. Éditions Anthropos, Paris. English edition: Lefebvre H (1991) The production of space (trans: Nicholson-Smith D). Blackwell, Oxford
Lesaffer R (2017) Mare clausum (the closure of the sea or the ownership of the sea) 1635 John Selden (1584-1654). In: Dauchy S et al (eds) The formation and transmission of western legal culture. 150 books that made the law in the age of printing. Springer, Cham, pp 190–194
Lovegrove R (2012) Islands beyond the horizon. The life of twenty of the world's most remote places. OUP, Oxford
Maluwa T (2016) Oil under troubled waters?: Some legal aspects of the boundary dispute between Malawi and Tanzania over Lake Malawi. Mich J Int Law 37:351–419
Marshall PJ (1999) Introduction. In: Marshall PJ (ed) The Oxford history of the British Empire, vol II, the eighteenth century. OUP, Oxford, pp 1–27
Marston SA (2000) The social construction of scale. Prog Hum Geogr 24:219–242
Matthews JA, Herbert DT (2008) Geography. A very short introduction. OUP, Oxford
Mentz S (2015) Shipwreck modernity: ecologies of globalization, 1550–1719. University of Minnesota Press, Minneapolis, MN
Merk W (1926) Wege und Ziele der geschichtlichen Rechtsgeographie G. Stille, Berlin
Miller D (2016) Boundaries, democracy, and territory. Am J Jurisprud 61:33–49
Nayak A, Jeffrey A (2011) Geographical thought. An introduction to ideas in human geography. Routledge, Abingdon and New York
Nicol S (2018) The curious life of Krill. A conservation story from the bottom of the world. Island Press, Washington et al
Núñez A et al (2017) From a geographical mountain to mountain geographies. A social geography analysis of the Chilean Andes. J Alpine Res/Revue de géographie alpine 105(4):1–12
O'Reilly J (2013) Antarctic climate futures: how Terra incognita becomes Terra clima. Polar J 3: 384–398
Ochoa Espejo P (2020) On borders. Territories, legitimacy, and the rights of place. OUP, Oxford
Pickles J (1992) Geography, GIS, and the surveillant society. Pap Proc Appl Geogr Conf 14:80–91
Pickles J (2004) A history of spaces: cartographic reason, mapping and the geo-coded world. Routledge, London
Pilch M (2017) The structure of Wittengstein's logical space. Wittgenstein-Studien 8:15–60
Pred A (1984) Place as historically contingent process: structuration and the time-geography of becoming places. Ann Assoc Am Geogr 74:279–297
Ratliff E (2007) The whole earth, catalogued. How google maps is changing the way we see the world. Wired 15:154–159
Reid S (2020) Solwara 1 and the sessile ones. In: Braverman I, Johnson ER (eds) Blue legalities. The life & laws of the seas. Duke UP, Durham and London, pp 24–44

Reidy MS, Rozwadowski HM (2014) The spaces in between: science, ocean, empire. Isis 105:338–335

Reiser A (2020) Clupea Liberum. In: Braverman I, Johnson ER (eds) Blue legalities. The life & laws of the seas. Duke UP, Durham and London, pp 201–218

Relph E (1976) Place and placeness. Pion, London

Rissanen M (2000) Standardisation and the language of early statutes. In: Wright L (ed) The development of standard English 1300-1800: theories descriptions conflicts. CUP, Cambridge

Rossi E (1988) I Giuristi alla Conquista della Marmolada. Foro Italiano I:3184–3191

Sammler KG (2020) Kauri and the Whale. Oceanic matter and meaning in New Zealand. In: Braverman I, Johnson ER (eds) Blue legalities. The life & laws of the seas. Duke UP, Durham and London, pp 63–84

Schuurman N (1999) Critical GIS: theorizing an emerging discipline. Cartographica 36:1–108

Selden J (1635) Mare clausum seu, De dominio maris. Pro Richardo Meighen, London. English edition: Selden J (1652) Of the dominion, or, ownership of the sea. William-Du Gard, London

Smith N (1992) Geography, difference and the politics of scale. In: Doherty J et al (eds) Postmodernism and the social sciences. Macmillan, London, pp 57–79

Stallard AJ (2016) Antipodes: in search of the southern continent. Monash UP, Melbourne

Steinberg P (2018) The power of water. In: Coleman M, Agnew J (eds) Handbook on the geographies of power. Edward Elgar, Cheltenham and Northampton, MA, pp 217–228

Stoddart DR (1981) Geography, ideology and social concern. Blackwell, Oxford

Symmons CR (1984) Who owns the Falkland Island dependencies in international law? An analysis of certain recent British and Argentinian official statements. Int Comp Law Q 33:726–736

Taylor P (1987) The paradox of geographical scale in Marx's politics. Antipode 19:287–306

Taylor PJ (1994) The state as container: territoriality in the modern world-system. Prog Hum Geogr 18:151–162

Thomas KA (2017) The river-border complex: a border-integrated approach to transboundary river governance illustrated by the Ganges River and Indo-Bangladesh border. Water Int 42:34–53

Thompson MM (1979) Maps for Americans: cartographic products of the U.S. geological survey and others. Department of the Interior, Washington DC

Thompson Ford R (2012) Law and borders. Ala Law Rev 64:123–139

Thomson J (1802) Britannia. In: The works of Mr. James Thomson: with his last corrections and improvements, vol 2. Printed for R. Baldwin, London

Tobler WR (1961) Map transformations of geographic space. PhD Thesis, University of Washington

Tobler WR (1966) Medieval distortions: the projections of ancient maps. Ann Assoc Am Geogr 56: 351–360

Tobler WR (2004) Thirty-five years of computer cartograms. Ann Assoc Am Geogr 94:58–73

Tobler WR (2018) A new companion for Mercator. Cartogr Geogr Inf Sci 45:284–285

Tombs R (2022) This sovereign Isle. Britani in and out Europe. Allen Lane, London

Tripathi D, Chaturvedi S (2020) South Asia: boundaries, borders and beyond. J Borderlands Stud 35:173–181

Valková I (2017) Claiming the Arctic: on the legal geography of the northernmost sovereignty dispute. Polar J 7:143–164

van Gent RH, Van Helden A (2007) Lunar, solar, and planetary representations to 1650. In: Woodward D (ed) The history of cartography, volume three (Cartography in the European Renaissance PART 1), Chicago UP, Chicago, pp 123–134

von Benda-Beckmann F, von Benda-Beckmann K (2014) Places that come and go: a legal anthropological perspective on the temporalities of space in plural orders. In: Braverman I et al (eds) The expanding spaces of law: a timely legal geography. Stanford UP, Stanford, pp 30–52

Von der Dunk FG (2020) Advanced introduction to space law. Edward Elgar, Cheltenham and Northampton

Westlake D (1985) High adventure publication. Mysterious Press, NYC

Whatmore S (2002) Hybrid geographies. Natures, cultures, spaces. Sage, London
Whatmore S (2014) Nature & human geography. In: Cloke P et al (eds) Introducing human geographies. Routledge, London and New York, pp 152–162
Williams G (1996) 'To make discoveries of countries hitherto unknown'. The admiralty and pacific exploration in the eighteenth century. Mariner's Mirror 82:14–27
Wittgenstein L (1921) Logisch-Philosophische Abhandlung. Annalen der Naturphilosophie XIV:185–262. English edition: Wittgenstein L (1922) Tractatus Logico-Philosophicus (ed Kegan P) Harcourt, Brace & Company, Inc.; Trench, Trubner & Co., New York and London
Wood D (1992) The power of maps. The Guildford Press, New York
Wood D (1993) The fine line between mapping and mapmaking. Cartographica 30:50–60
Wood D (1994) The traditional map as a visualisation technique. In: Earnshaw HM, Unwin DJ (eds) Visualisation in geographical information systems. Wiley, Chichester, pp 9–17
Wood D (2010) Rethinking the power of maps. The Guildford Press, New York and London

Chapter 4
Colonial Underpinnings: Spatiality of Law

4.1 Legal Geography, 'Triumphant'

In *Maps and Politics*, Jeremy Black formulates an intriguing statement on how maps are arranged in atlases. We do not usually pay particular attention to the 'disproportionate amount of space' they devote to 'maps of Europe and North America', which 'are also generally placed before maps of other continents.' We assume this as a fact. 'This very arrangement', Black continues, 'suggests a hierarchy of importance and a Eurocentric, or rather "Western", primacy' within geographical knowledge.[1]

Black grasps the point. This chapter further explores his predicament and the processes of spatial production underpinning such a Eurocentric depiction of the earth. The way continents and countries are arranged is a matter of hierarchy, which cannot be explained merely in cartographic terms. Meaningful places are not 'simply "there" waiting discovery and representation'; rather, their production is the outcome of a politico-legal discourse, with wider (geographical, political, and legal) implications.[2]

Hierarchy replicates the competition among different actors—in particular, Western political ones. Colonial geographies, for instance, reflected the 'politics of spaces' European powers had been pursuing since the dawn of the Age of Discovery.[3] Legal cartography was servient to these geopolitics: moving 'itself outwards, [Europe's] sense of cultural', as well as political, 'strength was fortified'.[4] European powers were thus able to shape African, American, Oceanian, and Asian legal cartographies. As noted, (see Sect. 3.1) this spatial production is labelled 'spatiality of law'; through it, Europe westernised the world and superimposed its own

[1] Black (1997), p. 40.

[2] Nayak and Jeffrey (2011), p. 242. See Sect. 3.1.

[3] Gordon and Klein (2001), p. 5. See Livingstone (1992), p. 32; Driver (2001) and Brotton (2018).

[4] Said [1978] (2003), p. 117.

M. Nicolini, *Legal Geography*, Ius Gentium: Comparative Perspectives on Law and Justice 105, https://doi.org/10.1007/978-3-031-19410-8_4

geopolitics of law onto it.[5] In so doing, it pointed to 'both opening and later closing' the territories acquired by Europeans.[6]

To this end, it was necessary to discover new territories. As well as to do something more: 'To plant Christian religion. To trafficke. To conquer. Or, to doe all thrée'. With these words Richard Hakluyt the elder (1530–1591) epitomised 'the ends of' colonial voyages in his *Inducements* (1585): 'TO plant Christian religion without conquest, will bée hard. Trafficke easily followeth conquest: conquest is not easie. Trafficke without conquest séemeth possible, and not vneasie. What is to be done, is the question'.[7]

The answer to this question was to make world open to competition; European countries contended with each other to project their own logical spaces onto the earth's surface. This had effects on the search for the *Terra Australis Incognita*. The mythical southern continent stirred up the imagination: 'men have been drawn to unknown regions . . . echoes of which ring in our ears . . . when on modern maps we see spaces labelled "unexplored", rivers shown by broken lines, islands marked "existence doubtful"'.[8] No wonder that this label was made up of 'the most promising words ever written on the maps of human knowledge' acting as the 'driving force behind the voyages of discovery' and scientific endeavour.[9] In *Heart of Darkness*, Conrad portrayed Europe's appetite for political prestige. Marlow, the main character, embodies the juvenile ardour of Europe's explorers. As a 'little chap', he 'had a passion for maps' and used to lose himself 'in all the glories of exploration'. At that time there were several 'blank spaces on the earth'. When one looked 'particularly inviting on a map', he 'would put [his] finger on it and say, When I grow up, I will go there'.[10]

Radiating from Europe, colonial powers reimagined spaces for new purposes. They surveyed and renamed them with their own nomenclatures. And filled the unexplored spaces: 'The blank has become a network of dark lines, the interspaces covered with the names of tribes and rivers and lakes', J. Scott Keltie highlighted as regards Henry Molton Stanley's surveys of Central Africa.[11] In appropriating these lands, they dispossessed their original inhabitants of their imaginative geographies and superimposed theirs. The process was functional to the second spatial step mentioned above, that of closing territories.

'Closing' these territories triggered two consequences. Firstly, it cut the ties between local communities and their territories, thus neglecting their legal geography (their revival will be addressed in Sect. 5.1.2). It then transplanted the territorial trap into them, whose thorough application replicated European geopolitics outside

---

[5]On 'Geopolitics of law' see Monateri (2013). See also Godlewska and Smith (1994), p. 2.

[6]Harley (1988).

[7]Hakluyt (1602), p. 30. See also Tomlins (2001), p. 320.

[8]Wright (1947), p. 1. See Sect. 3.6.3.

[9]Respectively: Boorstin (1983), p. xvi; Livingstone (1992), p. 35.

[10]Conrad [1899] (2008), p. 108.

[11]Keltie (1890), p. 51.

Europe. Hence, colonial competition was not confined to territorial acquisitions. Certainly, the more lands they discovered, the more territories they conquered, and the bigger their colonial empires would become. Competition also meant domination, i.e. the natural outcome of the cultural-legal superiority of Western civilisation and prestige. Imperial ambitions 'required a knowledge of the sea casts, fortifications, internal waterways', i.e. an 'Imperial policy to survey, chart, and map all the coasts of all the continents'.[12]

The law 'was ... the principal instrument by means of which newly acquired territories' and their resources met 'Europe's domination over the rest of the world'.[13] Conrad's *Geography Triumphant* is triumphant because it conveyed a cartographic visualisation of the legal systems of the world. Keenly interested in this process of legal mapping, at the end of the nineteenth century the nascent discipline of comparative law also reinforced the colonial attitude towards non-Western conceptions of the law, paving the way to the application of European legal paradigms (see Sect. 4.4). With the support of *Geography Triumphant*, the West gained the global politics of place; by virtue of its legal traditions, it ultimately imposed their allegedly superior *legitimacy* over other non-Western ones, rearranging colonial spaces through its own culture-specific 'legal frame' (see Sect. 4.3).[14]

In Conrad's words, that is why the spatiality of the law made legal geography 'triumphant'. Firstly, it disclosed the colonial underpinnings of legal geopolitics, with which this chapter will critically engage. Secondly, legal geography was triumphant because, unlike the law of spatiality, it did not merely reshuffle and normalise the meaning of the spatial furniture (see Sect. 3.2). By contrast, geographical attributes and politico-legal discourse upheld the superiority of Western spatial production. The interstate struggle was related to the prestige of Western legal systems, as well as to their ability to offer a pattern whereby 'peace, security and order, and therefore efficiency and progress' might be attained.[15]

The spatiality of law secures the superior *legitimacy* of the Western legal tradition through a set of circular justifications. The process of spatial production demonstrates the superiority of Western legal derivatives in arranging states as social containers even outside Europe. Such a superiority, it goes without saying, is traceable back to the superiority of Western cultural practices, which unsurprisingly favoured the fabrication of the global politics of space *also* through legal practices.[16]

The practice of ranking legal systems also characterises the imaginative geographies of globalisation. Transnational actors create their own geopolitics of law and spatial hierarchies. The difference lies in how the spatiality of law is represented: digital technology allows us to represent indexes and indicators whereby countries

---

[12] See e.g., Taylor (1994), pp. 3 and 7.

[13] Patton (2000), p. 26.

[14] Halpérin (2009).

[15] Delaney (2005), p. 19.

[16] On how superiority was displayed through colonial Exhibitions revolving around the opposition between 'British enterprise ... and African primitivism' see Driver (2001), p. 148.

are said to be more or less democratic than others, and systems ensure more elevated economic performances than others.

The 'interstate conflict and antagonism' of classical legal geopolitics has been replaced by a new geopolitical competition, i.e. the struggle for politico-legal dominance among legal families.[17] Economic and financial actors evaluate the efficiency of the legal systems and their attractiveness towards international investments: the more efficient the system, the more likely the investment, and the higher the economic returns (see Sect. 4.6).

4.2 Essentialising Space

4.2.1 On Place Names and Portolan Charts

'Seldom or never' do geographers nowadays have 'the possibility to enter literal *terrae incognitae*'.[18] From the fifteenth century onwards, the discovery of new continents marked the beginning of what Joseph Conrad labelled *Geography Militant*. Explorers and adventurers had an insatiable appetite for new lands; 'Nothing could satisfy their sense of the fitness of things but an enormous extent of solid earth'.[19] In discovering new lands, they added 'a bit of truth here and a bit of truth there' on maps and were often 'swallowed up by the mystery [that] their hearts were so persistently set on unveiling'.[20] Like in Antarctica, 'white spaces succumbed to the dominion of science century'.[21]

The manufacturing of the earth's last landmass, which is Antarctica, 'closed' both territories and the epoch of the politics of spaces. *Geography Triumphant* commenced. 'Condemned to make [their] discoveries on beaten tracks',[22] Europeans turned to territorial maps; they territorialised their imagined geographies and imposed the yoke of Western civilisation on indigenous populations. In the aftermath of the West African Conference in Berlin (1884–1885), European powers relinquished their informal empires, thus laying down the rules for the definitive partition of Africa. Its 'scramble' was concluded when the French Protectorate over Morocco was declared in 1912.[23] Exploitation of natural resources soon ensued. At

---

[17] Nayak and Jeffrey (2011), p. 235.

[18] Wright (1947), p. 1.

[19] Conrad (1924), p. 243.

[20] Conrad (1924), p. 243.

[21] Driver (2001), p. 3. See Sect. 3.6.3.

[22] Conrad (1970), p. 92. See Ho (2007).

[23] On the colonial partition, and on the conventional end-dates for the Scramble of Africa (Italy's invasion of Libya: 1911; the French Protectorate over Morocco: 1912; the Peace of Vereeniging ending the South African war: 1899–1902) see Sharkey (2013), pp. 153–154; Gjersø (2015); Griffiths (1986); Sanderson (1974). See also Griffiths (1986), pp. 204–205.

the same time, European powers entrusted their traders and agents with the civilising mission of the backward, i.e. non-Western, peoples.

Taking place in Africa, Asia, and Oceania, this mission was the main justification for both colonisation and world exploitation, unveiling a third type of 'closing of territories'; after territories and the politics of space, Europeans also closed the markets. Their ideology of domination had indeed an intrinsic mercantilist attitude. Colonial empires competed for the establishment of 'a coherent trading area subject ... to a relatively uniform system of law.'[24] Within such global geographies of power, politico-legal spaces were definitively caught in the trap elaborated by legal geography 'triumphant'; and 'the *raison d'être* of colonies [was] *to benefit the commerce of the Mother Country*.'[25]

Similar processes of spatial production characterise imaginative geographies. Said's *Orientalism*, for instance, depicts how the 'closing of territories' actually worked. Through its 'self-generated "expansion"', Europe (more generally, the West) 'shaped the modern world'.[26] 'European Modernity' also self-generated its own 'self' ...through a proliferation of spacings', which 'were all spacings *within* Europe', capable of 'set Europe off against its exterior "others"'.[27] A 'self-sufficient Europe' was able to 'essentialise', i.e. misrepresent, the rest of the world.[28] How did this occur? How was it possible to essentialise geographical knowledge and shape legal geographical spaces?

As a type of spatiality of law, essentialising non-Western physical and human geographies is a further application of the territorialised paradigm. In claiming *dominium* over newly discovered areas, Europeans re-*imagined* territorial spaces, turning them into their belongs of public law (see Sect. 2.4).

Essentialising requires physical and imaginative apprehension—as well as political appropriation.

Both apprehension and appropriation account for the 'linguistic impact of European imperial rule' on formerly colonised areas, even when these were situated in Europe.[29] Take, for example, the effects of the English rule over Ireland, which started in the Middle Ages with the Anglo-Norman conquest of the island (1169–1172). In his *Flores Historiarum*, Roger of Wendover commented that King John 'caused English laws and customs to be instituted in Ireland'.[30]

The English attempted to rearrange 'the histories of the rest of the British Isles and Ireland, at worst by ignoring them, at best by converting them into subsidiary appendages to its own central theme'—or by superimposing their own legal

---

[24] See Price (1998), p. 17.

[25] Egerton (1897), p. 2. See now Arewa (2021), pp. 4–5.

[26] Driver (2001), p. 7.

[27] Gregory (2004), p. 3.

[28] Driver (2001), p. 7.

[29] Black (1997), p. 42.

[30] Roger of Wendover (1886–1889), p. 56. See also Brand (1981).

geography onto the whole Irish island.[31] Such attempts are vividly portrayed in Gerald of Wales's *Topographia Hibernica* (1187) and *Expugnatio Hibernica* (1189). According to the rationale of Gerald's rhetoric, 'The Irish, once separate from the holy island on which they live, reveal themselves to be uncivilized and unchristian wild men, who lack a legitimate claim to the land they inhabit'.[32] The Welsh author eulogised Ireland but harshly criticised the nature of its inhabitants; this 'Celtic fringe' was indeed 'regarded by London elites as a sphere of cultural and legal primitiveness', thus upholding the need for British intervention.[33]

Intertwining with Conrad's *Geography Fabulous*, the Irish reacted to this topographic challenge prompted by Gerald of Wales and his place-name policy. In the aftermath of the English conquest of Ireland, the reaction was committed to the most relevant twelfth-century prosimetrum written in Middle Irish, i.e. the *Acallam na Senórach* (*Tales of the Elders of Ireland*). The text describes the encounter between Saint Patrick and the Fenian warriors Oisín and Caílte; the latter transmit the legacies of pre-Christian Ireland to Ireland's first saint and patron. The literary and cultural Irish reaction proved to be strategic. St Patrick and Cáilte mapped the Island and explained the Irish origins of the different Irish place names; the colloquies are a response to the English colonialist rhetoric attempting to sever the relations between the Irish community and its territory.[34]

By contrasting Gerald's rhetoric, *Acallam na Senórach* asserts the legitimacy and validity of Irish place names. It also confers legal significance upon the physical geography of Ireland as *the* central aspect of its identity, linking 'the land and the people ... through the narratives written and told about both.'[35] Territory and community are not separable and physical geography turns into *legal geography*.

Whereas Mediaeval England still regarded Ireland a series of '"magic" hill regions',[36] the transition from the Middle Ages to the Age of Discovery meant shifting from the 'extravagant speculation' and 'imaginary kingdoms' of *Geography Fabulous* to the discoveries of *Geography Militant*.[37] Take, for example, the first Atlantic expeditions the Portuguese undertook in the fifteenth century. What these early explorers had before them were open spaces and wide horizons, i.e. truly unexplored stretches of open waters. As Abulafia puts it, the Portuguese acquisition of the uninhabited Azores, Madeira, Cape Verde Islands, and the furthest island of São Tomé created 'an interconnected world ... beyond the familiar waters of the Mediterranean' disclosing 'open seas that had rarely been navigated before the fourteenth century'.[38] As commentators were 'reluctant to believe in the existence'

[31] Davies (1994), I, p. 3.

[32] Connell (2011), p. 16.

[33] Benton (2010), p. 228. Gerald of Wales (1978, 1982).

[34] See Nicolini (2016).

[35] Connell (2011), pp. 8–9.

[36] Benton (2010), p. 228.

[37] Conrad (1924), p. 241.

[38] Abulafia (2020), p. 483. See also Livingstone (1992), p. 41.

of continents distinct from Europe, Asia, and Africa, the Portuguese had to create a new logical space, within which they would reshuffle and normalise newly discovered physical, human, and imaginative geographies (see Sect. 3.3.2).[39]

The Portuguese discovered and reshuffled. New logical spaces were processed at the same time as they discovered new lands. The creation of logical space was coeval to the law of spatiality: discoveries, spatial reshuffling, and legal coding went hand in hand with new 'territories, monopolization of trade routes, and establishment of commercial centres through the coastal territories'.[40] It does not come as a surprise that the Portuguese equivalent for discovery, i.e. *descobrimento*, pointed to both logical space (i.e. 'the achievements of the Portuguese navigational developments') and spatial production (i.e. 'the territories to which such achievements provided access').[41] The same holds valid for the processing of Spain's logical space in the Americas. 'The "discovery" itself, even within the ideology of discovery, has no existence of its own'. It had to be imagined and processed through imagination, maps, and the law. Indeed, Spaniards 'symbolically invented a "New Spain" to be visualized, possessed, and controlled'.[42]

Cartography supplied an extremely functional tool in the form of portolan charts. Invented during the Middle Ages, these were accurate and extremely precise geographical maps. Based, as they were, on an unrivalled level of geometric measurement, they displayed the geographical data of the 'core area [consisting] of the Mediterranean, the Black Sea, and the Atlantic coast from Cap Drâa in present-day Morocco up to and including the south coast of England'.[43] As 'nautical aids' for navigating and trading in the Mediterranean and Black-Sea basins, they depicted 'key ports and basic coastal features of specific stretches of coastline' with the necessary accuracy.[44] Together with their partial mapping of the Atlantic Ocean, their accuracy was valuable during the progressive opening up of navigation and seaborne trade beyond the Pillars of Hercules. These charts, I assume, allowed explorers to progressively annotate new coasts and locales on the maps, thus storing concrete practical information and, at the same time, shaping a logical space that was closer to reality. To exit geography fabulous, adding 'a bit of truth here and a bit of truth there' was functional to the establishment of new imagined geographies.

By progressively charting new routes and territories, the Portuguese populated the earth's surface with a variety of legal spaces. The navigation along the coast of West Africa disclosed still unchartered Atlantic routes: Pedro Reinel's 1483 portolan chart, for example, added all the recently discovered areas: Cape Bojador (1434), the Cape Verde Islands (1460), São Tomé and Príncipe (1472–1475), with the Equator being crossed in 1473. In 1488 Bartolomeu Días rounded the Cape of Good Hope.

39 Brotton (2018), p. 29.

40 Brotton (2018), p. 29.

41 Brotton (2018), p. 72; Seed (1995), p. 129.

42 Craib (2000), p. 17.

43 See Nicolai (2015), pp. 517–518.

44 Brotton (2018), p. 29. See Campbell (1987), pp. 438–445; Nicolai (2015).

We shall never know what he felt when rounding it; nor shall we experience what, whilst crossing the isthmus of Panama in 1513, Vasco Núñez de Balboa felt when setting 'his eyes for the first time upon the ocean the immensity of which he did not suspect and which in his elation he named the Pacific'.[45] The same is valid for Vasco da Gama's landing in India in 1497 and for Ferdinand Magellan's navigation through his eponymous strait in South America.

Of these voyages, what we still enjoy is the legacy of their several imaginative legal geographies, that is, conceiving of the world after new global lines.

4.2.2 *Thinking Globally and Sanitising the World*

The Age of Discovery marked several legal geographical transitions: from geography fabulous to geography militant; from Mediterranean to Oceanic spaces; and from unchartered wide horizons to newly created logical spaces. This favoured the projection of new imagined geographies onto the earth's surface. The production of legal spaces soon ensued.

In his *The Nomos of the Earth*, Carl Schmitt likens the discovery of new territories to the advent of a new global spatial order. As the 'Age of Discovery' led to the expansion of European geographies of power, 'land-appropriation of a New World' required a new spatial order, which was arranged 'along global lines'.[46] Schmitt's analysis of these new geographies is restricted to the birth of international law. In so doing, he pays more attention to the outputs of spatial production than to spatial reshuffling itself. When stating that 'the struggle among European powers for land-appropriations made necessary certain divisions and distributions of spaces', he points to the division of spaces but does not explain why and how the criteria applied for land distribution and appropriation were elaborated. Schmitt terms these outputs 'global linear thinking'; meridians and parallels, however, merely grasp 'the external and superficial character of this thinking'.[47] The price to be paid for this was the flattening of the earth along the system of meridians and parallels that also saturated Ribeiro's and Mercator's projections, i.e. two of the most celebrated achievements of early modern cartography.[48]

European geography had to be rewritten, and the earth mapped again. This process had to be attained at a rapid pace. New routes were discovered. Magellan, Vespucci, da Gama, Tasman, and Cook—to mention just a few—opened new routes when navigating south. Being the major European powers, the Spaniards, the

[45] Conrad (1924), p. 243. See also Seed (1995), pp. 116–120; Brotton (2018), p. 74. On the progressive *descobrimento* of new coasts and the points reached by successive expeditions see Campbell (1987), pp. 411–414.

[46] Schmitt (2006), p. 86.

[47] Schmitt (2006), p. 86.

[48] Brotton (2013), pp. 186 and 218.

Portuguese, the English, and the Dutch had to accommodate newly discovered territories into a new political order becoming visible through maps. Amid such global confrontations, imperial powers projected their Western imaginaries onto the earth. Jeremy Brotton reminds us that, to commemorate his 'politico-legal' marriage with Catherine of Habsburg (that is, Charles's V sister), João III of Portugal commissioned a 'series of lavish tapestries entitled the *Spheres*' (1525). The last tapestry of the series epitomised the Eurocentric geography of the Age of Discovery: it indeed represents 'the sweep of territory and ocean stretching eastwards through Africa, the Indian Ocean and Southeast Asia, curving back towards Europe through the Red Sea and into the Mediterranean world and Central Asia'. On these territories, João 'claimed entitlement'.[49]

The process was not as linear as it might now appear. On its own, the spatiality of law is a composite process whereby space and law are mutually co-productive. As an admixture of several elements, it cannot be reduced to a linear and abstract thinking replicable irrespective of time and place. There are hints of universalism in linear thinking; spaces are unfettered from their societal contexts, and local variations are ignored. Quite unsurprisingly, this facilitated the imposition of Western, universal, and abstract legal derivatives on the human and physical geographies of the world.

By contrast, the production of concrete spaces requires an examination of all contingencies and variables that contribute to outlining their operational rules. As noted (see Sect. 2.3.2), the *acts of simplification* of legal cartography reduce the inherent complexity of the world to a picture by stripping away whatever is ideologically useless or dangerous for the purposes of political power.

A couple of examples will suffice.

The first of these relates to Portuguese achievements. The *Spheres* tapestry series is the tale of several encounters. The mysteries of Africa, the fascination of India, and the bonanza of the 'spice-producing islands of Moluccas' prompted immediate geographical assimilation. The tapestries convey the 'wonder and excitement of such exotic, commercially tantalizing distant territories', offering 'a reassuringly sanitized vision of long-distant travel, whose deleterious effects' were craftly occulted and tempered 'with the lavishly illustrated maps of the known world'—i.e. the world known and told after the European politico-legal discourse.[50]

Portugal was involved 'in changing the basic apprehension of the shape of the globe.'[51] Yet, the incorporation of these discoveries had several consequences; among the others, a close intermingling with places recently conquered. This changed Portugal itself, which did not apply the linear, abstract, and 'sanitised' approach to these new worlds. On the one hand, through discovery they gained new astronomical knowledge, which allowed them to 'establish the latitude of a new place' even under skies and stars completely unknown to other European powers. In Renaissance Europe, nobody 'regarded discovery as creating dominion'. Yet, the

---

[49] Brotton (2018), pp. 13 and 15. For more on the tapestries see Gschwend (1995), p. 8.

[50] Brotton (2018), pp. 13 and 15.

[51] Brotton (2018), p. 60.

Portuguese were able to discover and, through astronomical knowledge, to 'provide the central proof of their discovery', indicating its political apprehension.[52]

On the other hand, unlike other Europeans, they engaged in a dialogue with their new logical spaces. As mainstream historiographical chronicles of the collapse and decline of its empire uphold, Portugal had a close engagement with its colonies—and a concrete global thinking. These chronicles blame on Portugal the toleration of racial intermarriage, which led to a 'racial degeneration [eroding its] practical and administrative efficiency.'[53] The decline is directly linked to the dismissal of the sanitising function of linear thinking, which would have shielded the Portuguese from the social and cultural consequences of their seaborne empire. Unlike them, the other European powers manufactured their own spaces but occluded their politico-legal discourse beneath the lavishly (essentialising and sanitising) universal principles of the Renaissance.

The second example is drawn from early American colonial history. Linguists usually divide the history of U.S. English into three periods. The first of them 'extends from the settlement of Jamestown in 1607' to probably '1790, when the last of the colonies ratified [the Constitution] and the first census was taken'. And this is the most productive one in terms of legal geography.[54] In an effort to create colonial space, the examination of all the relevant contingencies outlines a clear operational rule. In those constitutive years, the intermixture between colonists and natives seems to 'overcome [their] traditional, essentialist notions of cultural identity in favor of a continuum of newly composed, mixed or even contradictory identities'.[55]

Chesapeake Bay hosted early settlements even before the foundation of Jamestown. Thomas Harriot 'visited Roanoke Island at the mouth of the Chesapeake' in 1585–1586; he gained some fluency in Algonquian and was able to secure a first settlement in 1587 at Roanoke. The intermixture between colonists and natives led to dismissing the sanitising function of essentialist European thinking. By 1590, the colony had vanished, though curiosity 'about the fate of the settlers led to suppositions that they had joined Native Americans and moved to the forests of the interior'.[56] Again, Europe's *acts of simplification* reduced the inherent complexity of the legal geography of the New World to an essentialised picture. Without 'the tradition of hospitality they encountered, the English settlers could not have survived, and to do so they had to identify the most powerful among the Native

[52] Seed (1995), p. 103. On the relevance of astronomy in overseas voyaging see Tomlins (2001), p. 322.

[53] Brotton (2018) 62. See Penrose (1952), Ch 4.

[54] See Baugh and Cable (2012), p. 341. See also Seed (1995), p. 16. The second era is said to close with the 1860 Civil War; the third period corresponds to the contemporary era.

[55] Kersten (2006), p. 75.

[56] Bailey (2012), p. 18. See also Seed (1995), p. 17. On the socio-economic conditions of Chesapeake's colony see Rutman (1987) and Carr (1992).

Americans and persuade them to help'.[57] In this reductionist narrative, the way the English were helped by Powhatan, i.e. the Algonquian monarch and Pocahontas's father, was stripped away because of its being dangerous for the white-dominated U.S. political epic.

4.2.3 Partitioning the Earth Along Global Lines

The last example demonstrates how linear global thinking was not able to completely evacuate concrete legal geographies from the contexts where new claims were being processed.

As is well-known, the discovery of the Americas led to the first partition of the globe between Spain and Portugal. Soon after Alexander VI had issued the Papal Bulls *Inter Caetera* in 1493, the two Catholic powers signed the 1494 Treaty of Tordesillas.[58] This represented the earth along global lines: 'a boundary or straight line be determined and drawn north and south, from pole to pole, on the said ocean sea . . . This boundary or line shall be drawn straight at a distance of three hundred and seventy leagues west of the Cape Verde Island'.[59]

Schmitt tells us that this line was labelled *raya*, inasmuch both parties shared a common logical space: 'in the background these princes still shared the authority of a common *ordo*'—which means a shared international legal system—that 'distinguished between the territory of Christian and non-Christian princes and peoples.'[60] Besides partitioning the world between Portugal and Spain, the pontiff entrusted them with the propagation of the Gospel. Mission, navigation, and trade were closely intertwined and supported the projection of global linear thinking at the expenses of non-Western cultures.[61]

Again, however, the process of spatial abstraction had to be reconciled with the concreteness of the spatiality of the law. Where did the antemeridian *raya* to the Tordesillas line have to be drawn in the Pacific? In the sixteenth century, this question was related to the possession of Moluccas, the 'spice-producing islands'. It was not just a matter of claiming *dominium* over their *imagined* territorial spaces. The Moluccas dispute gave rise to a geopolitical struggle over new 'belongs of public law'. In his *A History of the World in Twelve Maps*, Jeremy Brotton lists the array of arguments and contextual variables used by Spanish and Portuguese delegations to draw the antemeridian. The dispute related to the creation of new legal

---

[57] Bailey (2012), pp. 19–22. See Miller (2004/2005).

[58] See Vander Linden (1916).

[59] Brotton (2018), p. 94. The text in is Davenport (1917), p. 97. See also Brotton (2018); Diener and Hagen (2012), p. 47; Korosy (2020), p. 228. For the exegesis of both Bulls see Vander Linden (1916), p. 9.

[60] Schmitt (2006), p. 91.

[61] See Seed (1995), pp. 154–160.

spaces was settled by the 1529 Treaty of Saragossa—at the cost of flattening, i.e. commodifying, the earth around parallels and meridians. To prevent future clashes between the spatialities of Spanish and Portuguese laws, Portugal and Spain entered into a bargain; Spain renounced the Moluccas in return for financial compensation. Both powers created a common logical space, which reproduced a 'normalised' world that had been previously agreed. The normalisation of the earth's surface is described, with hints of irony, by Brotton: 'Portugal and Castile finally agreed to place the Moluccas on a world map.'; the antemeridian to the *raya* fell '17 degrees (which equal 297 ½ degrees) east of the Moluccas'.[62] The politico-legal discourse thus complemented the creation of empires in what, from a cartographic perspective, could not be considered a pure geographical project.

Likewise, the 'Treaty between France and Spain, concluded at Chateau Cambrésis' (1559) drew the amity lines. An oral agreement concerning the Indies was attached thereto. The lines were not formally mentioned in the Treaty, but the oral agreement was made 'to the effect that west of the prime meridian and south of the Tropic of Cancer might be made right, and violence done by either party to the other should not be regarded as in contravention to the treaties.'[63] Schmitt reminds us that, unlike the *raya*, these lines separated the old from the new world. There was an 'overseas zone in which . . . only the law of the stronger applied. Unlike the *rayas*, then, these lines did not partition the earth but created a sphere of conflict between contractual parties seeking to appropriate land, precisely because they lacked any common presupposition and authority'.[64] There were two logical spaces; beyond the amity lines, the creation of logical space was coeval to the production of the law of spatiality. Treaties would lose their legal force beyond the lines, and ships could be captured and deemed to be a good prize. In sixteenth and seventeenth century, the amity lines implied that 'great areas of freedom were designated as conflict zones in the struggle over the distribution of a new world'.[65]

The lines drew a distinction between justice and injustice. On this side of the lines there was the civilised, logical space of the Christian world; beyond them, chaos, darkness, unlimited war, and freedom of land occupation. In legal geographical terms, several spatialities of the law were thus established. Each European power could create its own legal regime and adapt it to its colonies. Their Eurocentric attitudes made use of these lines for a sanitising function, which would protect them from the inferior socio-cultural contexts of their colonies. As a result, they essentialised their own spaces and possessions and applied a clear policy of domination over them (see Sect. 4.4).

---

[62] Brotton (2013), pp. 186–211 and 212.

[63] Davenport (1917), pp. 219–220. On the amity lines see Williamson (1949), p. 46; Mattingly (1963); Schmitt (2006), pp. 86 et seq.; Netzloff (2019).

[64] Schmitt (2006), pp. 93–94.

[65] Schmitt (2006), p. 97.

4.3 'The Earth Is the Lord's'. British Integrated Geography of Land and Sea

4.3.1 *The 'Character' of the Sea*

In *The Nomos of the Earth*, Carl Schmitt extends linear global thinking to the processes of land- and sea-appropriation. According to him, *global* encompasses the 'planetary', 'external', and 'superficial character of this type of thinking, based on the equation of land and sea surfaces'.[66] His references are the fifteenth- and sixteenth-century discovery of oceanic spaces and the Selden-Grotius debate on the freedom of the seas.[67] But, he leaves the following questions unanswered: how does the process of sea-appropriation work? Is it really possible to apprehend it materially?

Schmitt introduces the idea of common and integrated processes for the manufacturing of both earthly and maritime spaces. Yet, this reasoning is underpinned by the binary opposition between land and sea. The text constantly swings back and forth between the 'earth' (which extends over both land and sea) and the 'new World' (which is only made up of land). To certain extent, maritime spaces, discoveries, and navigation are still functional to land appropriation. Lauren Benton reminds us that unbounded oceans paved the way to further territorialised spaces. For example, the 'new earth' beyond the seas turned out to be 'firm land'. Besides oceanic corridors, Europeans 'focused on rivers as the points of entry and pathways to trade and settlement'. Especially in West Africa and America, 'as they moved through river regions, expeditions [became] embroiled in internal, i.e. terrestrial, "disputes"'.[68] This resulted in a thorough application of the territorialised paradigm in new discovered lands. Certainly, hints of territoriality were present in the oceanic enterprises. Despite this, colonial policies preferred to conceive of watery logical spaces as functional paths to land, i.e. the earth's element that we, as humans, know better and with which we are more familiar.

Schmitt's arguments revolve around the interconnectedness of land and sea as 'free' spaces. Their discovery turns them into vast logical spaces in which spatial reshuffling and occupation were freely exercisable by Europeans. The advent of geography triumphant, however, only saturated terrestrial spaces. The sea, on the contrary, fell (and still falls) under international law, which applies the territorialised paradigm when it comes to demarcating, say, the territorial sea, the EEZ, and the seabed.[69] The '*nomos* of the sea' means its territorialisation, flattening, partition, and

[66] Schmitt (2006), p. 88.

[67] Schmitt (2006), p. 94. See also Campling and Colás (2021), p. 71 and Sect. 3.6.

[68] Benton (2010), pp. xii–xiii and 43 et seq.

[69] See Sect. 3.5.2. See Reid (2020).

exploitation; related to state interests 'in using spaces in desired ways', it continues the politics of spaces triggered by *geography militant*.[70]

Applying the territorialised paradigm to the territorial seas reflects the spatiality of the law. The high seas and their resources are 'things which cannot be occupied'; things that can never be occupied 'can be proper to none . . . no part of [them] can be accompted in the territory of any people'.[71] As Schmitt puts it, 'the sea has no *character*, in the original sense of the word, which comes from the Greek [χαράσσειν] *charassein*, meaning to engrave, to scratch, to imprint . . . On the waves there is nothing but waves'.[72] The impossible physical apprehension of its 'undifferentiated surface'[73] led to an alternative spatial construction; maritime logical space has become a 'geo-economic battleground' for the appropriation of resources and a geopolitical battlefield 'conditioned by the capacity of one state or another to control those waters'.[74]

4.3.2 Manufacturing Imperial Spatialities

An alternative process of oceanic production eludes the binary opposition between land and sea as practised in international law and human geography. It navigates different routes and tries to answer the questions Schmitt left unanswered. At the same time, it searches for a new paradigm through which marine logical spaces might be apprehended. The result is a spatiality of the whole earth that crosses the divide between the geographies of land and sea and outlines a holistic, i.e. ecological, paradigm, which 'globally' captures Schmitt's planetary equation between them.

We examined holistic paradigms when considering the ecological spatial normativies of chthonic legal traditions (see Sect. 2.1.2). They are present in discrete cultures. Within the Māori cosmogony, for example, the term *whakapapa* discloses the 'connection, lineage, or genealogy between humans and ecosystems and all flora and fauna.'[75] The world is more 'a sea of islands' than a pattern of 'islands in the sea' sizeable according to 'the dominant hegemonic view of oceans.'[76] Ecological paradigms are present in the Bible, where narratives related to the earth point to both land and sea—but the sea takes precedence over the land.[77] In *Psalm* 24:1–2, processes of land- and sea- appropriation are reversed:

---

[70] Campling and Colás (2021), p. 70.

[71] Campling and Colás (2021), p. 72.

[72] Schmitt (2006), pp. 42–43.

[73] Steinberg (2001).

[74] Campling and Colás (2021), p. 73.

[75] Harmsworth and Shaun (2013), p. 275. See also Sammler (2020), p. 66.

[76] Respectively: Hau'ofa (1994), p. 153; Sammler (2020), p. 68.

[77] See e.g., *Ge* 1.1-31; *Jonah* 1:9; *Ps* 95:5.

[1]THE earth is the Lord's, and all that therein is :

the compass of the world, and they that dwell therein.

[2] For he hath founded it upon the seas :

and prepared it upon the floods.

Likewise, the validity of Southern American claims over the Antarctic region displayed a divine provenance (see Sect. 3.6) grounded in the 1493 Bulls partitioning the world between Portugal and Spain (see Sect. 4.2.3). The Argentinian government has also appealed to geographical titles, such as 'geographical proximity, and geological continuity' to support its 'natural right' to the so-called *Antártida Argentina.*[78] The term 'Antarctandes' was indeed used 'to refer to the geological extension of the Andes Mountains underneath the Drake Passage into America'.[79]

The English 'were late in the field compared to Spain and Portugal'. In partitioning the world between these two Catholic powers, the 1493 Papal Bulls *Inter Caetera* must have 'greatly discouraged exploration'.[80] The English Reformation gave impetus to navigation mainly for geopolitical reasons. It was necessary to 'alert the nation and the state to the advantage that their Spanish counterparts had gained over them through transoceanic exploits'. Continental trading networks had to be circumvented through 'direct trading connections with Asia and to exploit whatever opportunities existed in America'. A 'spectacular expansion in trade' soon ensued.[81] A group of geographers, country gentlemen, and militant Protestant stimulated the English navigational enterprise. This resulted in their appropriating heathen 'sundry rich and unknown lands ... fatally, and it seemeth by Gods providence, reserved for England and for the honor of her Majesty'.[82] The Anglo-British reversed these processes of maritime appropriation, devising an ecological and holistic approach to both the geographical and the legal. Like in the Bible, its validity depended on its divine provenance.

In the aftermath of the Reformation, the English explored the earth upon direct divine mandate without the geographical limitations set by the Bulls. On the one hand, the Anglo-British shared the common Catholic feature of their Spanish and Portuguese counterparts, i.e. the colonising (and civilising) mission aimed at propagating the Gospel: 'Go ye into all the world, and preach the Gospel to every creature'.[83] On the other, the English had neither the pontiff nor any other representative of God nor earthly mediator. Since they were *Defensores Fidei*, English monarchs were 'as entitled as any Pope to make universal concessions of

[78]See Collis (2010), p. 389.

[79]See also Howkins (2017), pp. 9–10.

[80]Egerton (1897), p. 14. See also Tomlins (2001), pp. 318–319.

[81]Canny (2001), p. 4.

[82]Petition of divers gentlemen of the west parts of England to the Queen. To allow of an enterprise by them conceived, and with the help of God, ... to be performed for discovery of sundry rich and unknown lands, in Sainsbury (1893), p. 1.

[83]*Ma* 16:5.

sovereignty so as to promote the spread of the Gospel around the world'.[84] Particularly westwards and northwards, i.e towards North America, according to both the Bible[85] and George Herbert's *The Church Militant* (1625, lines 296–297):

> Then Shall Religion to America flee:
>
> They have their times of Gospel, ev'n as we.

The Catholic-but-Reformed England had directly inherited the title from God. This mandate is state in more general terms in *Psalm* 2:7–8:

> 7 I will preach the law, whereof the Lord hath said unto me : Thou art my Son, this day have I begotten thee.
>
> 8 Desire of me, and I shall give thee the heathen for thine inheritance : and the utmost parts of the earth for thy possession.

The legal creation of oceanic spaces was possible due to some intrinsic qualities of the English common law. Acting upon direct divine mandate depended on the English self-identification as God's chosen people. As John Aylmer (1521–1594), Bishop of London under Elizabeth I, put it, 'God is English': 'For first you have God, and al his army on angels on your side; you have right and trouth, and seeke not to do them wronge, but to defend your own right'.[86] The divine provenance of the British titles is reflected in the 'Common Laws of England', which 'are grounded upon the law of God, and extend themselves to the original Law of Nature, and universal Laws of Nations, and . . . are not originally *Leges Scriptae*' (Dugdale 1680, 3b). Being 'the law of nature part of the law of England', it follows that, as Coke put it in *Calvin's case*, the English common law was also written 'with the finger of God',[87] thus being 'immemorial in literal sense'.[88]

The unmediated divine title endowed the British with a 'well-earn'd empire of the deep' under God's protection.[89] Such self-identification was highly productive in terms of discoveries, expeditions, and connections between science and sovereignty. The rise of Puritanism reinforced the navigation-trade nexus; Oliver Cromwell's *Navigation Act* of 1651 transformed England into a mercantile power. The English became 'a people . . . that are to God as the apple of His eye'; in commercial terms, they were 'under the shadow' of mercantilist 'wings' of God.[90]

[84] Pagden (2001), p. 37. See Edwards (1992), p. 274; Tomlins (2001), p. 319.

[85] See *Is* 43:5 and *Mt* 24:27.

[86] Aylmer (1559), p. 36; Ward (2020), pp. 30–37; Goodrich (2021), pp. 15 and 29.

[87] *Calvin's Case*, 7 *Reports* (1608) 4: 12b.

[88] Baker (2016), pp. 85 and 86, where it is stated that 'evidence of a legal culture among the ancient Britons was found in the tales of the legendary King Brut and King Lucius'.

[89] Thomson (1802), line 166.

[90] Quoted in Molyneaux (2014). See *Ps* 18: 8.

Having sent His 'word of command to rebuke the raging winds, and the roaring sea', God preserved His Chosen People 'from the dangers of the sea, ... from the violence of the enemy', and the dragons of the deeps:[91]

> 14 Thou didst divide the sea through thy power :
>
> thou brakest the heads of the dragons in the waters.
>
> 15 Thou smotest the heads of Leviathan in pieces :
>
> and gavest him to be meat for the people in the wilderness.[92]

Consequently, he granted the English

> 6 [. . .] dominion of the works of thy hands :
>
> and thou hast put all things in subjection under his feet;
>
> [. . .]
>
> 8 The fowls of the air, and the fishes of the sea :
>
> and whatsoever walketh through the paths of the seas.[93]

The self-identification with God's inheritance was therefore highly productive in terms of discoveries, expeditions, and spatial production. The English were able to secure the validity of their title over dryland and oceans because their navigation was firmly grounded in open waters: 'Thy way is in the sea, and thy paths in the great waters: and thy footsteps are not known' (*Ps* 77:16). We saw that this was possible because of some intrinsic qualities of the English common law, which allowed the Anglo-British to reframe the processes of maritime appropriation, thus founding the colonial enterprise, as Psalm 24 puts it, 'upon the seas'. Again, with *Psalm* 24, I may say that the act of impressing the common law into colonial soil was firstly 'prepared ... upon the floods'. The integrated Imperial legal geography was founded on a remarkable process of politico-legal invention taking place in early modern England; the Reformation and the immemorial antiquity of the common law were reimagined in the light of the religious fervour of militant Protestantism, the Age of Discovery, and the transformation of the English nation from a community of farmers into a society of sailors and traders.

The leap is particularly apparent in terms of imaginative geography. Take, for example, the frontispiece of John Dee's *General and Rare Memorials pertayning to the Perfect Arte of Navigation*. The staunch supporter of English imperialism we have encountered throughout the essay is adamant in assuming that 'God [has taken] up the cause of the expansionist English'. As verses 8–9 of Psalm 139 confirm, God will never abandon His Chosen People:

> 8 If I take the wings of the morning : and remain in the uttermost parts of the sea;

[91] 'Praise the Lord upon earth: ye dragons, and all deeps' (*Ps* 148:7).

[92] BCP [1662] (2004), p. 439 (*Psalm* 74: 14–15).

[93] BCP [1662] (2004), p. 353 (*Psalm* 8:6, 8).

> 9 Even there also shall thy hand lead me : and thy right hand shall hold me.

Indeed, the frontispiece depicts God's 'sword-wielding angel, Michael. By weaking its enemies and strengthening its friends, God has shown England that it is now time to act'.[94] Perhaps most striking of all, *Michael* was the only flagship that had survived 'the Arctic ice' during the expedition to the Hudson Bay in 1578 and came back home safe.[95]

Divine inheritance secured the validity of this holistic, i.e. ecological, title, which extended over dryland and oceans. Instead of territorialising it, the land was firmly grounded in open waters. At the beginning, 'the spirit of God moved upon the face of the waters'.[96]

The consequences are threefold. Firstly, the divine provenance of the British titles is reflected in the English legal tradition, in existence since the creation of the world.[97] Secondly, when navigating the South Atlantic Ocean, the 'path in the great waters' led to 'a cascade of islands' and 'chains of islands and reefs'.[98] Stretching from Saint Helena, Tristan da Cunha, and Ascension, this insular constellation is the 'essential staging post for flighting to and from the Falkland Islands' and its former dependencies.[99] As *Psalm* 97:1 reminds us,

> 1 THE Lord is King, the earth may be glad thereof :
>
> Yea, the multitude of the isles may be glad thereof.

With its inherited faith, Britain governed 'a sea of islands' and not a pattern of 'islands in the sea'. This leads us to the third consequence of being the English God's Chosen People, i.e. their indissoluble connection with their own Established Church of England, which is, needless to say, *catholick* and universal. This section has abundantly drawn quotations from the *Book of the Common Prayer* enacted by the 1662 *Act of Uniformity*,[100] which had a pivotal role in propagating the Gospel and the British title to the word. With the words of *Acts* 1:8,

> 8 But ye shall receive power, after that the Holy Ghost is come upon you: and ye shall be witnesses unto me . . . unto the uttermost part of the earth.

Intriguing as it is, this portion of scripture is *legally* appointed for Ascension Day,[101] thus contributing to upholding the validity of the British *catholick* paradigm over

---

[94] Cormack (2001), p. 50.

[95] Abulafia (2020), p. 664.

[96] *Ge.* 1:2.

[97] See *Wallyng v Meger* (1470) 47 SS 30, *per* Catesby Sjt.; Sir Edward Coke, *Posnati. Calvin's Case*, 7 *Reports* (1608) 4: 12b.

[98] See, respectively, Bottomley (2020), p. 18; Mentz (2015), p. 52.

[99] Lovegrove (2012), p. 68 (Ascension); 104 (Tristan da Cunha), and 136 (Georgia). See Fichter (2008).

[100] 12 &14 Char II c 4 cl 10.

[101] BCP [1662] (2004), p. 145.

physical, human, and imaginative geographies of the earth. In so doing, it confirms the interweaving of the politico-legal, the religious, and the geographical as these existed and were effectively practiced at the onset of the Anglo-British Empire. The common law was able to impress its character into the oceans and colonial soils, thus testifying to English superiority in mastering both land and sea. In legal geographical terms, the dawn of the English Imperial enterprise was an intriguing cross-disciplinary achievement, probably regarded with bitterness and awe by the actors competing in the global production of space. *Psalm* 65:8 vividly illustrate this:

> [8] They also that dwell in the uttermost parts of the earth shall be afraid at thy tokens :
>
> thou that makest the outgoings of the morning and evening to praise thee.

4.4 Processing Spatial Superiority

I have followed the Europeans navigating the oceans in search of new logical spaces. In reimagining them for radically new purposes, they adopted a linear, abstract, and 'sanitised' approach. In so doing, they tried to avoid the mistakes made by the Portuguese, whose close engagement with their colonies led to a swift decline of its imperial ambitions. Or, at least, this is what mainstream historiography traditionally relates (see Sect. 4.2.1).

This sanitising posture concealed a policy of conquest, exploitation, and domination of newly discovered territories. And, for our purposes, of spatial production. Certainly, linear-abstract thinking prevented colonisers from contracting the same colonial malaise to which the Portuguese had fallen prey. Like a vaccine, the lavish *Sphere* tapestries commissioned by João III of Portugal, their globes, and maps complemented the universal (i.e. abstract) principles of Western civilisation, shielding them from the malaise, but also disguising a concrete production of logical spaces. Accordingly, Europeans avoided any direct contact with non-European, i.e. inferior, cultures, which facilitated the manufacturing of new legal spaces.

Unlike the Portuguese, many European powers chose alternative paths of spatial production. Their outcomes were as concrete as those achieved by the Portuguese, in that they led to the physical appropriation of new territories. The process did not end there, but required a further step, i.e. their physical-cum-cultural appropriation. To be naturalised, 'the natural content of the landscape [had to] be culturalized *into* existence'. Daniel Wood terms this 'a labor of culture ... of identifying, of bounding, naming, inventorying, ... it is a labor of mapping'. Indeed, 'land is not *systematically* divided into plains and hills that are subsequently delimited, named and inventoried'. Quite the opposite: they are brought into existence as 'a human landscape', which in turn 'is brought into being historically, sometimes in a rush, but usually bit by bit in a patient dance of disjoined incrementalism'.[102]

---

[102] Wood (1992), p. 78.

Europeans achieved this goal by superimposing their own nomenclatures and imaginative geographies onto newly acquired territories. Through surveys and by renaming territories, they appropriated them and 'culturalised' their natural landscapes. As a result, they made these territories meaningful spaces for them, both culturally and legally.[103]

Halpérin reminds us how these processes of adaptation to Western law took place in several non-Western contexts. In terms of legal geography, the American logical space was completely reshuffled. Barely inhabited, Halpérin reminds us, the continent was 'populated by Indians weakly organized in polities without proper rules and likely to be Christianized'. The presence of highly civilised groups in Asia, by contrast, made the Europeans understand 'rather quickly that they could not evangelize all these people'; the subjugation of these societies was thus to be attained through legal coding of their human geography. This meant transforming 'social norms, which were not binding rules before the local courts, into law', and therefore fabricating legal traditions.[104] In India, in particular, Matthew H Edney demonstrated that the use of cartography by British geographers 'came to define the empire itself, to give it territorial integrity and its basic existence'.[105] In Africa, European colonial law promoted 'social engineering', i.e. the economic development, modernisation, and transformation of indigenous African societies.[106] Colonial agents forged new institutions whereby African societies could be both governed and 'civilised': chiefs, tribes, and customary courts are 'invented traditions', which 'became in themselves realities through which a good deal of colonial encounter was expressed.'[107]

I noted that the colonial enterprise was manufactured through a politico-legal discourse. To this extent, not only was colonialism a necessary condition for the 'closing of territories', but it also supplied a further tool that proved functional to secure Europe's domination over the rest of the world. Processing the spatiality of law also meant securing the cultural (and legal) superiority of Western civilisation; indeed, the land was materially encoded with European legal paradigms. This was made possible because landmasses beyond the lines were considered uninhabited lands—uninhabited, we add, by Christian peoples. Political appropriation justified physical apprehension, inasmuch as non-Christian peoples were 'savage both by dint of their paganism but also by dint of their failure . . . to cultivate or "improve"'.[108]

No wonder, therefore, that, according to the rule in *Calvin's case*, 'if an uninhabited country be discovered and planted by English subjects, all the English laws are immediately there in force. For as the law is the birthright of every subject, so wherever they go they carry their laws with them'. An exception was created for

[103]Blomley (2015). See also Patton (2000), p. 173.

[104]Halpérin (2009), p. 342.

[105]Edney (1997), p. 1.

[106]Allott (1967), p. 262; Eisenstadt (1965), p. 453.

[107]Ranger (1983), pp. 211 and 212. See also Nicolini (2019).

[108]Tomlins (2001), p. 335. See also Bennett (1996), pp. 65–66.

those tracts of lands already inhabited, conquered, or occupied by Christian peoples under the rule in *Campbell v Hall*: 'in conquered or ceded countries, that have already laws of their own, the king may indeed alter and change those laws; but, till he does actually change them, the antient laws of the country remain, unless such as are against the law of God, as in the case of an infidel country'.[109]

Henry Jones explains how the rule in *Calvin's case* operated in 'uninhabited' Canada, where the land was turned into territory and then into private property, which meant not only claiming a title to it, but also legally infusing it with the principle and superior *legitimacy* of the Western cultural legal tradition.

In 1583, at the onset of the colonial enterprise, Sir Humphrey Gilbert landed at St John's Harbour, Newfoundland, with a royal charter. Issued by Elizabeth I, it entrusted 'all such lands, countries and territories to be discovered' to Gilbert, 'with full power to dispose thereof ... in fee simple or otherwise, according to the Laws of England'.[110] The land was encoded with the principles of English land law; Gilbert had brought four surveyors to Canada. Jones tells us that they 'immediately began to measure and map the area'.

The same holds valid for the French territories in North America. Along the St Laurence River, which was the 'heart' of French colonisation in Canada, the Ancient-Regime seignorial system was introduced in the eighteenth century. It survived until 1854, well beyond its cession to the United Kingdom (1763) and its abolition during revolutionary France.[111] Perhaps most striking of all, this also meant transplanting into Québec the adage *Nulle terre sans seigneur*—which styles the radical idea of legal geography (see Sect. 2.6.2).[112]

The same occurred in the recently discovered New England: from a public-law perspective, colonial charters allowed colonisers to 'describe and pursue claims to American spaces in detail'; and 'America ... came predefined, "produced" as English territory by legal documents that created jurisdictions in bounded spaces'.[113] The first charter of Virginia (1606), for example, established that 'all the lands, Tenements, and Hereditaments' were held of the Crown, 'as of our Manor at East-Greenwich, in the County of Kent, in free and common Socage only, and not in Capite'.[114] This meant that the holder of was not a King's tenant-in-chief (such as in the case the land was held *in capite*); rather, the feudal tenure of land required the holber pay charges of other non-military service to a superior. In 1663 Charles II issued a charter making 'a group of eight confidants ... proprietors' of Carolina,

[109] *Calvin's Case* (1608), 77 ER 377 and *Campbell v Hall* (1774) 1 Cowp 204, 98 ER 1045. See also *Phillips v Eyre* (1870) LR 6 QB 1; *Sammut v Strickland* [1938] AC 678 (PC). Quotations are drawn from Blackstone [1765] (2016), p. 75.

[110] Jones (2019), p. 187.

[111] Grenier (2004).

[112] Grenier (2004).

[113] Tomlins (2001), p. 316.

[114] Quoted in Tomlins (2001), p. 319.

i.e. a colony 'encompassing what is now most of North Carolina, South Carolina, and Georgia'. Consequently, they were enfeoffed as 'absolute lords'.[115]

Not only did this originate politico-legal apprehension of new tracts of land, but it also resonated with Schmitt's νόμος (the law) and νέμειν ('to divide'). Again, not only did this point to the transplant of English private property into the colonies, but it also assigned a new legal-geographical meaning to the new world. A new political, social, and proprietary order became visible. Bhandar terms it 'abstraction', since it 'turns the land into fungible property, commodifies it'.[116] The idea also reflects Conrad's insatiable appetite of blank spaces of the map, 'a deeply ingrained colonial mentality [also] entrenched in seventeenth-century New England.' The 'world was full of empty spaces ready for taking by Englishmen'.[117]

In suffusing these territories with Western legal principles, the process of legal cartographic encoding also infused the land with 'spatial hierarchies'.[118] And the process of land appropriation impressed the seal of Westernisation into the fertile grounds of the colonies. These became the lands of colonisers; 'by racializing the land holding of the indigenous people abstracted into savages', they 'enabled the dispossession of indigenous communities'.[119]

4.5 The Law of Spatiality Beyond the Lines

4.5.1 Ethnocentrism: The Spatiality of Legal Systems

As explained in the previous section, newly discovered territories were encoded and infused with the superior principles of Western legal civilisation. This process still has some bearing on how we display legal systems on maps. It is also the continuation of the politics of spaces that took place in the Age of Discovery. In legal geopolitical terms, it triggers a competition among superior Western legal practices. The struggle reflects the fabrication of the global politics of space through legal practices, thus conveying the idea that legal systems are rank-able depending on their performativity (see Sect. 4.6).

Impressing the seal of Westernisation into the fertile grounds of the colonies also explains why comparative law has traditionally focused (and still focuses) on the similarities between former colonies and Western legal systems. By emphasising

[115] The 1663 Charter contained the 'Durham Clause' giving the same proprietary rights held by 'any Bishop of Durham', within the County Palatine of Durham. Proprietors were thus 'granted directly from the king as freehold', which gave them far more power and freedom then any traditional manorial lord or knight service in capite': Jones (2019), p. 199. For other cases of 'Durham Clauses' appended to colonial charters see Tomlins (2001).

[116] Jones (2019), p. 191; Bhandar (2014).

[117] Wood (1992), pp. 45 and 205.

[118] Bennett (2008), p. 17.

[119] Bhandar (2014), pp. 211–212. See also Patton (2000), pp. 27–28.

legal-colonial links, scholars rank French, Spanish, Portuguese, Dutch, and Italian former colonies among the civil-law legal systems; former British colonies and protectorates are ascribed to the common-law legal tradition, whereas countries with a mixed colonial past join the mixed jurisdictions.[120] This perpetuates the ethnocentric approach of nineteenth-century geography and comparative law, when only the limited set of civilised, i.e. 'European', systems were considered generalisable and exportable. Located in the Global North, these have set the benchmark in relation to which all systems must be measured.

By encoding colonial landscapes Europeans superimposed their own spatiality of law onto them; peoples, communities, and territories, still bear the consequences of colonial domination. The thorough application of legal encoding had the 'material effect' of outlining 'depopulated', 'often void of human traces', and visually 'empty' spaces,[121] which would be subsequently arranged so as to comply with 'Western beliefs'.[122] With sour irony, we may say that the process of legal sanitation was impeccable; once cordoned off, the non-Western world was essentialised and reduced to Harley's 'cartographic silence'.[123]

Silence is a clear (albeit implicit) statement of the inferiority of these legal geographies; it then discloses an ethnocentric approach to the spatiality of colonial laws. The new 'spatial model' was firmly grounded on a political application of legal cartography. Sixteenth-century achievements in mapmaking framed the world according to a geometrical frame. Thus, the 'point of view was elevated, to the point of grasping in a single instant the convexity of the terraqueous globe'. Yet, this geometrical (and sanitising) cartographic discourse still needed a '*centre*, an *omphalos*, in order to describe, in degree of civilisations, its difference from [the] diffuse *periphery*' which was the uncivilised, i.e. inferior, 'rest of the world'.[124]

This centre was the West. Take, for example, comparative law. Several handbooks dedicate chapters to, say, African and Asian legal systems;[125] though, their analyses are still affected by methodological biases, such as the Ethnocentric, and colonial, attitude towards non-Western conceptions of the law. Ethnocentrism advocates the superiority of European legal paradigms; non-Western, i.e. inferior, systems do not have anything to teach them. This results in the limited extension of chapters on non-Western law in these handbooks; references to Asian, African, and Pacific legal traditions are usually limited in range and scope and are confined within either classifications of the legal systems or micro-comparative analyses.

The bias is apparent even when scholars suggest the adoption of new taxonomies. For example, Patrick Glenn advocated the establishment of chthonic legal traditions,

[120] Zimmermann and Visser (1996), pp. 7–8.

[121] Crampton (2003), p. 47. See also Jones (2019), p. 203.

[122] Gordon and Klein (2001), p. 1. See also Korosy (2020), pp. 227–288.

[123] Harley (1988), p. 57.

[124] Gordon and Klein (2001), p. 3; Lestringant (1994), p. 5.

[125] See, among others, Kischel (2019); Rambaud (2017); David et al. (2016); Bussani and Mattei (2012); Reimann and Zimmermann (2006); Menski (2006).

into which several pre-colonial legal traditions (i.e. African, Asian, Polynesian, and Inuit) coalesce.[126] However, the legal descriptor does not account for the rich variety of 'non-Eurocentric conceptions of the law': within the chthonic milieu, all beyond-the-West conceptions of the law lose their own legal-specific features.

The way comparative legal cartography is arranged in manuals still reflects late nineteenth-century ethnocentric attitudes in the study of the law. The approach exhibited by comparative law towards the mapping of legal systems and their grouping into a family is due to our specific cultural habits. Europeans considered 'mapping the law' servient to their gaining control of new territories: their laws were classified by reducing them to a limited number of families. Besides ensuring a better understanding of the world, these classifications silenced the non-Western legal traditions and, in so doing, superimposed the European ones onto them in accordance with the abovementioned cultural habits. Again, this is a legacy of the colonial attitude in the realm of politics and law.[127]

Monateri provides us with clear geopolitics of law when listing some of these classifications. In his presentation to the 1900 Congress of Paris, for example, Adémair Esmein referred to the legal families to which the countries of the 'Concert of Europe' belonged. The Roman, German, Anglo-Saxon, Slavic, and Islamic families reflected Europe's balance of power between France, Austria, Prussia, the United Kingdom, Russia, and Turkey. The rest of the world is relegated to politico-legal irrelevance. Sanitised and essentialised, it is located in a peripheral *département d'outre-mer*.[128]

Likewise, in his 1922 comparative-law treatise, Lévi-Ullman categorised the legal systems into three groups: the Continental, the Anglophone, and the Islamic systems, essentialising both the Western and the Mediterranean legal cultures to the exclusion of the rest of the world. The same holds true for Sauser-Hall, whose controversial classification around linguistic criteria reduced the law to three groups, i.e. the Semitic, the Indo-European, and the Mongol ones.[129]

These classifications resonate with legal applications of Mackinder's *Geographical Pivot of History*, as they revolve around European and Asian landmasses. This was the centre from which all legal systems of the world radiated and were arranged upon a hierarchical scale. Even in the post-colonial era classifications, this ethnocentric approach to legal cartography was preserved.[130] Yet, the seminal comparative-law treaties that René David dedicated to the legal systems of the World still focus on the three major legal Euro-Asiatic legal traditions (Romano-Germanic; Common law, and Soviet – then Russian – law), essentialising Hindu, Eastern Asian, Islamic, and African laws under the banner 'Other Conceptions of the

[126] Glenn (2014), p. 60. See also Sect. 2.1.1.

[127] See Pargendler (2012).

[128] Esmein (1900). See Monateri (2013, 2021), pp. 23–41.

[129] Levy-Ullman (1923); Sauser-Hall (1913), pp. 59–63.

[130] Indeed, Arminjon et al. (1950) and Zweigert and Kötz (1996) devised classifications that broadened the scope of traditional legal cartography but still had Europe as their centre.

Law and the Social Order'. When proposing a classification grounded in the deep-rooted differences between all models of social organisation, even Ugo Mattei[131] did not escape Western-driven methodological biases. The way he arranges systems according to the rules of *professional law*, *political law*, and *traditional law* clearly resonates with the hierarchical distinction between: (1) superior (i.e. Global-North) systems based on separation of powers and division between morality and the law (rule of professional law); (2) systems which either are experiencing a transition from a colonial, undemocratic regime to democracy or make the law servient to politics (rule of political law), such as China, Russia, and several, allegedly Third-World, countries; finally, (3) Western systems and jurisdictions where non-Western (and therefore backward) conceptions of the law are still dominant.

Last but not least, U.S. scholars also arranged Ethnocentric legal cartographies. The most relevant U.S. classification was devised by Wigmore in 1928.[132] His *A Panorama of the World's Legal Systems* lists five cultural contexts, which might evolve through various stages of development leading to the same stage of superiority envisaged by U.S. and European comparative legal traditions: the primitive, ancient, Euro-American, religious, and Afro-Asian contexts. Perhaps the most intriguing of all classifications was devised by Berman in *Law and Revolution.*[133] Berman's assumption is that the Western world has elaborated a unique legal tradition, into which the basic features of Western legal culture have coalesced, i.e. the 'sagesse' of English law, the principles of the French Enlightenment, and the German 'genius'. It goes without saying that these features have come to blend in the USA, which now represents the universal model of Western legal culture.

4.5.2 Spatialities Beyond the Lines: Americas ...

Perhaps most paradoxical of all is that Berman essentialised the European legal traditions. The ideological intent is apparent: he backed U.S. geopolitics by applying legal cartography. Not only does the U.S. law embed the principles inherited by European colonisation, but it is also the dominant Western tradition. Within spatial production, Berman's reasoning supplements the U.S. politico-legal discourse with a sort of 'counter-pivot of history'. The American continent is the land where individuals enjoy their freedoms; in the classification devised by the Argentinian comparative lawyer Enrique Martínez Paz, this continent was also home to several liberal democracies, such as the Latin American ones in the 1930s.[134] As a land of freedom, the American continent, in general, and the USA, in particular, sharply

[131] Mattei (1997).

[132] Wigmore (1928).

[133] Berman (1983).

[134] Martínez Paz (1934), p. 155.

contrasted with Europe, i.e. the land of monarchies. Not only did the American continent inherit the superiority of the Western law, but it also ranked first within it.

This normalised the U.S. politico-legal discourse as regards its logical space. Perhaps forgetting how the practice of slavery was deeply embedded in U.S. history, Berman add a note to Blackstone's *Commentaries*: 'the spirit of liberty is so deeply implanted in our constitution, and rooted even in our very soil, that a slave . . ., the moment he lands in England [or, in the USA] [. . .] becomes *eo instanti* a freeman'.[135]

How this has happened is, again, a consequence of American logical spaces arranged along global lines. As said above (Sect. 4.4), the continent was barely populated in its Northern part; in addition, Spaniards and Portuguese undertook a process of subjugation (not to say extermination) of pre-Columbian civilisations in its Southern areas. In facilitating spatial reshuffling and normalisation of the legal-geographic features of the continent, this resulted in impressing the seal of Westernisation into these fertile grounds, where democracy was established by virtue of the American Revolution and the nineteenth-century Independence Wars in the vice-realm of New Spain and Brazil. In terms of legal geopolitics, the Americas are completely comprised within the Western hemisphere, which meant cordoning off the whole continent to prevent its being corrupted by the belligerent and non-democratic European monarchies. The decline was avoided by applying the sanitising function of the linear and abstract thinking along hemispheres. 'All of the states of the New World' thus had 'their independence [secured] against the ambitions of Europe'. This argument is also known as the 'Monroe Doctrine', as it reproduces the statement released by the U.S. President Monroe in 1823 and confirmed by the American International Congress which met in Panama in 1826: 'no European country may acquire, no matter what the motive, any portion of the New World'.[136]

The consequences of U.S. global thinking are manifold. On the American side of the line, it justified two policies deeply rooted in U.S. logical space, i.e. hegemony and imperialism. The first one allowed the United States to impose 'the opinions of her State Department upon America at large' and to intervene 'in the internal and external affairs of any country of Latin-America to protect the interests of the United States'. Imperialism aimed to expand its commerce towards 'new markets in different continents until there shall be established a political and economic supremacy'.[137] This happened during the Cold War. The victory in WWII upheld the primacy of the USA within the Western legal tradition, as well as the validity of this strategic global thinking, which definitively relegated Europe as the continuation of the Western hemisphere in the Eastern one.

---

[135] Blackstone [1765] (2016), p. I.123. See also Nicolini (2018).

[136] See Alvarez (1911), pp. 24 and 25.

[137] See Alvarez (1911), pp. 25–26.

4.5.3 . . . *And Africa*

It is not really a coincidence that the ‘American line’ resonates with the fifteenth-century amity lines. These had already drawn a distinction between two logical spaces. On this side of the lines there was the civilised, Christian world; beyond them, chaos, darkness, unlimited war, and freedom of land occupation. Not only did this trigger the progressive marginalisation of non-Western legal systems, but it also led to the establishment of separate new legal regimes for the colonies, which also required tailor-made political structures. These were deliberately asymmetrical and reflected the inequality between Europeans and the backward Africans; they were eventually settled in the aftermath of the 1884–1885 Berlin Conference (see Sect. 4.2.1).

The British fostered the creation of Crown Colony governments—where indirect rule or ‘government from above’ was deemed essential ‘in those lands whose populations [were] politically and economically backward, and where in many cases society [was] organised on a tribal basis.’[138] The UK also resorted to indirect rule, a method of administration which associated tribal chiefs with colonial governance; indeed, the British accepted that ‘*some* Africans could become members of the governing class of colonial Africa’.[139] Other European colonial powers prompted different colonial policies. Hence, France’s *mission civilisatrice* aimed to assimilate African natives, as well as ‘to propagate the best of French culture along with the rationalist and libertarian values deriving from the Enlightenment and French revolution.’[140] As for French colonial governance, the revolutionary principles of *Liberté , egalité, fraternité* may well have been of relevance only at home: ‘None but Frenchmen should go to the colonies of “Liberty, Equality, and Fraternity”; for there is Little Liberty, less Equality, and no Fraternity in the French colonies for Whites or Blacks.’[141]

Colonial approaches towards African law varied enormously. France’s colonial policy, for example, pursued social and legal uniformity in Africa, Asia, and Oceania. It also forged the *indigénat*: originally established in Algeria in 1881, this policy was applied across French colonies and eventually abolished in 1946. The *indigénat* neither recognised indigenous legal systems nor recollected customary law; it was ‘a regime of exception’, an ‘unspoken and roughly sketched domain of “non-law”’. By merely defining ‘the very status of “native”’, it listed the ‘offenses that “by definition” only “natives” could commit’.[142]

In South Africa, the Boer Republics of Transvaal and Orange Free State recognised customary law in 1885. The creation of the Union of South Africa

---

[138] White and Hussey (1958), p. 194.

[139] See Ranger (1983), pp. 220–221: On direct rule see Lugard (1922), pp. 192–213; Mann and Roberts (1991), p. 20. Oplot (1992), pp. 91–92.

[140] Sharkey (2013), p. 156.

[141] Sir J. Harris (1912), p. 97.

[142] Mann (2009), p. 336.

(1910) resulted in even more conflicting approaches: there was 'complete non-recognition in the Cape, limited application in the Transvaal and full recognition and application in Natal and the Transkeian territories'.[143] Odd as it may seem, African customary law obtained full recognition with the implementation of apartheid; the *South African Native Administration Act* (Act No 38 of 1927) recognised customary law and established a separate system of courts for Africans with the main purpose of fostering separateness among the different races living within the Dominion.

The limited recognition of non-Western legal systems is usually traced back to their intrinsic features, such as oral transmission and law-making as a communal performance, the output of which is collective legal wisdom. These features hardly squared with Western legal mentality and policies. Consequently, local variations in customary law were reduced through legislative action and restatement, whereby customary rules were recollected in written form and accommodated to the colonial legal framework. In addition, European colonial authorities established legal dualism, within which customary law and European law coexisted. Their mutual interactions were arranged upon a hierarchical scale: according to the repugnancy clauses appended to restated law, customary law was applied to the extent that it was not 'contrary to justice and humanity'.[144] In the event of inconsistency between European law and local law, the former prevailed.

Ethnocentrism complemented global linear thinking in arranging the concrete legal geographies beyond the lines. Europeans reshuffled spatial furniture in a strategic way, either silencing or cordoning off the application of non-European legal traditions. The infusion of colonial lands with the principles of Western (civil- or common-) law reflected the narrative of property, economy, and progress secured by the civilising mission; and this proprietary narrative represented the 'just' title for the acquisition of colonial lands and the 'just' indiscriminate dispossession of natives' legal geographies. Colonial policies disregarded ancestral and non-territorial arrangements based on borderless logical spaces. Divided among different states, communities were arbitrarily separated and subsequently merged with other groups with the aim of creating political entities based on territorial jurisdictions.

That two spatialities of law took place beyond the lines. The first one deprived natives of their legal geographies and superimposed the Western legal geography onto them. The second one assumed that even the marginalised non-Western legal traditions might progressively evolve through various stages with the Western legal paradigm as the natural end point. This entails that it is feasible to legally process even the lower stages of legal development so that they may reach the same stage of superiority envisaged by European legal traditions.

Modernisation was also achieved by backing 'official' customary law, the development of which required the 'unification' of native customary law, that is, the

[143] Gant (2006), p. 13. See Himonga and Nhapo (2014), pp. 9–13.

[144] See, among others, Sect. 12(1)(a) *Local Courts Act* 1966, Act No. 20 of 1966 (Zambia).

progressive amalgamation of its local variations.[145] The results are interesting: in Madagascar (1957), Senegal and Tanganyika (1961), Kenya (1968–1969). and Malawi (1970–1971), restatement altered, i.e. modernised, 'native' customary law. The first attempts to modernise it date back to the early twentieth century: Germany started restating Tanganyikan family law in 1907—and the process was subsequently carried on by the United Kingdom in the 1940s.[146] The 'School of Oriental and African Studies' (SOAS) in London fostered its own *Restatement of African Law Project* in 1959: this was a comprehensive pattern for the study and restatement of African customary law of 16 Anglophone countries in the fields of land tenure, succession, family law, and status of women. The colonial legacy is apparent, because the project was externally driven, that is, elaborated and implemented in London. In the aftermath of decolonisation, the task of modernising African law was resumed by the Law and development movement, whereby European and U.S. legal and economic assistance aimed to develop African countries by imposing their own legal paradigms.[147]

Pervasive changes in native customary law altered social geographies and local communities. To mitigate the strictures of native customary law, European values, such as individualism and liberalism, were infused into traditional systems, leading to the relaxation of group-centred traditional societies as in the case of African land law. African land tenure system was mainly communal and governed by both supernatural entities and the group; there was no room left for Western possessive individualism. The rise of trade pushed for its suppression—or, at least, reduction—, 'because the land market could not fit with ideas regarding the communal nature of African land tenure'.[148] In the aftermath of decolonisation, Western, i.e. individual, land titles were retained and colonial laws regarding customary lands adapted to the African context: land legislation transformed former communal lands into public lands, such as in Tanzania and Ghana. Like in England, Tanzanian legislation assigns the land to the Head of State (the President), who acts as trustee on behalf of all citizens: the latter 'cannot own land, but they can own rights over the land,' which 'may be bought or sold, and inherited, and can thus be seen as (limited) decision-making rights.' The 1992 Constitution of Ghana does the same: 'All public lands in Ghana shall be vested in the President on behalf of, and in trust for, the people of Ghana'; whereas 'stool lands', which the communal soul (the stool) granted to its own people, are vested 'in the appropriate stool on behalf of, and in trust for, the subjects of the stool in accordance with customary law and usage'.[149]

---

[145] On such processes see Prinsloo (1987), p. 411.

[146] Sippel (1998), p. 378.

[147] Merryman (1977), p. 457.

[148] Joireman (2010), p. 298.

[149] Sect. 255(1) and 167(1) of the 1992 Constitution (Ghana); *Loi* no 034-2009/AN du 16 juin 2009 *portant régime foncier rural* (Burkina Faso); URT, *Land Act* (No. 4), sec. 7 and URT, *Village Land Act* (No. 5), sec. 8(1), 12(1) (Tanzania). See Locher (2016).

4.6 Mapping and Ranking the Spatiality of Global Law

4.6.1 The Logical Space of Global Actors

In the wake of decolonisation, we might be persuaded that dismantling colonial Empires has put an end to the geopolitics of law. *Geography Triumphant* led to what Harley termed as the 'closing of territories'; at the same time, Europeans filled all the blank spaces spotted on maps. It was thought that the next phase might well herald the application of the territorial trap on a global scale (see Sect. 4.1).

It is no coincidence that the current state of world affairs is labelled 'globalisation'. This phenomenon has generated its own spatiality of law and the inception of a new global politics of space (see Sects. 5.3.1, 8.2, and 8.3). Like in the previous epochs of spatial appropriation, the territorialised paradigm still haunts, as a 'hidden geography',[150] the legal geography of the global politics of space. The process of spatial production still points to 'both opening and later closing' territories in a time of globalisation.

The dismantlement of colonialism in the 1960s and 1970s, the end of the Cold War, and the advent of 'new world order' announced by the US President George Bush led to the reopening of territories. The post-Cold-War, the 1994 Uruguay Round, and the World Bank advocated for a new logical space which had to be governed under the 'good governance' scheme. A true Neoliberal era begun. As well as a new spatiality of law. The ambition was that this would promote 'peace and security, freedom and the rule of law' alongside economic development: indeed, the global village had already been established 'from the economic ground up'.[151] 'Civilisation [still] follows trade', we may say paraphrasing Joseph Conrad's final lines of *An Outpost of Progress*.[152] We have got used to equating trade and civilisation within the context of globalised markets and to praising global law, which is the outcome of the 'historical process of the worldwide generalization of Western legalism'. Embedding it triggers 'a kind of impending radiant future for humanity ... characterized by open markets that ensure (sustainable) development and (environmentally-friendly) prosperity'.[153]

With such a liberating force, globalisation opens up new cultural possibilities. It is impossible to resist the fascinating trends of universalism, which facilitate harmonisation of laws and stimulate economic development. Indeed, global markets need the reopening of territories to favour the smooth flows of goods, capital, and services across frontiers and legal systems, which act as 'striations and discrepancies that interrupt [the] monolithic spaces' of globalised normativities.[154] Consequently, the globalisation of financial flows requires new legal spaces. The practice of ranking

---

[150] Agnew (1995).

[151] Rosow (2000), p. 29.

[152] Conrad [1897] (2008), p. 25.

[153] Xifaras (2016), p. 216.

[154] Carpi (2020), p. 225. On the normativities of globalisation see Husa (2018).

legal systems also characterises the current state of affairs. As a part of a new process of spatial production, it reflects the imaginative geographies of global and transnational actors, their geopolitics of law, and spatial hierarchies. The difference lies in how the logical space is arranged and the spatiality of law is manufactured.

4.6.2 The Legal Geography of Indicators

In this section I shall examine how globalisation has favoured new waves of universalism. Take, for instance, the dissemination of European constitutionalism around the world: 'the basic principles of constitutional law' are now deemed to be 'the same around the world',[155] their 'universal respect' and 'observance' is dependent on their being 'a common standard of achievement for all peoples and all nations'.[156] Global law is supposed to 'offer everyone the benefits of peace, human rights, representative democracy, and moderate government'.[157] Over the past few decades, we have also been experiencing a considerable degree of legal convergence: the comparative-law discourse is oversaturated with 'transplants', 'borrowings', and so on.[158] These changes have paved the way to several cartographic experiments. In our globalised world, recipient countries usually adopt Western legal paradigms as their benchmark, against which it is possible to assess 'whether governance comports with universal human rights norms'.[159] In addition, the use of quantification in comparative law allows us to measure, inter alia, 'global-transnational influences on constitutional human rights'.[160]

Indexes, such as the Global Democracy Index, 'illustrate the point that an instrumental part of cartography (and more broadly of scientific practice) is ideology'; and the joint application of thematic mapping and GIS creates a hierarchy of colours and jurisdictions within maps.[161] Each country is coloured to reflect its rank and democratic outlook in terms of shades of blue and red. The redder the country, the less democratic it is; the darker blue the recipient, the more democratic it is. The possibility of clicking on this map to ascertain whether a country comports with further five indicators (i.e. electoral process and pluralism, functioning of government, political participation, democratic political culture, and civil liberties) also reflects an additional pattern of normalisation within the Global Democracy Index. Its logical space is objectivised both in geographical and comparative-law terms. As regards legal cartographic discourse, the logical space embeds the 'various

---

[155] Beatty (1995), p. 10.

[156] See Arnold (2013).

[157] Xifaras (2016), p. 216.

[158] For more on these processes see e.g., Nicolini (2020).

[159] Udagama 148; Davies 542.

[160] Beck et al. (2019).

[161] Crampton (2010), p. 70.

"interests" of those who make and use them [i.e. the global holders of capital], rather than reflections of the environment'.[162] In economic terms, the global convergence of laws has twisted the rule of law into a measurable value-neutral legal concept that secures predictability on investment returns.

Whatever projection is chosen for representing the world, the cartographical representation is no longer limited to technicalities for recording and storing information. It is intuitive that legal imaginative geographies of transnational actors promote the change in national logical spaces, whose decision-making processes must adapt to the interests of transnational actors to code their own interests through the law. The relationship between 'spatial technologies' (cartography, remote sensing, and GIS, just to quote a few) and globalisation marks a new 'defining moment' for cartography 'coming to grips with' a new function of maps—that is, the propaganda of global interests relegating the representation of the earth to a purely decorative role.[163] The GIS represents indexes whereby countries are said to be more or less democratic than others; it also outlines systems according to their economic performances. And, probably, economic and financial actors now advocate for new representational tools.

This has contributed to the steady rise of 'calculative practices' and fostered the 'use of quantification' in the form of both indicators and rankings, through which 'the most economical solutions have been chosen and . . . are permanently in control'.[164] Indeed, quantitative methods allow scholars to assess the 'quality' of the legal systems and rank them, say, on how they actually implement the rule of law and protect foreign financial investors. For instance, the world-famous essay *Law and Finance* measured and ranked forty-nine national legal systems: the benchmark was the quality of the shareholder and creditor protection regulation.[165]

Such a *quantitative frame* has 'Turn[ed] Law into Numbers' and made the shift in logical space possible.[166] Like in the spatiality of law, we may use the verb 'to naturalise' when pointing to the processes whereby the measurement of legal frameworks becomes 'common sense'. In particular, the logical space of globalisation has used this process of naturalisation as regards the meaning of the legal lexicon, as well as a qualitative change in manifold legal systems. The shift is functional to both common financial language and a global market regulation in the form of the Law of the Buyosphere (see Sect. 8.5). How this occurs is due to pressure from economic models, which aim to make laws converge towards economic semantics: the substitution of meaning is thus complemented by the substitution of words, and economic lexemes replace legal jargon. Linguistics and legal

[162] Crampton (2010), p. 17 quoting Wood (1992).

[163] See Pickles (1992) and Crampton (1994).

[164] Rottenburg and Merry (2015), p. 3.

[165] La Porta et al. (1998). An assessment of quantification in numerical comparative law is in Siems (2022), p. 208; Michaels (2009). On how global actors (the World Bank, and the EU Commission) use of such methods when promoting legal change see Nicita and Benedettini (2012).

[166] Siems (2011).

change are a projection of global financial dominance, and the rule of law expresses a performance measure, which evaluates the efficiency of the legal systems and their attractiveness to international investments: the more efficient the system, the more likely the investment, and the higher the economic return.[167]

The link between law and economic performance has a huge impact on classifications.[168] 'Power and authority' shifts from countries to non-democratically accountable global actors;[169] the ranking of legal systems depends on their performativity, which is rooted in their legal origins. The *Western legal tradition* is dominant, and, within it, the *common law* prevails over the *civil law*.[170] Although civil-law systems have demonstrated their ability to secure a shareholder and creditor protection regulation higher than common law jurisdictions, the latter are said to ensure more elevated economic performances. Measured, as they are, through numerical indicators, legal systems now fall under 'numerical comparative law'. Doing comparative law by number presupposes that 'legal institutions have an impact on economic growth'—and 'in tune with neoclassical law and economics'—the dominant paradigm is based 'essentially on the idea that law should be measured by the incentives it sets for welfare-maximizing conduct'.[171] The holders of capital have created spatial hierarchies through 'cross-country comparisons including rankings of the attractiveness of different legal systems for doing business'.[172]

Indicators share the biases of traditional legal cartography; indeed, 'through numerical representation, one always loses some aspects of the reality'.[173] Their ability to reconcile scientific objectivity and measurement with real-world tests is attained by leaving out of the picture the striations that may disturb the application of transnational derivatives wherever economic interests deem it fit to do so.[174] Like in traditional cartography, transnational actors superimpose their own nomenclatures and imaginative geographies onto the earth's surface. Indicators thus deliberately omit data and variables, notwithstanding their relevance in real-world assessments. And such omissions may also trigger a strabismus in metrics, which unavoidably affects the reliability of their outlook. Odd as it may seem, this type of strabismus may prevent them from scratching the surface of the data and realising that there are concrete signs of dangers, if not of a catastrophe. But indicators also point to dangers, which we may either ignore or take seriously. The strabismus of indicators occurred, for instance, just before the 2007 financial crisis, when 'all the standard financial indicators declared that financial system was safe'. It could have been sufficient to scratch the surface and realise that real economy exhibited signs of

[167] Oto-Peralía and Romero-Ávila (2017), pp. 85–119 and 121–135.

[168] See Siems (2016); Oto-Peralía and Romero-Ávila (2017), pp. 57–83.

[169] Cheng (2005), p. 469.

[170] Grosswald Curran (2009).

[171] Michaels (2009), p. 768.

[172] Michaels (2009), p. 766.

[173] Rottenburg and Merry (2015), p. 8.

[174] Davies (2015), p. 285.

dangers; and yet, there was a 'collective declaration of safety being issued by the financial market indicators'.[175]

Made up of indicators and data, logical space holds a particular fascination because it is *consociational*. Coined by Arend Lijpart, the expression points to those federalist processes whereby highly conflictive societies, usually divided along ethnolinguistic cleavages, are accommodated and the maintenance of territorial integrity is secured (see Sect. 7).[176] The consociational argument arranges different sectionalities around a power-sharing organising principle. It assumes the unequal distribution of wealth; but this does not prevent each member of the community from gaining access to it.

The consociational argument is also useful for propagating global comparative law as it favours its generalisation. Egalitarian in its commitment, it facilitates the harmonisation of laws to stimulate business and economic development. It also proposes a transnational, borderless legal geography into which heterogeneous legal systems coalesce. Again, this occurs under pressure from economic models, which aim to make laws converge towards a common economic semantics. The alleged egalitarianism of transnational logical space is thus a projection of global financial dominance, which evaluates the efficiency of legal systems and their attractiveness provided that they positively discriminate in favour of international investors.

References

Abulafia D (2020) The boundless sea: a human history of the oceans. Penguin, London

Agnew J (1995) The hidden geographies of social science and the myth of the 'geographical turn'. Environ Plan D: Soc Space 13:379–380

Allott A (1967) Law in the new Africa. Afr Aff 66:55–63

Alvarez A (1911) The Monroe Doctrine at the Fourth Pan-American Conference. Ann Am Acad Polit Soc Sci 37:24–30

Arewa OB (2021) Disrupting Africa. Technology, law, and development. CUP, Cambridge

Arminjon P, Nolde B, Martin W (1950) Traité de Droit Comparé. Tome I. LDGJ, Paris

Arnold R (2013) Reflections on the universality of human rights. In: Arnold R (ed) The universalism of human rights. Springer, Dordrecht, pp 1–12

Aylmer J (1559) An harborowe for faithfull and trewe subiectes agaynst the late blowne blaste. London. Printed by John Day

Bailey RW (2012) Speaking American: a history of English in the United States. OUP, New York

Baker JH (2016) The reinvention of Magna Carta 1216–1616. CUP, Cambridge

Baugh AC, Cable T (2012) A history of the English language, 6th edn. Routledge, Abingdon and New York

Beatty D (1995) Comparative law in theory and practice. University of Toronto Press, Toronto

Beck CL et al (2019) Constitutions in world society: a new measure of human rights. In: Schaffer G et al (eds) Constitution-making and transnational legal order. CUP, Cambridge, pp 85–109

[175] Farlow (2015), pp. 221–222.

[176] On consociational federalism see Lijpart (1979), p. 505.

Bennett TW (1996) African land – a history of dispossession. In: Zimmermann R, Visser D (eds) Southern cross: civil law and common law in South Africa. Clarendon, Oxford, pp 65–94

Bennett A (2008) What is geography? SAGE, London

Benton L (2010) A search for sovereignty: law and geography in European empires, 1400-1900. CUP, Cambridge

Berman HJ (1983) Law and revolution: the formation of the western legal tradition. Harvard UP, Cambridge, MA

Bhandar B (2014) Property, law, and race: modes of abstraction. UC Irvine Law Rev 4:203–218

Black J (1997) Maps and politics. The University of Chicago Press, Chicago

Blackstone W [1765] (2016) Commentaries on the Laws of England, I, of the rights of persons. In: Prest W (ed) The Oxford edition of Blackstone. OUP, Oxford

Blomley N (2015) The ties that blind: making fee simple in the British Columbia treaty process. Trans Inst Br Geogr 40:168–179

Boorstin DJ (1983) The discoverers. A history of man's search to know his world and himself. Random House, New York

Bottomley A (2020) Between islands: colonial legacies and cultural imaginaries. Pólemos 14:1–25

Brand P (1981) Ireland and the literature of the early common law. Irish Jurist XVI:95–113

Brotton J (2013) A history of the world in twelve maps. Penguin, London

Brotton J (2018) Trading territories: mapping the early modern world. Reaktion Books, London

Bussani M, Mattei U (eds) (2012) The Cambridge companion to comparative law. CUP, Cambridge

Campbell T (1987) Portolan charts from the late thirteenth century to 1500. In: Harley JB, Woodward D (eds) The history of cartography, volume one (Cartography in prehistoric, ancient, and Medieval Europe and the Mediterranean). Chicago UP, Chicago, pp 371–462

Campling L, Colás A (2021) Capitalism and the sea. The maritime factor in the making of the modern world. Verso, London and New York

Canny N (2001) The origins of empire. An introduction. In: Canny N (ed) The Oxford history of the British Empire, vol I, The origins of empire. British overseas enterprise to the close of the seventeenth century. OUP, Oxford, pp 1–33

Carpi D (2020) The island metaphor in literature and law. Pólemos 14:225–235

Carr LG (1992) Emigration and the standard of living: the seventeenth century Chesapeake. J Econ Hist 52:271–291

Cheng T-H (2005) Power, authority, and international investment law. Am Univ Int Law Rev 20: 465–520

Collis C (2010) Critical legal geographies of possession: Antarctica and the International Geophysical Year 1957–1958. GeoJournal 75:387–395

Connell S (2011) Writing on the land of Ireland: nationality, textuality, and geography in the Acallam na Senórach. Hortulus 7:1–27

Conrad J (1924) Geography and some explorers. Natl Geogr Mag 45:241–356

Conrad J (1970) Travel. In: Last essays. Doubleday, New York, pp 84–92

Conrad J [1899] (2008) Heart of darkness. In: Watts C (ed) Heart of darkness and other tales. OUP, Oxford, pp 103–187

Cormack LB (2001) Britannia rules the waves?: images of empire in Elizabethan England. In: Gordon A, Klein B (eds) Literature, mapping and the politics of space in early modern Britain. CUP, Cambridge, pp 45–68

Craib RB (2000) Cartography and power in the conquest and creation of new Spain. Latin Am Res Rev 35:7–36

Crampton J (1994) Cartography's defining moment: the Peters projection controversy, 1974–1990. Cartographica: Int J Geogr Inf Geovis 31:16–32

Crampton J (2003) The political mapping of cyberspace. Edinburgh UP, Edinburgh

Crampton JW (2010) Mapping: a critical introduction to cartography and GIS. Blackwell, Oxford

Davenport FG (ed) (1917) European treaties bearing on the history of the United States and its dependencies, IV vol. Carnegie Institution of Washington, Washington, DC

David R et al (2016) Les grands systèmes de droit contemporains, 12th edn. LGDJ, Paris

Davies RR (1994) Presidential address: The Peoples of Britain and Ireland 1100-1400. I. Identities. Trans R Hist Soc 4:1–20

Davies W (2015) Spirits of neoliberalism: 'competitiveness' and 'wellbeing' indicators as rival orders of worth. In: Rottenburg R et al (eds) The world of indicators: the making of governmental knowledge through quantification. CUP, Cambridge, pp 283–306

Delaney D (2005) Territory. A short introduction. Blackwell, London

Diener AC, Hagen J (2012) Borders. A very short introduction. Oxford UP, Oxford

Driver F (2001) Geography militant. Cultures of exploration and empires. Blackwell, Oxford

Edney MH (1997) Mapping an empire: the geographical construction of British India, 1765-1843. University of Chicago Press, Chicago

Edwards P (1992) Edward Hayes explains away Sir Humphrey Gilbert. Renaiss Stud 6:270–286

Egerton H (1897) A short history of British Colonial Policy. Methuen & Co., London

Eisenstadt SN (1965) Social change and modernization in African societies South of the Sahara. Cahiers d'Études africaines 19:453–471

Esmein A (1900) Le droit comparé et l'enseignement du droit. Revue historique de droit français et étranger XXIV:489–499

Farlow A (2015) Financial indicators and the global financial crash. In: Rottenburg R et al (eds) The world of indicators. The making of governmental knowledge through quantification. CUP, Cambridge, pp 220–253

Fichter J (2008) The British Empire and the American Atlantic on Tristan da Cunha, 1811–16. J Imp Commonw Hist 36:567–589

Gant E (2006) Human rights, cultural diversity and customary law. South Africa. J Afr Law 50:2–23

Gerald of Wales (1978) Expugnatio Hibernica or the conquest of Ireland, ed with translation and historical notes by Scott AB, Martin FX. Royal Irish Academy, Dublin

Gerald of Wales (1982) Topographia Hibernica or the history and topography of Ireland (trans: O'Meara JJ). Penguin, London

Gjersø JF (2015) The scramble for East Africa: British motives reconsidered, 1884-95. J Imp Commonw Hist 43:831–860

Glenn HP (2014) Legal traditions of the world: sustainable diversity in law, 5th edn. OUP, Oxford

Godlewska A, Smith N (eds) (1994) Geography and empire. Blackwell, Oxford

Goodrich P (2021) Advanced introduction to law and literature. Edward Elgar, Cheltenham and Northampton, MA

Gordon A, Klein B (2001) Introduction. In: Gordon A, Klein B (eds) Literature, mapping, and the politics of space in early modern Britain. CUP, Cambridge, pp 1–12

Gregory D (2004) The colonial present. Blackwell, Oxford

Grenier B (2004) "Nulle Terre Sans Seigneur"?: Une Étude Comparative de la Présence Seigneurial (France-Canada), XVIIe-XIXe Siècle. Fr Colon Hist 5:7–24

Griffiths I (1986) The scramble for Africa: inherited political boundaries. Geogr J 152:204–216

Grosswald Curran V (2009) Comparative law and the legal origins thesis: '[N]on scholae sed vitae discimus'. Am J Comp Law 57:863–876

Gschwend AJ (1995) In the tradition of princely collections: curiosities and exotica in the Kunstkammer of Catherine of Austria. Bull Soc Renaiss Stud 13:1–9

Hakluyt R [1585] (1602) Inducements to the liking of the voyage inten|ded towards Virginia in 40. and 42. degrees of latitude, written 1585. by M. Richard Hakluyt the elder, sometime student of the Middle Temple. In: Breton J. A briefe and true relation of the discouerie of the north part of Virginia being a most pleasant, fruitfull and commodious soile: made this present yeere 1602. London, [Printed at Eliot's Court Press] impensis Geor. Bishop, pp 25–36

Halpérin J-L (2009) The concept of law: a western transplant? Theor Inq Law 10:333–354

Harley JB (1988) Victims of a map: New England Cartography and the Native Americans. Paper read at the Land of Norumbega Conference, Portland, Maine, December 1988

Harmsworth GR, Shaun A (2013) Indigenous Māori knowledge and perspectives of ecosystems. In: Dymond JR (ed) Ecosystems services in New Zealand – conditions and trends. Manaaki Whenua Press, Lincoln, NZ, pp 274–286
Harris J Sir (1912) Dawn in darkest Africa. Smith, Elder & Co., London
Hau'ofa E (1994) Our sea of islands. Contemp Pac 6:148–161
Himonga C, Nhapo S (eds) (2014) African customary law in South Africa. Post-apartheid living law perspectives. OUP Southern Africa, Cape Town
Ho J (2007) The spatial imagination and literary form of Conrad's colonial fictions. J Mod Lit 30:1–19
Howkins A (2017) Frozen empires. An environmental history of the Antarctic Peninsula. CUP, Cambridge
Husa J (2018) Advanced introduction to law and globalisation. Edward Elgar, Cheltenham-Northampton
Joireman SF (2010) Entrapment or freedom: enforcing customary property rights regimes in common-law Africa. In: Fenrich J-M et al (eds) The future of African customary law. CUP, Cambridge, pp 295–311
Jones H (2019) Property, territory, and colonialism: an international legal history of enclosure. Leg Stud 39:187–203
Keltie JS (1890) What Stanley has done for the map of Africa. Science 15(364):50–55
Kersten H (2006) America's multilingualism and the problem of the literary representation of 'Pidgin English'. Amerikastudien/Am Stud 51:75–91
Kischel U (2019) Comparative law. OUP, Oxford
Korosy Z (2020) Whales and the colonization of the pacific ocean. In: Braverman I, Johnson ER (eds) Blue legalities. The life & laws of the seas. Duke UP, Durham and London, pp 219–235
La Porta R et al (1998) Law and finance. J Polit Econ 106:1113–1155
Lestringant F (1994) Mapping the renaissance world: the geographical imagination in the age of discovery. Polity Press, Cambridge
Levy-Ullman H (1923) Observation générales sur les communications relatives au droit privé dans les pays étrangers. In: Les Transformations du droit dans les principaux pays depuis cinquante ans (1869-1919). Livre du cinquantenaire de la Société de Législation comparée, Paris, pp 85–111
Lijpart A (1979) Consociation and federation: conceptual and empirical links. Can J Polit Sci 13: 499–516
Livingstone DN (1992) The geographical tradition. Blackwell, Oxford
Locher TM (2016) 'How come others are selling our land?' Customary land rights and the complex process of land acquisition in Tanzania. J East Afr Stud 10:393–412
Lovegrove R (2012) Islands beyond the horizon: the life of twenty of the world's most remote places. Oxford UP, Oxford
Lugard FD (1922) The dual mandate in British tropical Africa. William Blackwood and Sons, Edinburgh
Mann G (2009) What was the 'Indigénat'? The 'Empire of Law' in French West Africa. J Afr Hist 50:331–353
Mann K, Roberts R (1991) Law in colonial Africa. Heinemann-James Currey, Portsmouth (NH)-London
Martínez Paz (1934) Introducción al estudio del derecho civil comparado. Imprenta de la Universidad, Córdoba
Mattei U (1997) Three patterns of law: taxonomy and change in the world's legal systems. Am J Comp Law 45:5–44
Mattingly G (1963) No peace beyond what line? Trans R Hist Soc 13:145–162
Menski WF (2006) Comparative law in a global context: the legal systems of Asia and Africa. CUP, Cambridge
Mentz S (2015) Shipwreck modernity: ecologies of globalization, 1550–1719. University of Minnesota Press, Minneapolis, MN

Merryman JH (1977) Comparative law and social change: on the origins, style, decline & revival of the law and development movement. Am J Comp Law 25:457–491
Michaels R (2009) Comparative law by numbers? Legal origins thesis, doing business reports, and the silence of traditional comparative law. Am J Comp Law 57:765–795
Miller HM (2004/2005) The archaeology of colonial encounters along Chesapeake Bay: an overview. Revista de Arqueología Americana 23:231–290
Molyneaux G (2014) Did the English really think they were god's elect in the Anglo-Saxon Period? J Eccles Hist 65:721–737
Monateri PG (2013) Geopolitica del diritto: genesi, governo e dissoluzione dei corpi politici. Laterza, Rome-Bari
Monateri PG (2021) Advanced introduction to comparative legal methods. Edward Elgar, Cheltenham and Northampton, MA
Nayak A, Jeffrey A (2011) Geographical thought. An introduction to ideas in human geography. Routledge, Abingdon and New York
Netzloff M (2019) Lines of amity. The law of nations in the Americas. In: Sowerby TA, Craigwood J (eds) Cultures of diplomacy and literary writing in the early modern world. OUP, Oxford, pp 54–68
Nicita A, Benedettini S (2012) Towards the economics of comparative law: the *Doing Business* debate. In: Monateri PG (ed) Methods of comparative law. Edward Elgar, Cheltenham and Northampton, MA, pp 291–305
Nicolai R (2015) The premedieval origin of portolan charts: new geodetic evidence. Isis 106:517–543
Nicolini M (2016) Narrators of fables or framers of the constitution? The Acallam na Senórach beyond time, place, and law. In: Carpi D, Leiboff M (eds) Fables of the law. Fairy tales in a legal context. de Gruyter, Berlin and Boston, pp 137–163
Nicolini M (2018) Writing for the "scholar and the gentleman." Language, society, and legal education in Blackstone's commentaries. Cardozo Electron Law Bull 24:1–32
Nicolini M (2019) African legal traditions. In: Mousourakis G (ed) Comparative law and legal traditions. Springer, Cham, pp 281–297
Nicolini M (2020) Methodologies of comparative constitutional law: universalist approach. In: Max Planck Encyclopaedia of Comparative Constitutional Law [MPECCOL]. OUP, Oxford
Oplot JSE (1992) The resilience of the British colonial police legacies in East Africa, Southern Africa, and West Africa. Police Stud: Int Rev Police Dev 15:91–99
Oto-Peralía D, Romero-Ávila D (2017) Legal traditions, legal reforms and economic performance. Theory and evidence. Springer, Wien
Pagden A (2001) The struggle for legitimacy and the image of Empire I the Atlantic c. 1700. In: Canny N (ed) The Oxford history of the British Empire, vol I, The origins of empire. British overseas enterprise to the close of the seventeenth century. OPU, Oxford, pp 34–54
Pargendler M (2012) The rise and decline of legal families. Am J Comp Law 60:1043–1074
Patton P (2000) The translation of indigenous land into property: the mere analogy of English jurisprudence. Parallax 6:25–38
Penrose B (1952) Travel and discovery in the Renaissance, 1420–1620. Harvard UP, Cambridge, MA
Pickles J (1992) Texts, hermeneutics, and propaganda maps. In: Barnes TJ, Duncan JS (eds) Writing worlds: discourse, text, and metaphor in the representation of landscape. Routledge, London and New York, pp 193–230
Price JM (1998) The imperial economy, 1700-1776. In: Marshall PJ (ed) The Oxford history of the British Empire, vol II, The eighteenth century. OUP, Oxford, pp 1–27
Prinsloo MW (1987) Restatement of indigenous law. Comp Int Law J South Afr 20:411–420
Rambaud T (2017) Introduction au droit comparé, 2nd edn. PUF, Paris
Ranger T (1983) The invention of tradition in colonial Africa. In: Hobsbawm E, Ranger T (eds) The invention of tradition. CUP, Cambridge, pp 211–262

Reid S (2020) Solwara 1 and the sessile ones. In: Braverman I, Johnson ER (eds) Blue legalities. The life & laws of the seas. Duke UP, Durham and London, pp 24–44
Reimann M, Zimmermann R (2006) The Oxford handbook of comparative law. OUP, Oxford
Rosow SJ (2000) Globalisation as democratic theory. Millennium 29:27–65
Rottenburg R, Merry S (2015) A world of indicators: the making of governmental knowledge through quantification. In: Rottenburg R et al (eds) The world of indicators: the making of governmental knowledge through quantification. CUP, Cambridge, pp 1–33
Rutman AH (1987) Still planting the seeds of hope: the recent literature of the early Chesapeake region. Va Mag Hist Biogr 95:3–24
Said E [1978] (2003) Orientalism. Penguin, London
Sainsbury N (ed) (1893) Calendar of state papers colonial, America and West Indies. 1893. Volume 9, 1675-1676 and Addenda 1574-1674. Majesty's Stationery Office, London
Sammler KG (2020) Kauri and the whale. Oceanic matter and meaning in New Zealand. In: Braverman I, Johnson ER (eds) Blue legalities. The life & laws of the seas. Duke UP, Durham and London, pp 63–84
Sanderson GN (1974) The European partition of Africa: coincidence or conjuncture? J Imp Commonw Hist 3:1–54
Sauser-Hall G (1913) Fonction et méthode du droit comparé. Sirey, Paris
Schmitt C (1950) Der Nomos Der Erde: Im Volkerrecht Des Jus Publicum Europaeum. Duncker & Humblot, Berlin. English edition: Schmitt C (2006) The Nomos of the Earth in the international law of the Jus Public Europaeum (trans: Ulmen GL). Telos, New York
Seed P (1995) Ceremonies of possession in Europe's conquest of the new world, 1492-1640. CUP, New York
Sharkey HJ (2013) African colonial states. In: Reid R, Parker J (eds) The Oxford handbook of modern African history. OUP, Oxford, pp 151–170
Siems M (2011) Measuring the immeasurable: how to turn law into numbers. In: Faure M, Smits J (eds) Does law matter? On law and economic growth. Cambridge, Intersentia, pp 115–136
Siems M (2016) Varieties of legal systems: towards a new global taxonomy. J Inst Econ 12:579–602
Siems M (2022) Comparative law, 3rd edn. CUP, Cambridge
Sippel H (1998) Customary family law in colonial Tanganyika: a study of change and continuity. Comp Int Law J South Afr 31:373–383
Steinberg PE (2001) The social construction of the ocean. CUP, Cambridge
Taylor IC (1994) Official geography and the creation of 'Canada'. Cartographica 21:1–15
The Book of the Common Prayer And Administration of the Sacraments, And Other Rites and Ceremonies of the Church, According to the Use of The Church of England: Together with the Psalter or Psalms of David, Pointed as they are to be sung or said in Churches [1662] (2004), CUP, Cambridge [BCP]
Thomson J (1802) Britannia. In: The works of Mr. James Thomson: with his last corrections and improvements, vol 2. Printed for R. Baldwin, London
Tomlins C (2001) The legal cartography of colonization, the legal polyphony of settlement: English intrusions on the American mainland in the seventeenth century. Law Soc Inq 26:315–372
Vander Linden H (1916) Alexander VI. and the demarcation of the maritime and colonial domains of Spain and Portugal, 1493–1494. Am Hist Rev 22:1–20
Ward I (2020) English legal histories. Bloomsbury, London
Wendower, of R (1886–1889) Flores Historiarum, ed. Henry Gay Hewlett, vol II, Printed for H. M. Stationery Office, London
White LW, Hussey WD (1958) Government in Great Britain, the empire, and the commonwealth. CUP, Cambridge
Wigmore JH (1928) Panorama of the world's legal systems. West Pub. Co., Saint Paul
Williamson AJ (1949) Hawkins of Plymouth. Black, London
Wood D (1992) The power of maps. The Guildford Press, New York

Wright JK (1947) Terrae Incognitae: the place of the imagination in geography. Ann Assoc Am Geogr 37:1–15

Xifaras M (2016) The global turn in legal theory. Can J Law Jurisprud XXIX:215–243

Zimmermann R, Visser D (1996) Introduction. South African law as a mixed legal system. In: Zimmermann R, Visser D (eds) Southern cross: civil law and common law in South Africa. Clarendon Press, Oxford, pp 3–30

Zweigert K, Kötz H (1996) Einführung in die Rechtsvergleichung, 3rd edn. J.C.B. Mohr Siebeck, Tübingen

Chapter 5
Critical Legal Geographies

5.1 The Revival of Marginalised Legal Geographies

5.1.1 Buried, but Still Alive

Within the spatiality of law, European powers superimposed their legal-geographical nomenclatures onto newly discovered territories, of which they claimed full possession. Colonisers squared non-Western legal geographies with their 'more familiar' legal taxonomies, 'either to reject them as radically unlike, or to analogise them as unsettlingly similar'.[1] In reflecting Western logical spaces, these practices silenced the voices of native communities and their cartographic traditions. Those of the Aztecs, for instance, had impressed Cortés, whose 'encounter with native mapmaking was of sufficient moment to be mentioned in two of the letters he wrote to Emperor Charles V in 1519–20'.[2] This undoubtedly matches the process that, with Harley, I termed 'closing of territories'. Before 'opening' them to spatial reshufflings, Europeans erased their pre-colonial past and buried native logical spaces under the 'sanitising' effects of linear thinking (see Sect. 4.2.2). This erasure resonates with a type of colonialism labelled 'settler colonialism'. Unlike its 'traditional' counterpart, it 'seeks to eliminate the native presence from the land upon which it aims to build a settler society'; in so doing, it replaces the former belongs of public law with new processes of legal (and spatial) production (see Sect. 6.2).[3]

The burial process was not so simplistic as colonisers had supposed. A couple of quotations help us to explain why. The first one is literary. Thomas Gray's *Elegy Written in a Country Churchyard* 'movingly capture[s] that life-wish that refuses to surrender easily to death' and burial rites: 'Ev'n from the tomb the voice of nature

[1] Blomley (2015), p. 173.

[2] Harley (1992), p. 525.

[3] Bhungalia (2018), p. 314. See also Wolfe (1999), Veracini (2014), and Correja (2021).

M. Nicolini, *Legal Geography*, Ius Gentium: Comparative Perspectives on Law and Justice 105, https://doi.org/10.1007/978-3-031-19410-8_5

cries, / Ev'n in our ashes live their wonted fires'.[4] The second quotation is legal. In his *The Forms of Action at Common Law*, Maitland reminds us that interring cultural and legal artefacts means burying neither their memory nor their performative force, which remains impressed in our logical space: 'The forms of action we have buried, but they still rule us from their graves'.[5]

Law and geography are populated with processes where spatial burying reflects policies of domination and dispossession. At the same time, both experience the resurgence of pre-colonial legal spatial settlements whose performative force Europeans assumed to have silenced. In former colonial contexts, attempts to replace Irish placenames with English nomenclatures became a 'topographic' challenge to Irish identity, triggering a reaction in the *Acallam na Senórach*.[6] Likewise, outside Europe, post-colonialism led to removing 'the linguistic impact of European imperial rule', which meant reviving marginalised legal geographies. When looking at maps, 'readers outside India' usually 'wish to find Bombay, Delhi and Madras', but also Calcutta and Pondicherry, i.e. placenames that have traditionally stirred our imagination as regards the Orient.[7] Maps now bear 'native' names: Mumbai, Dilly, Chennai, Kolkata, and Puducherry. The same occurs in Asia (such as Myanmar), Africa (such as the former British colonies and protectorates in Southern Africa),[8] and Oceania (e.g. New Zealand/Aotearoa; for federal systems see Sect. 6.4.1). The revival of historically marginalised toponyms is of the utmost legal importance; its impact on both human and physical geographies alike makes it *the* central aspect of territorial identity, and logical spaces turn into *legal geography*.

Renaming territories gave new impetus to offset colonial geographies. In a time of *Geography Militant*, colonised people were deprived of their lands. They were 'rightless' as they could not demonstrate to possess 'a recognizable legal system'.[9] In many cases, Europeans held that such a requirement was absent; in common law (and its Anglo-British colonies), the land was (and still is) vested in the Crown. A recognisable legal title to the land was therefore impressed into the grounds of the colonies; as 'tribal occupation did not rest upon a crown-derived basis', it was 'ungranted land'; tribes did not have any type of 'land right' of which 'a common law court might take cognizance'.[10] The feudal system of tenure thus deprived natives of their land rights.

---

[4] Respectively: Siebert (2013), p. 123; Gray (1751), lines 91–92.

[5] Maitland (1936), p. 1.

[6] See Sect. 4.2.1; see also Sect. 6.4.2.

[7] Black (1997), p. 42.

[8] See e.g., Nyasaland/Malawi, Rhodesia/Zimbabwe-Zambia, and Bechuanaland/Botswana. The last country that replaced its colonial denomination with a native one was Swaziland/Eswatini (2015).

[9] Bennett (1996), pp. 66, and 68 et seq (negation of aboriginal titles under the British rule in nineteenth-century Cape Colony).

[10] McHugh (2011), p. 111.

In post-colonial environments, the 'question of land rights' gives rise to the 'issue of restoring' them.[11] But: which title must we consider? The pristine native title or that processed (hence, Westernised) under the politico-legal order imposed by the Europeans? Canadian and Australian courts have developed what has been termed the 'aboriginal title doctrine'—or, *mutatis mutandis*, the 'native title doctrine'.[12] As a common-law derivative, it is applied in several former British colonies and protectorates. To be restored, native titles must precede the advent of the common-law logical space and must not have been 'completely extinguished', i.e. they must have survived the conquest.[13]

This doctrine aims to reconcile native and colonial legal geographies. This amicable conciliation is not the whole truth Take, for example, the Canadian 'Comprehensive Land Claim' (CLC), which was established in 1973 to settle disputes related to aboriginal titles by adopting Crown-Indigenous group agreements.[14] There is something treacherous in this approach; native groups must indeed 'cede, release and surrender' all Aboriginal, and 'pre-existing', titles and rights on native lands. Alternatively, natives must concede 'to never assert these rights and titles'; if titles 'other than those [contemplated] in the agreement' are discovered after entering into it, these 'must be immediately ceded to Canada'. Whether surrendered or non-exercised, lands and rights become Crown lands, which First Nations are supposed to hold in fee simple 'as a collective'.[15] Reconciling Crown and Aboriginal titles processes the land under the spatiality of European 'imposed law',[16] replaces native logical spaces with Western nomic settings, and definitively buries the former ones but does not kill them. The spatial fabric of the common law thus invests the Crown with the radical title, triggering the ultimate dispossession of Aboriginal lands. An estate (i.e. an interest in land) is assigned to native groups and fills the land 'with white legal fictions'.[17]

Likewise, in New Zealand, the 1840 *Te Tiriti o Waitangi* (Treaty of Waitangi) signed by the Crown and the Māori people directly secures the persistence of Māori law. It also gave the Crown 'the exclusive right of pre-emption over such lands as the proprietors thereof may be disposed to alienate'; and such lands were subsequently sold to settlers. The objective of processing Native titles as Western ones was pursued under the Native Lands Act 1862. It aimed to 'greatly promote the peaceful settlement of the Colony and the advancement and civilization of the Natives' by

[11] Gilbert (2007), p. 585. See also Patton (2000), pp. 28–38; Attwood (2020).

[12] See the topical judgement *Mabo v Queensland (No. 2)*, 175 CLR. See Kedar (2003), p. 414.

[13] Such as in Malaysia, Belize, Botswana, Kenya, and RSA. See Gilbert (2007), pp. 585 and 586–592 for a legal comparative survey of its application in these jurisdictions.

[14] Samson (2016).

[15] Blomley (2015), p. 172. According to Kedar (2003), p. 413 ethnocratic legal regimes use these practices to 'reproduce and reinforce . . . social stratification', and 'control most land resources'. On Indigenous environmental jurisdictions and fee simple property see Van Wagner (2021).

[16] Samson (2016), p. 104.

[17] Blomley (2015), p. 171.

ascertaining, defining, and assimilating their rights on such lands 'as nearly as possible to the ownership of land according to British law'. The Native Land Act 1865 added fuel to the fire, by encouraging 'the extinction' of Native (i.e. customary) 'modes of ownership into titles derived from the Crown'. To this end, it established the Māori Land Court (a Māori Appellate Court was then created in 1894), which acted as a 'veritable engine of destruction for any tribe's tenure of land, anywhere'.[18] and dismantled tribe's tenure of land. As Richard Boast states, this 'policy of individualisation of customary tenures as reflected in the Native Lands Acts has some affinities with parliamentary enclosure in Britain and the abolition of customary tenures in Ireland and Scotland, but also with trends in the United States, Hawaii and Latin America'.[19] The communal titles were impressed with the seal of Western individualism, giving rise to the so-called Māori freehold land,[20] which the most recent *Te Ture Whenua Māori*/Māori Land Act 1993 limits. The Act recognises that 'land is a taonga tuku iho [heritage] of special significance to Māori people', thus promoting its retention 'in the hands of its owners, their whanau [family group], and their hapu [tribe], and to protect wahi tapu [sacred place]'. The Act now makes it 'very difficult to alienate Māori freehold land to persons outside the group, to devise by will to persons outside the family, and to change its status from Māori freehold land to general land'. It retains the Court, whose 'paternalistic' attitude towards Māori land is arguably 'outdated given current economic trends',[21] thus making 'occupation, development, and utilisation of that land for the benefit of its owners, their whanau, and their hapu' even more complicated. As is evident, the Act is a 'linear descendant of the original statutes of 1862 and 1865'.[22] Concealed under their paternalistic attitudes, Aboriginal legal geographies 'continue to be ... erased and ignored'.[23]

5.1.2 *Digging Out Native Legal Geographies*

Aboriginal voices still cry from the decomposing residues of these titles. Things are changing, though. The native title doctrine accounts for an 'intensification of resistance' to Western legal geographies that started in the 1970s. The 'un-ceded indigenous interest' to their lands entails a 'challenge' and a 'refusal'—in sum, an open disagreement—as regards Western legal spatialities.[24]

---

[18] Kawharu (1977), p. 15.

[19] Boast (2013), p. 164.

[20] On conversion of customary land into freehold in Pacific states, such as Papua New Guinea, see Kwapena (2021), pp. 101–102.

[21] Boast (2017), pp. 110–111.

[22] Boast (2017), p. 133.

[23] Blomley (2015), p. 171; Comaroff (2001), p. 306.

[24] Blomley (2015), p. 171; Borrows (2002), p. 306.

In some cases, digging out native legal geographies relies on GIS, GPS, and remote sensing, thus boosting the 'scientific' character of cartography (see Sect. 3.3.4). In a field dominated by technical issues, they have facilitated their 'naturalisation',[25] which, though, does not secure their unbiased use. As Denis Wood states, GIS 'focuses our attention on the new technologies with their satellites, their gazillion of optical fiber, their computer hardware'.[26] Not only do these means allow politico-economic powers to capitalise their investments in this field, but they also trap us in their logical space. Hence, we a-critically accept that their view is the real world. Consequently, our lived geographies take place within a reproduction of the world that has been previously arranged by the holders of geo-technology.

Countering this mainstream narrative takes legal geography 'in a whole new direction, a direction with the *potential* to free maps [and geography] at least from the tyranny of the state'.[27]

Forest resource mapping in Indonesia provides an example. Remote sensing and GIS have a pivotal role in augmenting 'our understanding of environmental phenomena', as well as developing 'innovative solutions to ever more complex social, environmental, and economic challenges'.[28] They help us to monitor land exploitation by humans, to survey water supply reservoirs (as well as their water quality), and, especially in tropical regions, to clear cut forests under economic pressure and complacency of governmental oversight.[29] Forests are strategic places: apart from being fragile ecosystems, they are rich in timber and other valuable resources meaningful for the Indonesian state and trade. In cartographic terms, the simplest way to benefit from them is to silence customary legal geography, evacuate it from governmental maps, and make room for their exploitation. Several projects now challenge such omissions; the 'superseding of customary forest rights through official planning and mapping efforts' is actively opposed through 'an alternative or "counter" mapping movement'; activists formalise claims to forest traditionally managed by native communities 'matching their sketch maps to points on the [GPS] and the official Indonesian forest planning maps'.[30]

Again, politico-legal geographies are confronted through 'geographical resistance' in the interest of Indonesian indigenous peoples. GIS and GPS are extremely useful for countering neo-colonial attitudes, gaining recognition of land rights, and demarcating traditional territories.[31] Management of traditional land boosts community awareness and, at the same time, protects their knowledge. Participatory GIS and crowdsourcing outline counter-legal geographies as processes of spatial production that unwrap the ideological underpinnings of the spatiality of state law. They

[25] See also Crampton (2010), pp. 103–104.

[26] Wood (2010), p. 111.

[27] Wood (2010), p. 111.

[28] Khorram et al. (2012), p. 63.

[29] Khorram et al. (2012), pp. 69–72.

[30] Peluso (1995), pp. 384, 393, and 398. See also Kusters et al. (2013).

[31] See Calyx et al. (2020).

contribute to planning and policy-making processes, especially in the context of environmental management. The judiciary has complemented this participatory spirit. Indonesian legislation originally declared customary forests 'state forests'. In 2013 the Constitutional Court of Indonesia declared this piece of national legislation unconstitutional; the Court bracketed the word 'state', turning customary forests into 'forests located in the areas of custom-based community'. Which meant opening 'opportunities for greater documentation of Indigenous Forest governance'.[32] Ultimately, this contrasted the neo-colonial attitudes of GIS and GPS.

5.2 Law as Mode of Spatial Production

5.2.1 Between Domination and Critique

The aforesaid histories of dispossession highlight the dialectic between domination and resistance. Lying at the heart of legal-geographical production, this dialectic is examined both in 'critical geographies' and 'critical legal studies'; like legal geography, they address the interactions between space, power, and the law. Both approaches, however, assess it from a purely academic angle; 'legal-geographical resistance' is an academic speculation that unpacks the law as an ideologically motivated device implementing policies of domination.

Within counter legal geographies, the domination-resistance dialectic is not an academic speculation. Quite the opposite: it is a variety of spatial production, i.e. 'critical legal geography'. Certainly, it has traits in common with the colonial approach of the 'spatiality of law', as it examines the spatialities of domination in territorialised paradigms. Processing Aboriginal titles as interests in land (or gaining control of forests through GIS) unmistakably points to logical spaces made up of physical apprehension and political appropriation. Yet, the relation between logical space, ideology, and power departs from both the law of spatiality and the spatiality of law and displays an alternative attitude towards domination and resistance.

The law of spatiality addresses how ideology normalises geographical features through politico-legal discourse; these are legally coded to provide competing agencies with a common geopolitical grammar (see Sect. 3.3.3). In gaining control of world knowledge, agencies engage in a geopolitical struggle; and 'juxtaposition' denotes a dialectic among actors, not between domination and resistance. The competition is between 'ideologies' that try to normalise their own representation of the world.

Within the spatiality of law, 'juxtaposition' also refers to geopolitical competition. Western cultural superiority created a politico-legal framework where colonial powers competed in what we labelled the geopolitics of law (see Sect. 4.1). This was

[32] See *Indigenous People Alliance of the Archipelago v Indonesia* (2013), as quoted and discussed in Calyx et al. (2020), p. 37.

functional to their cultural (and legal) domination over the rest of the world. Likewise, spatial reshuffling of globalisation superimposes its own spatiality of the law onto the earth's surface and replace local legal geographies with normativies rooted in global markets.

In sum, both the spatiality of law and the law of spatiality focus on competing 'ideologies' that process normative spatialities functional to their policies of domination. In both processes there is no room left for resistance, which is slyly kept in the background stage. Marginalised and silenced, resistance bears the consequences of domination, pillaging, and environmental disasters.

By contrast, critical legal geography examines processes of spatial production dependent upon the juxtaposition between domination and its critique. Human agencies normalise their ideology through an agreed politico-legal discourse; but their sanitised 'nomic settings' are challenged. Take, for example, cartography. Traditionally conceived of as 'sources of power for the powerful',[33] its maps and GIS have been 'democratised' (see Sect. 3.4.1). Thus, they are appropriated by counter legal geographies to defy the monopoly of power over them. The way deforestation in Indonesia has been addressed by 'democratised' GIS is a case in point, where counter legal geographies display their *potential* to free maps from domination.

Legal geographies of domination assign their 'preferred' meaning to the production of space. Though, as Sarah McFarland Taylor highlights, what they 'intentionally *encode* into cultural communications is not necessarily what is *decoded* by consumers of those same communications'. Normalisation and politico-legal discourse are 'open to *any* possible meaning'; there are, of course, 'dominant' meanings; but addressees 'can decode' them 'in a variety of ways, including engaging in negotiated and oppositional readings'.[34] The dialectic between domination and critique lies in this exchange of meanings; domination enforces its legal geographies encoding 'them in the landscape as facts on the ground',[35] whereas critique contests 'the spaces we inhabit and produce'.[36]

Lexical items such as 'resistance', 'challenge', 'refusal', and 'resurgence' evoke the open disagreement as regards the meanings of spaces. Like in anti-geopolitics, it gives voice to 'subaltern (i.e. dominated) positions'; actors within civil society and outside the state challenge mainstream ideologies, such as the 'military, political and economic and cultural hegemony of the states and its elites'.[37] Now, also of transnational elites.[38]

[33]Peluso (1995), p. 385. See also Harley (1988), pp. 278–279; Harley (1989); Wood (1992).

[34]McFarland Taylor (2019), p. 76.

[35]On how legal meaning is implied in spatial production see e.g., Delaney (2010), p. 5.

[36]Blomley (2015), p. 168.

[37]See Nayak and Jeffrey (2011), p. 233.

[38]Allen (2018).

5.2.2 *Ideo-logical Spaces as Spaces of Domination*

The strategy of open disagreement has often been used in literature. Readers have to 'reassess almost everything hitherto taken for granted', Wordsworth stated in his *Lyrical Ballads* (2nd edn, 1802). In her introduction to the Oxford Classics edition, Fiona Stafford reminds us that, in the 1790s, Wordsworth's poetry highlighted how 'the homeless, the uneducated, the abandoned, the poor, the very old or very young should not only matter, but have things to *teach* those whose lives were rather more secure'. This 'idea ... was decidedly unsettling'.[39] Yet, Wordsworth overturned poetic style and focus, concentrating on 'common life', describing it 'throughout, as far as was possible, in a selection of language really used by men'.[40] Poetry had to proceed from the 'suffering of those unprotected by wealth and rank', which meant opening the reality (and the very idea of disagreement) against the 'measures designed to contain any internal threat to national stability'. In Wordsworth's time, domination and critique already outlined the critical legal geographies of acquisitiveness and social exclusion.[41]

Likewise, post-colonial discourse revises history by overturning the 'location of culture'.[42] History must be examined through the lenses of the peripheries (the 'margins') of colonial empires.[43] Take how André Brink criticised apartheid in his literary production (1935–2015). An Afrikaner professor of English literature, he was a worldwide known dramatist and novelist, whose literary production stands out for the intelligent criticism of apartheid. Brink epitomises the very idea of the dissenter 'fighting from within' the dominant and overtly racist Afrikaner regime.[44]

Instead of adopting a purely oppositional attitude, Brink bypasses Afrikaner history giving voice to the disregarded by 'official' colonial historiography. In *Reinventing a Continent* (1996), he confirms this change. After realising the brutality of apartheid, he decided to 'know the place [*his* Africa] for the first time'. Brink reminds us that what fired his writing 'was no longer what I had in common with writers in Europe, but what tied me to Africa'. For the first time, he continues, 'I was possessed by the passionate need to define my roots and invent my sub-continent'.[45]

This is evident in *Die Ertse Lewe van Adamastor* (1988),[46] which he had written in Afrikaans before apartheid was dismantled. The *Boer* perspective is totally reversed. The narrator, a South African native pertaining to the Khoikhoi ethnic group, describes events traditionally referred to by Europeans, such as 'the first

---

[39] Stafford (2013), p. xix.

[40] Wordsworth [1802] (2013), p. 96.

[41] Stafford (2013), p. xix.

[42] Bhabha (1994). See also Sidaway (2000).

[43] Nayak and Jeffrey (2011), p. 260.

[44] See Brink (1976). For more on Brink's critique to apartheid see Nicolini (2015).

[45] Brink (1996), p. 20.

[46] Brink [1988] (1993).

contact' between them and the colonial 'other'.[47] The narrator is Adamastor (from ancient Greek, 'salvage'), i.e. the mythical giant described in Luiz de Camoes' *Lusiads* (1572)—the Portuguese epic where there is also an account of the first encounter between Europeans and Africans.[48]

Brink overturns the perspective and narrates the first 'constitutional history' of South Africa through the eyes of black people. Prior to him, a similar attempt had been undertaken by John Purves (professor of English and Italian at the Transvaal University College of Pretoria) on August 28, 1909, during a lecture on *Camoens and the Epic of Africa*. In footnoting 'much of [Camoens's] material from a South African point of view',[49] he concluded 'The *Lusiads* is . . . not only the first but the greatest of South African poems. It is our portion in the Renaissance'.[50] Brink goes above and beyond Purves's predicament, sets aside the 'official version of Afrikaner history', adopts the angle of the marginalised, and directs the narrative 'against an established master-narrative valorising the European "discovery" of Africa'.[51] His counter legal geography is remarkable: Adamastor, not the Europeans, narrates the history of their arrival. This totally reverses how the history of South Africa has been told. Brink challenges the Afrikaner narrative of how the Dutch colonised the Cape, which Stephanus Jacobus du Toit narrated in his 1897 epic poem *Hoe die Hollanders die Kaap ingeneem het* (How the Dutch people took the Cape).[52]

Re-reading pre-colonial South African history in the light of post-colonialism is an example of the 'normative spatialities of dissent'. The territorialised paradigm receives ideological application, and the law plays a relevant role in spatial construction. We may say that *the law acts as a mode of spatial production* making it possible 'to track the effect that a given law or ruling has on some specific structure'.[53]

Traditionally considered the precinct of geographers, spatial construction deserves the attention of legal geographers, too.

Within the spatialities of dissent, the law has a highly ideological connotation. As a mode of spatial construction, it enforces the policy of domination 'on bodies'; the spaces it creates are 'saturated with legal meanings' which are 'complex, ambiguous, and potentially unstable'; their meaning depends on how ideologies vary and change. In spatialising itself, the law also spatialises them, penetrates space, and arranges its own hierarchies.[54] If power is legally 'distributed, relational, networked,

---

[47] Petzold (2000), p. 48.

[48] Vaz de Camões (1572).

[49] Gray (1977), p. 1.

[50] Purves (1909), p. 745.

[51] Petzold (2000), p. 52.

[52] du Toit [1897] (1966).

[53] Blomley (1989), p. 515. See also Mitchell (1997, 2003).

[54] Delaney (2003), pp. 73 and 69. See also Wright et al. (2018). On postcolonial geographies and their implications in terms of race, space, and resistance see Ryan (2004), p. 475.

and immanent in all relationships', spatial production 'does not exist apart from objects and people'. These are constantly reshuffled through 'situated legal practices [that] contribute to the shaping and reshaping of the spatialities of social life'.[55]

My point is this: domination has its own logical spaces. Or, as Lefebvre put it, 'What we call ideology only achieves consistency by intervening in social space and its production'.[56] We may term them *ideo-logical spaces.* Ideology is immediately inferred from contexts where these highly ideologised and 'situated legal practices' are enforced. These are present in several hierarchical geographies, where the territorialised paradigm entrenches the ideology of domination.

I shall discuss how the forces of global capital implement their desired legal geographies and create their ideo-logical spaces (see Sect. 5.3). Suffice it here to consider how urban centres are made 'attractive to both footloose capital and to the footloose middle classes'. In search of places to gentrify, these desired geographies adopt a 'sanitised' approach to urban spatial reshuffling. The more 'real estate prices escalate', and the more 'speculation and elite constructions of escape or congestion' are required, the more important 'the clearing of land for capital becomes'.[57] Capital requires a 'purified public space' evacuated by those who are not affluent, such as the homeless (see Sect. 5.3.6).

Threatening 'so many neighbourhoods' around the world, such a clearing triggers 'annihilation of space by law',[58] i.e. legal remedies that 'cleanse the streets of those left behind . . . by simply erasing the space in which they', as homeless people, 'must live'.[59] And that now must leave.[60] Likewise, the law creates racialised nomic settings, enforcing 'a wider political economy of domination', say, in prisons, immigration detention centres, borders, and inner-city black neighbourhoods.[61]

Legal geography is crucial when examining ideo-logical spaces. They become manifest through the law, permeate legal systems with their performative character, and reshuffle the relations between society, power, law, and geography.

5.2.3 *Stirring Ideo-logical Law*

Thinking in terms of legal geography discloses the dialectics between domination and critique.

---

[55] Respectively: Coleman and Agnew (2018), p. 7; Delaney (2003), p. 68.

[56] Lefebvre (1991), p. 44. See also Delaney (2010), p. 101.

[57] Jeffrey et al. (2012), p. 1251. On urban sanitised places see Black (1997), p. 115.

[58] Jeffrey et al. (2012), p. 1251. See Mitchell (1997). On such clearings in, say, New York, Medellín, and Shanghai see Sennett (2019).

[59] Mitchell (1997).

[60] Nayak and Jeffrey (2011), p. 240.

[61] See e.g. Delaney (2003), pp. 70–73; Wright et al. (2018), p. 75.

To this end, geographers must be complemented by legal scholars, in general, and comparative legal scholars, in particular. Albeit instructive, geographical critique is limited to unveiling the violence of ideo-logical spaces and situated legal practices. This certainly grasps the effects of legal geographies of domination. Yet, this assessment is purely political: 'legal power is concentrated' in the geographies of 'corrections'; the 'violence of the law is overt and ever present'; its interpretation is 'violent and puts bodies on the line'. Under this 'alleged neutral, abstract, race-blind' attitude, the law is used, say, in immigration policy, 'as a white supremacist standpoint' describing immigrant bodies as 'brown/immoral/law breaking' and its enforcement against the poor as 'a horrible vengefulness' of the bourgeoisie.[62]

A second risk to be avoided relates to 'bracketing'. I have already examined the unfettered law, which is insulated from contexts where it is applied (see Sects. 2.3.1 and 5.1.1). Bracketing arranges the domination-resistance dialectic within the realm of the law and constructs 'a boundary within which interactions' between ideo-logical spaces and the law taking 'place more or less independently of their surrounding context.'[63]

Confining resistance to a purely political critique within the brackets of the law, we play the game of formalism according to the rules unilaterally laid down by ideology. Within ideo-logical spaces, by contrast, law and performance constantly engage in a dialogue with the outer world. Bracketing prevents us from assessing the law as a situated legal practice. As hierarchies, values, and priorities change, ideology also alters law-making processes adapting them to its new desired legal geographies. Bracketing also favours the normalisation of ideo-logical spaces and promotes the generalisation of its own political purposes.

Comparative law is therefore crucial. We must not play the game according to the rules laid down by ideology. Quite the opposite: in complementing geographical critique, legal scholars set the game on a level playing field with ideology.

I term this legal competence 'stirring the law'. A typical Anglo-British legal device, it describes the relation between equity and its formalistic counterpart, i.e. the more 'classical' common law moulded in Westminster Hall. Equity offers us a paradigm applicable to cross-disciplinary research. The way it stirs the law reminds us that equity (and resistance) is like 'a door having one side within the law and one side without,' whilst ideo-logical spaces prefer 'to keep the door closed' and let geographers bracket it and 'see only their side of it'.[64] As comparative lawyers, we challenge the formalistic and self-referential attitude of legal studies; as legal geographers, we disclose the ideological underpinnings of our situated legal practices and open them to mutual interference and interdisciplinarity. In stirring the law, we disclose the domination-resistance dialectic, which is internal to it.

---

[62] See respectively Delaney (2003), p. 72; Wright et al. (2018), p. 75; Mitchell (1997). See also Boyce (2018), p. 352.

[63] Blomley (2015), p. 169; Callon (1998), p. 249. For more on bracketing see Blomley (2014), pp. 135–136. On the 'curious silence of law' on these topics see Barkan (2011), p. 591.

[64] Watt (2009), p. 1.

Again, this evokes the open disagreement as regards the dominant meanings of spatial normativies. The comparison with equity assists us in redeeming the interests of our communities and in challenging strategic formalism incrementally. By progressively introducing a set of 'comparatively novel legal principles', we can 'override the older jurisprudence' that advocates transnational legal derivatives as the new organising themes within our societies.[65] Counter legal geographies, in sum, mobilise the most active forces of our societies. This perfectly matches 'the deep-rooted Anglo-American practice of the common law, according to which the law involves an unceasing exercise of imagination'.[66]

5.3 Varieties of Ideo-logical Spaces

I shall now examine ideo-logical spaces based on the domination-resistance dialectic to explore how ideology arranges spatial reshuffling and establishes its own desired legal geographies. I shall also assess how the law enforces them through situated legal practices. Far from being an alleged neutral, abstract, and race-blind force, it becomes, as Ian Ward vividly puts it, a 'political instrument wielded to political purposes'.[67]

If, as legal geographers, we unveil how domination is implemented and legally enforced in spatial production, as *critical* legal geographers we are called to unpack the ideological underpinnings of the law. Un-bracketing the law also discloses how ideology processes its own nomic settings.

The domination-resistance dialectic opens up the law to a more contextualised analysis. This entails reconsidering how logical spaces (and their normative spatialities) organise societies. When the former become 'meaningful', they develop into places. We may say that the 'violence of the law' (see Sect. 5.2.3) is nothing other than spatially situated practices converted into societal 'repositories' by an 'overt and ever present' policy of domination.

The reshuffling prompted by ideo-logical spaces lies at the heart of the following subsections. Owing to their impact on the organisation of political communities, ideo-logical spaces have infiltrated the public sphere and contributed, globally and domestically, to a constellation of 'mortgaged democracies.' I shall examine them in due course (see Sect. 5.3.2). Suffice it here to say that a 'mortgaged society' is a political community governed by situated legal practices reflecting the 'mutual support for capitalism and for the nation-state'.[68] At the same time, it perpetuates the idea of the 'well bounded order of politics' (see Sect. 1.1) against the 'fear' of terrorism (see Sect. 5.3.2) and unbridled immigration (see Sect. 5.3.4).

[65] Sumner Maine [1861] (1908), p. 45.

[66] Bennett et al. (2020), p. 9.

[67] Ward (2021), p. 12.

[68] Wallace-Wells (2019), p. 192. See also Mann and Wainwright (2018).

5.3.1 *Global Ideo-logical Spaces*

That the law may be used to fulfil political purposes is evident as regards the ideo-logical space of globalisation, which holds a particular fascination for us. We have got used to localising trade and civilisation within globalised markets; as legal scholars, we also praise global law, as well as its 'aesthetic consequences . . . [and] sublime effects'.[69] Our Western minds describe them in accordance with a narrative of *superiority* and *progress*. Global law possesses a liberating force that facilitates the harmonisation of laws, stimulates economic development, and promises an 'impending radiant future' to humanity.[70] It does not come as a surprise that global markets need the reopening of territories (and new legal spaces) to favour the smooth flows of goods, capital, and services across frontiers and national legal systems.

The reopening of territories does not mean rubbing borders out of maps. It solely requires new legal geographies to enforce the ideo-logical space of globalisation, i.e. the outcome of spatial production stimulated by transnational actors. Henri Lefebvre highlighted that, as it is produced by socio (and legal) practices, even space is not immune from the exercise of the hegemony of the holders of global capital. They make 'use of it . . . on the basis of an underlying logic and with the help of the knowledge and technical expertise, of a "system"'.[71] To this end, they create several cross-border regulatory systems that match their 'private ordering in the name of party autonomy'.[72] They have also found ways to generalise their interests by promoting changes in national spaces. As a result, their ideo-logical space impacts on how political communities arrange their organising themes.

But: how do these processes of spatial production occur? How is it possible that the holders of global capital have infiltrated the public sphere and contributed to manufacturing domestic spaces in accordance with their own interests? In *The Code of Capital* (2019), Katharina Pistor provides us with a possible answer. She argues that globalisation holds a particular fascination *also* for states, which 'are not neutral when it comes to whose interests in an asset shall be given priority'. The prospect of benefitting from capital gains is 'more likely to find their blessings than claims that assert self-governance or seek to ensure environmental sustainability.'[73]

Per se, the allure of globalisation does not trigger changes in domestic normative spaces. Something more is required. The forces of capital seek a connection with the public sphere, which is usually drawn by suggesting a renovated 'morally eligible foundation on which' communities might 'organise their practical affairs'.[74] What is shared is the maximisation of profits for shareholders, capital investors, and assets

[69] Hugues (2007), p. 696.

[70] Xifaras (2016), p. 216.

[71] Lefebvre (1991), p. 11.

[72] Muir Watt (2014), p. 9.

[73] Pistor (2019), p. 23.

[74] Mullender (1997), pp. 23–34. On how economic actors penetrate the public sphere see e.g., Braithwaite (2005), p. 9.

owners. Indeed, 'the control and accumulation of resources . . . was and remains' the organising theme within political communities, as well as a 'key element of political tension.'[75] In the public sector, private investors' concern 'meets that of public finance and the income and solvency of state'.[76] In so doing, the interests of transnational actors turn out to be the interests of the whole community.

The connection revolving around public finance concerns took place at the very onset of the modern nation state. In modern constitutional history, for example, the establishment of the Bank of England (1694) tackled the 'dramatically increased rate of growth in public finance' by resorting to national debt and other 'means of state funding through new kinds of long-term and short-term loans.'[77] Also, public and private concerns interweaved with the commons. To increase the production of commodities and maximise profits, several tracts of lands, which had been the major source of sustenance for commoners, were enclosed. Enclosure became the new organising theme of England: turning lands into capital, it allowed their owners 'to extract monetary values.'[78]

The case of enclosure is particularly fitting, as it demonstrates how private interests were (and are still) able to infiltrate the public sphere. Although it reflected the interests of the affluent sectors of English society, it became part of the public policy and was pursued by pieces of legislation. By authorising it, Parliament did not establish 'equality of goods and lands;' but, as the Digger William Walwyn put it in 1649, freeholders legally appropriated commons making 'themselves thieves by act of Parliament.'[79]

Whether politically correct or not, Walwyn's predicament outlines a pattern that we also find in the enforcement of the globalised ideo-logical space. Like free-holders, the holders of global capital have 'found ways to utilize the law' by turning their interests into *the* organising theme of communities.[80] Fascinated by globalisa-tion, state authority shows a 'readiness' to allow transnational actors—which are not democratically accountable—to 'make a considerable part of the law' in national communities.[81]

The consequences are threefold. The first one regards the legislative process, through which the holders of capital now pursue their interests by creating a national, but-globally-driven, 'safe normative space'.[82] This nomic setting allows private actors to gain the 'simplification of the conveyancing process' and the global 'possessive market society' to flourish.[83] When states 'create the law that was

[75] Capra and Mattei (2015), p. 53.

[76] Hoyle (1992), p. 1.

[77] Rudolph (2013), p. 131.

[78] Pistor (2019), p. 29. See also Turner (1984), p. 36; Blomley (2007).

[79] Walwyn (1649). See Hill (2001), p. 139; Clark and Clark (2002).

[80] Pistor (2019), p. 154.

[81] Watson (1988), p. 87.

[82] Mullender (2018), p. 697.

[83] Respectively: Gravells (2010), p. 43; Macpherson (2010), p. 48.

wanted' by the holders of capital,[84] they enforce externally imposed values, thus disregarding the interests of national communities.

The second consequence refers to the purpose of the law. We have been taught that it possesses a strong 'egalitarian commitment' and promotes substantive equality.[85] As lawyers, we teach that it is instrumental in protecting individuals' interests from private groups whose aim is to amass affluence and control 'the levers for the distribution of wealth in society.'[86] Yet, this predicament scarcely squares with the current reality. Public finance concerns have caused a shift from the legal to the economic sphere, particularly apparent in the case of the rule of law. Within global contexts, this promotes efficient institutional changes and acts as the driver of good governance, as well as the 'guiding light' ensuring elevated economic performances. Before it became the shibboleth of global governance, it had had a different legal meaning: it set constraints on political power to protect the rights of the individuals. Instead of putting bridles on financial actors, the rule of law now evaluates the efficiency of the legal systems and their attractiveness towards international investments (see also Sect. 4.6.1).[87]

The third consequence relates to the role of the law in the redistribution of wealth. Legislation implements its egalitarian commitment by preventing private actors from benefitting from an uneven distribution of wealth.[88] They aim to gain control over the distribution of wealth by placing 'the law ... firmly in the service of [their] capital.'[89]

This reveals a disjuncture at the heart of legal taxonomies. The law acts inconsistently when it comes to addressing ideo-logical spaces of globalisation. Holders of global capital infiltrate the public sphere and formally act as legislators of national communities; formalism probes a *strategic* use of the law, inasmuch as it enforces their interests as if they were *the* organising theme of communities. In other words, they act *strategically*, and code, i.e. generalise, their own interests through the law.

This kind of formalism traps the law in a set of circular justifications. It accounts for the generalisation of private actors' interests through the law; in coding such interests, it favours the preservation of the capital-nation state nexus, which acts as a powerful, 'strategic' device. In shaping our communities, strategic formalism tries to convince us that there are no real alternatives to the narrative of acquisitiveness of global ideo-logical space. 'The trick', Sarah McFarland Taylor upholds, is to convince every citizen to 'simply adjust [their] behavior' to this lack of alternatives—and modify their 'outlook in keeping with the system, not to challenge

[84] Watson (1988), pp. 87 and 96.

[85] For further references see Nicolini (2020).

[86] Pistor (2019), pp. 17, 19.

[87] On this shift in the meaning of the rule of law see Nicolini (2019).

[88] Husa (2018), p. 74.

[89] Pistor (2019), p. xi.

or to change it'. Strategic formalism assumes that adaptation to global ideo-logical spaces is the 'progress' of our times.[90]

5.3.2 *The Advent of Mortgaged Societies*

At the domestic level, the ideo-logical space of globalisation has changed our perception of communitarian bonds. As finance has 'taken in contemporary economy and society',[91] we have become part of mortgaged societies. As a variant of Stanley Fish's 'interpretive communities,' their members share a 'sense of relevance as to what is at stake in the fields that concern them.'[92] But, which are the relevant themes that concern us all?

As said, such transformative effects have paved the way to the so-called 'mortgaged societies' (see Sect. 5.3.1). In broader terms, the expression refers to the new role of the law in allocating the world's scarce resources. The debates about inequality, democracy, and global law now take place within the ideo-logical space of transnational actors. Notwithstanding the current level of global wealth, economic inequalities still exist. And thrive. Half of the world wealth is in the hands of an insignificant percentage of the world's population: 'In the twenty-first century, a small handful of global corporations perpetuate the conditions that flourished' in the commercial environment of the late seventeenth century.[93]

In a narrower sense, 'mortgaged society' refers to how private interests reframe the organising themes around their ideo-logical space. The laws of the market percolate through the political system; the new foundation of the political obligation allows communities to reconcile the variety of conflicting economic interests giving primacy to private actors' 'legal coding'. The organising theme of transnational interests is the promotion of efficient institutional legal paradigms. I term it the 'things-as-wealth principle'. It affects political societies turning them into acquisitive communities, within which individuals become consumer-borrowers. This change has also been favoured by a change in the perception of charges applicable to real property, in general, and of mortgage conditions, in particular. Instead of responding to the needs of consumer-borrowers, the law encourages them to take out mortgages to an extent that our democracies are now 'real-property-mortgaged-to-a-building-society-owning' democracies.[94]

This organising theme has turned the political bond in acts of conveyance, whereby consumers borrow money from lenders, which are granted a security by transfer of proprietary rights.

---

[90] McFarland Taylor (2019), p. 92.

[91] See Dixon (2018), p. 123.

[92] Mullender (2019), pp. 549–550. For more on interpretive communities see Fish (1980).

[93] Athreya (2011), p. 5.

[94] *Pettit v Pettit* [1970] AC 777, HL, *per* Lord Diplock, at 824.

The mortgage metaphor has been borrowed by the common law. English law has traditionally showed attentiveness to private interests. Common lawyers have always been used to 'treat things as things' and, at the same time, 'as wealth.'[95] This has allowed different interests—namely, those of debtors and those of creditors—to cohabit in the same land. Take, for example, the Law of Property Act (LPA) 1925, which crowds it with them: 'legal estates', 'equitable interests,' and 'equitable interests "capable of subsisting as a legal estate"' coexist.[96] If they conflict—co-ownerships entitlements, estates upon condition, charges by way of legal mortgage—, the law assumes that the mortgagee's 'legal coding'[97] is 'more equal' than the mortgagor's.

The practice of conveyancing estates in fee simple into mortgagee's estates in the land is long-standing. Mortgagor's and mortgagee's interests in the same land has always diverged, and the interests of the latter were always ranked as superior. Common law and equity consented to such a multiple coexistence of distinct-and-often-conflicting interests in the same land.

The legal coding whereby the common law incorporates creditors' interests in land proves to be useful when assessing how the ideo-logical space of globalisation establishes 'mortgaged societies'. English common law has historically constituted the backbone of global law; the same U.S. common law, which is supposed to be the hegemonic legal system, owes its features to English law. Even if we assume that globalisation prospers without 'a global state or a global law,' we must concede that transnational legal modules are derivatives of a process of legal enculturation promoted by the world-wide dominant common-law legal profession.[98] Consequently, common-law paradigms are applied globally for processing things as wealth and turning them into capital. English legal language is now the language of the global society. Odd as it may seem, English legal language (now the global legal language) was created when 'the bulk of capital was land.' As the processes of capital coding turned out to be transnational and loosely tied to territories and communities, English legal paradigms adapted to 'stocks, shares, bonds and the like ... crossing oceans at the touch of a key-pad in the search for a fiscal utopia.'[99] Like in English land law, 'the feudal calculus lives and breeds, but its habitat is wealth not land.'[100]

Such a mortgage-effect gives rise to several concerns. The hierarchical order of interests gives preference to the mortgagee's. As a result, the new political bond does not consist of symmetric acts of conveyance. Quite the opposite: it conceals asymmetric relations mimicking the balance of politico-economic power between the

---

[95] Rudden (1994), p. 82.

[96] LPA, s 205(1)(x).

[97] Pistor (2019), p. 5. See LPA, ss 34 and 87. See Rudden (1994), p. 95: 'the lender's interest is merely in the thing as wealth,' whereas 'the family member's interest is in the thing' as such.

[98] Pistor (2019), pp. 176, 158.

[99] Rudden (1994), p. 82.

[100] Clarke and Kohler (2005), p. 55.

mortgagor and the mortgagee.[101] In the Middle Ages, when mortgage was 'invented', this order of interests was not morally eligible. Although it was 'not forbidden by the court of the lord king, the mortgages were deemed to be a kind of usury,' as well as 'unjust and dishonourable.' If the mortgagee died 'seised of such a gage and after his death this [was] proved, his property [was to] be disposed of as the property of a usurer'.[102]

In politico-constitutional terms, the shift is remarkable. States are accountable *vis-à-vis* financial markets and international investors if capital is allocated at their level. As global financial dominance causes a shift from the political to the economic sphere, nation states rely on a new form of confidence between political power and sovereign financial markets, where 'distressed sovereign debt can be sold . . . and the debtor [is] subjected to the harsh economics of private law'.[103] Transnational actors aim to dominate, not to be governed; interests conveyed by legislation do not have to be accommodated but articulated as if they were arranged upon a hierarchical scale. Transnational actors are the owners of the levers of wealth.

The process whereby conveyancing is accepted by the political community is a deliberative process. To be convincing and gain the broadest consent possible on new organising themes, arguments must be persuasive and, at least, 'incipiently egalitarian in orientation.'[104]

The argument is persuasive because it is *consociational*. I have already used the term before (see Sect. 4.6.2). This argument arranges the different sectionalities around a power-sharing organising principle, assuming that there is an unequal distribution of wealth, which does not prevent members of communities from gaining access to it. Although it does not reflect an egalitarian commitment, the consociational argument is egalitarian in its commitment. The LPA holds this assumption: a legal mortgage of land may be created 'by a charge by deed expressed to be by way of legal mortgage', i.e. without transferring the legal title to the land to the mortgagee. s 87(1) of the LPA gives us the illusion of taking part in the global distribution of wealth.

Consequently, and if the deed so establishes, the mortgagor has also right of redemption, including an option to repurchase.[105] In mortgaged societies, however, the 'terms and conditions' of the political obligation require unceasing acts of conveyancing—and this makes the political liens often irredeemable. 'The current packages on offer', in other words, 'could have the effect of removing or rendering illusory' this very basic right.[106] For when the 'mortgagee in possession' has entered into and is in possession of the mortgaged property,' as Thomas More already

[101] For more on this asymmetry see Sect. 5.3.1 and Nicolini (2020), p. 29.

[102] Glanvill (ed. GDG Hall 1965), Book x, ch 6 and 8: See also Co. Litt. 206, 253; Bl. Comm. III, 158.

[103] Muir Watt (2012), p. 286.

[104] Mullender (2018), p. 697.

[105] LPA 1925, ss 5.1 and 205.1.xvi.

[106] Houghton and Livesey (2001), p. 167.

realised, 'landlords as well as tenants are turned out of possession by tricks or main force;' or 'wearied by ill usage they sell at last.'[107]

5.3.3 *Safe Ideo-logical Spaces*

When referring to globalisation, I used the expression 'safe normative space' (see Sect. 5.3.2). Coined by Richard Mullender,[108] it identifies the nomic settings created by the global holders of capital pointing to 'a normatively appealing, enduring, but not entirely safe politico-legal space'.[109] Rooted in an economic organising theme shared by public sphere and private interests, space is relatively safe, inasmuch as it is functional to the maximisation of profits. As it reflects the desired geographies of globalisation, its logical space is *absolute*; within it, forces of capital normalise their politico-legal discourse about the world as a comfort zone for commerce and trade (see Sect. 3.3.2). It is also *safe* since the maximisation of profits is secured (and enforced) through the law. At the same time, it is *relatively safe*, because financial risk is inherent to investments and business ventures.

Mullender elaborates this concept when discussing Stanley Fish's *Winning Arguments* (2017). Fish explains 'that arguments may sweep away the politico-legal frameworks, or normative words, we make and inhabit and that invest our lives with a sense of significance and security'.[110] This reading is fitting for our purposes. Ideo-logical spaces are erected as 'something that will neutralize arguments' with 'the desire for another world'; they take the form of a 'safe oasis' from which we benefit 'by fastening on a body of "truth" capable of bringing argument to an end'.[111] Safe spaces are construed as politico-legal discourses with a good dose of 'political motivation'.[112] In critical geopolitics, Geraoid O'Tuathail calls their construction 'geo-graphing',[113] which emphasises 'the creativity inherent in the process of using geographical reasoning in the practical service of power'. Every logical space shares this quality; its absolute-ness is enforced through a political-legal discourse, which aims to normalise 'geo-graphing' as 'an objective account rather than one written from a position grounded within the events being discussed'.[114]

This connotation is found in another ideo-logical space. Take, for example, the *motto* 'the world is safer'. Nick J Sciullo reminds us that it was uttered by the then

---

[107] More (1908), 24-2).

[108] Mullender (2019).

[109] Mullender (2019), p. 550.

[110] Mullender (2019), p. 550 referring to Fish (2017), pp. 3 and 120.

[111] Respectively: Fish (2017), p. 3; Mullender (2019), p. 551.

[112] Sharp (2014), p. 535. See also Dodds (2019), p. 35.

[113] O'Tuathail (1996).

[114] Sharp (2014), p. 535.

U.S. President Barack Obama in the aftermath of Osama bin Laden's death.[115] Surfing the internet, we find it used in several sites, blogs, and reports. The expression is part of a broader politico-legal discourse revolving around the state of the world 'out there'. *That* world 'out there', though, 'is not a collection of incontrovertible facts but it is instead about power'. The world 'out there' is *geographed* and spatially reshuffled by 'individuals and institutions' who superimpose their aspirations, interests, and representations onto geographical facts and relationships. These become meaningful because their ideological discourse provides the practical knowledge policymakers need to manufacture their own legal spatialities.[116]

Sciullo terms this ideo-logical space 'spatio-thought'.[117] It is an absolute 'where' providing us with a comfort zone and a 'safe normative space' both made meaningful in geopolitical terms. To build a spatio-though, we need both physical space and symbols. The former is usually a country, i.e. the nation state with its well bounded order of politics (see Sect. 1.1). At the same time, physical space relates to territoriality, i.e. a territory made politically meaningful through spatial production and territorial demarcation (see Sect. 2.4). Borders become associated with state jurisdictions, protect national legal systems, and separate their communities from the 'out there' and the mayhem of the international community.

There is a sense of cultural-legal superiority in spatio-thought. Within the West, this ideo-logical space prioritises securitisation and domination. The state as a bounded space justifies a variety of legal measures to monitor its borders: immigration controls, expulsion of illegal aliens, customs and border protection, border patrolling, and so on. As Peter J Taylor taught us, 'by controlling access to a territory through boundary restrictions, the content of a territory can be manipulated', and 'its character designed'.[118] Such a spatial production certainly turns spaces into meaningful places but make them neither 'repositories' of shared knowledge nor 'profound centres of human experience'.[119] Quite the opposite: they are safe only for those dwelling in them and gathered around a 'sort of collective place-consciousness', an ideo-logical *mentalité* which may be termed *cité*, i.e. the 'feelings people harboured about neighbours and strangers and attachments to place'.[120]

To create such consciousness, spatial production also requires symbols. These have a legal and geopolitical force and are, on their own, the outcomes of spatial reshuffling, which has maked them meaningful for state communities. Indeed, we usually 'define countries in relation to their capital city', Sciullo affirms. He mentions the 'area' of Washington D.C., which 'defines the United States not in terms of

[115]Sciullo (2012), p. 317 quoting from Wilson and Kornblut (2011) A1.

[116]Sharp (2014), p. 535. See also Dodds (2019), p. 35.

[117]Sciullo (2012), p. 330.

[118]Taylor (1994), p. 151. On the transformation of border control in immigration issues see Burridge et al. (2017).

[119]See Relph (1976), p. 43 as discussed in Sect. 3.1.

[120]Sennett (2019), pp. 1–2.

its boundaries, but in terms of its seat of government'. States do not confine their discourse within their precincts; in legal terms, extraterritoriality allows them to carry out several undertakings outside, or across, borders to protect and pursue their own interests. Geopolitical prestige is then another symbol: 'a country does not exist in relation to other countries, but in relation to its national identity'.[121]

Yet, spatio-thought does not end here. Particularly in the Western world, the creation of spatio-thought rests on the essentialising effects of our thinking along global lines. To avoid the decline of the West, ideo-logical safe spaces must not dismiss the sanitising function of such linear thinking. The necessity of protecting our 'collective place-consciousness' probably lies in the undisciplined 'openness' of the Portuguese to non-Western (i.e. inferior) communities located beyond the lines (see Sect. 4.2.2). Probably, we unconsciously preserve the memory of the socio-cultural consequences of such an undisciplined 'openness'. Likewise, Edward Said's *Orientalism* explains how Europeans manufactured new spaces by concealing their politico-legal discourse beneath the essentialising principles of Western civilisation.

Western safe spaces trigger further consequences. As Sciullo caustically states, when Western politicians 'say that the world is safer', probably they do not realise how this 'revelation' may be 'startling … for those in the tornado ravaged South, those in Southeast Asia at constant risk of tsunami, and those who fear earthquakes, like the Haitian people'. The world beyond the lines has 'largely fallen out of Western consciousness'.[122] The production of such ideo-logical space has indeed made us lose control of the reality.

Because of our living in the Global North, we think that 'wealth [is] a shield against the ravages' of exploitation, wars, and climate change. Consequently, we keep on 'mortgaging' the social and 'ecological future of the planet' and, at the same time, allow our behaviour to have 'cruelest impacts' on the less-developed countries.[123]

Our ideo-logical space reflects policies of domination promising Western-like progress, which involves 'simplifying the look and construction of a place so that [it] fits one kind of person, but not others'.[124] It is also part of the narrative of acquisitiveness proposed by the global ideo-logical space, which we find particularly fascinating. It is safe, hyperregulated, and monitored; it grants us protection through the enforcement of securitisation policies; it offers us the possibility to share our lived geographies through free-but-not-gratis mashups. It asks us, in sum, to be consumers-citizens, letting the forces of capital influence our behaviours.

The outcomes of ideo-logical spaces of acquisitiveness lie in this: the forces of capital normalise their politico-legal discourse about the world and convince us that our safe space coincides with the comfort zone for commerce and trade. Perhaps most striking of all is that the *Oxford English Dictionary* defines 'commodity' as 'A

[121] Sciullo (2012), p. 329.
[122] Sciullo (2012), p. 329.
[123] Wallace-Wells (2019), pp. 1 and 54.
[124] Sennett (2019), pp. 1–2.

thing which is beneficial to or advantageous for a person'. And for enterprises. As individuals, we peruse it for our 'private or selfish interest', especially in the tranquillity of our safe places; as entrepreneurs, we turn it into 'something *useful* that can be turned to *commercial advantage*' and traded in the comfort zones of capitalist economies.[125] Ideo-logical spaces favour a change in our perception of spatiality: commercial comfort zones are functional to safe spaces of consumption, thus maximising financial returns. The quieter the space, the safer it is, and the higher the returns are. Paraphrasing Wordsworth, within *safer* spaces we enjoy 'incidents and situations of common life', 'the essential passions of the heart ... are less under restraint', and 'our elementary feelings co-exist in a state of greater simplicity'. In safe places, we have our commodities shipped and 'recollected in tranquillity'.[126]

To conclude, safe normative places are processes of spatial construction where there is no room left for resistance. This is slyly kept in the background so as not to cause a threat to comfort zones of commerce, upon which our mortgaged communities build their safe places. Evidently, a safe space is not a 'space of doubt'; resistance is silenced and marginalised, together with the socio-climatic consequences of domination, pillaging, and environmental disasters.[127]

5.3.4 *Ideo-logical Spaces of Doubt and Terror*

Despite its being kept in the background, the idea of resistance resurfaces (and challenges) our 'safer' places. These embed a politico-legal discourse, whose ideo-logical space consists of reassuring imagined geographies. The discourse is a convincing one, whose texture does not accept conflicting arguments.[128] The presence of winning arguments, though, does not automatically exclude loopholes within the texture of spatio-thought. Following in the footsteps of Fish and Mullender, we understand that 'language [also] gives us the power to unsettle existing certainties: e.g., by opening up a "space of doubt" that' resistance may try to 'fill with "alternative readings" of our circumstances'.[129] In discussing Fish's arguments, Mullender suggests the legitimisation of safe spaces be considered 'contingencies (rhetorical achievements) vulnerable to the assaults' of new (winning) arguments. In this event, new arguments will trigger a 'change' and 'reversal' within the bounded argument space.[130] The idea of resistance thus insinuates itself into safe places and creates a space of doubt.

[125] Respectively: Oxford English Dictionary (2021), s.v. commodity, and Watts (2014), p. 391.
[126] Wordsworth [1802] (2013), pp. 96–97.
[127] Fish (2017), pp. 16 and 23.
[128] Macintyre (1985), p. 17.
[129] Mullender (2019), pp. 551–552 quoting Fish (2017), p. 24.
[130] Sennett (2019), p. 268.

That is why, with Mullender, we label our safe space as a 'not entirely safe politico-legal space'.

This is relatively safe since an ideo-logical space is always *relative*: economic comfort zones are spaces of financial risk, which is inherent to every investment (see Sect. 5.3.3). Hence, Fortuna—who, in ancient times, was the goddess of chance—may cast her lots and work against our financial returns, comfort zones, and safer spaces, making our transaction run amuck and even more vulnerable to the assault of climate change.[131]

There are other contingencies that make the 'accents of infallibility' of safer spaces 'vulnerable to the assault of arguments.' The first contingency is *hostility*. We must bear in mind that the whole field of geopolitics is based on the notion of security, i.e. 'the maintenance of the state in the face of threats, usually from external powers'.[132] Likewise, in legal geography, owing to their natural geographical features, 'certain spaces are seen as either easier or harder to defend'; distance has 'effects on them ... and certain topographical features promote security or lead to vulnerability. Suffice it here to remember the Roman conception of 'natural' borders, whose *arcifinious* nature makes them politically and legally a natural defence (see Sect. 3.6.2).[133]

The second contingency lies in the shift from rhetorical to *physical vulnerability* is apparent when spatio-thought meets the hostility of terrorism. We saw that the absolute 'where' of the former becomes a concrete comfort zone, as well as a safe normative space, through physical space and symbols (see Sect. 5.3.3). Particularly in the West, the well bounded order of politics with its culturally and legally superior spatio-thought is dependent upon the powers of its symbols. When targeting the Pentagon and the Twin Towers in New York, the 9/11 terrorist attacks enforced their own ideo-logical space of resistance. In so doing, they defied the ideo-logical safe space of Western domination. 'Certainly, the target of 9/11 was not coincidental'.[134] Indeed, terrorists 'have attacked the United States because of *where it is, what it is, and what it does* ... It is clear that America is detested by many people the world over for what it is – a successful and dynamic modern society'.[135]

Thus, we have experienced ominous terrorist attacks all around the world: in Madrid, London, Mumbai, Bali, and Paris, to name just a few. We have witnessed the wars in Iraq and Afghanistan, as well as the insurgence of Islamic States in Syria-Iraq, North Niger, Mozambique, and now in Afghanistan again.[136] The overall idea is that we must make our normative safe spaces safer and safer.

---

[131] Watts (2014), pp. 405–406. On the economic impact of climate change see Wallace-Wells (2019), p. 115.

[132] Sharp (2014), p. 531.

[133] On the interactions between geographical studies and terrorism see Bahgat and Medina (2013), pp. 40–44.

[134] Ward (2009), pp. 5–6.

[135] Hashim (2001), p. 31. See Sciullo (2012), p. 317; Gregory (2004), p. 20.

[136] Smith (2015).

Terrorism is triggering a new 'politics of spaces' (see Sect. 4.1), which is causes a third contingency, which is made up of *antagonistic and competing ideo-logical spaces*. The idea is rooted in the so-called 'No where' of terrorisms, whose ubiquity makes our comfort zones and terrorists' imagined geographies interweave and overlap. This is reflected in how we label such a 'No-where'; and the response is implicit in the phrase 'global war on terror'.[137] Although 'imprecise', the expression has clear legal geographical implications. Indeed, 'Language and geography intersect in this phrase, implicating that important legal standards and notions of legality and space are applicable'.[138] This expression constitutes a further application of the territorialised paradigm: as Carmody argues, Islamic terrorist movements 'have a different vision for the type of global state they want'. In this 'global Islamic state . . . sovereignty belongs to God', which implies physical apprehension and political organisation.[139] It may also 'serve ends of political and capital accumulation', thus requiring the creation of comfort zones of trade and commerce functional to the establishment of a global Islamic ideo-logical space.[140] Ideo-logical spaces of terrorism—and, it goes without saying, counterterrorism—aim to physically apprehend and politically appropriate concrete legal geographies. They compete to defeat each other. As a 'complex battle against an unknown', the global war on terror has used 'geographical imaginations and abstraction . . . to construct' their foes 'as "legitimate" targets for military . . . violence', also relying on (GIS) for the security of safer spaces.[141]

These ideo-logical spaces manufacture counter legal geographies following logics of 'concatenation' and 'juxtaposition' (see Sect. 3.6.2). The former is patent in the implementation of their respective imagined geographies. Both tend to establish their own ideo-logical spaces of domination. Unlike the law of spatiality, though, these spaces are not interested in knowing the world through an agreed lexicon. Here, 'juxtaposition' does not presuppose a common lexicon; and ideological struggles between agencies (and their logical spaces) take place in a global clash.

In their respective normative spaces, there is no room left for alternative arguments; nor do they accept doubt or resistance. The ideology of terrorism is immediately inferred from the contexts where its situated legal practices are enforced: extermination and subjugation for the infidels and traitors; establishment of a relatively unsafe normative space for its acolytes. Again, the space is 'relatively' safe, because terrorists unceasingly redetermine the 'accents of infallibility' of their doctrine, and there is no due process of law when it comes to depriving the subordinates of life, liberty, or property, if it arbitrarily seems right to do so.

---

[137] Sciullo (2011).

[138] Sciullo (2012), p. 317.

[139] Bahgat and Medina (2013), p. 44.

[140] Carmody (2018), p. 305.

[141] Respectively: Sciullo (2011), p. 564; Bhungalia (2018), p. 313. See Gregory (2004). On GIS and terrorism see Bahgat and Medina (2013), p. 40.

By contrast, Western ideo-logical spaces preserve their traditional, liberal, and democratic common lexical index. And the law enforces them through a set of situated legal practices, which reveal that, far from being a neutral force, the law becomes a political instrument for political purposes. Safe ideo-logical spaces prioritise securitisation and domination. As a bounded space, it justifies a variety of legal measures to monitor state borders for preventing terrorist attacks. Within the order of politics, the response is enforced through pieces of legislation enacted by allegedly democratically elected and accountable political branches. It is true that counter-terrorism law is enacted according to our democratic law-making processes; in so doing, though, they are steeped in Western ideology. In normalising ideo-logical spaces, they promote their generalisation, as well as their hierarchies, values, and priorities. 'Ours is an "Age of terror", Ian Ward warns us, 'because we are told it is and because we tend to believe the rhetoric'; it is a rhetoric that 'lays waste to all the cosy complacencies of liberal democratic politics, and its law'.[142]

Albeit particularly tough, Ward's predicament grasps the point when he examines the law as a political instrument for political purposes. In shaping our communities, it enforces what I have termed strategic formalism. To defeat the enemy, the legal securitisation of our ideo-logical space requires pieces of legislation limiting individual rights according to the due process of law. Our safe ideo-logical space convinces us that there are no real alternatives to the limitation of rights, which means adopting 'preventive measures against any perceived "threat"'.[143]

The idea of legally excluding the enemy from our comfort zones by using preventive measures reflects the process of spatial production through which our places are made safer for those who identify with its *mentalité*. To create its *cité*, the ideo-logical production of space requires, as said, symbols. These are traditionally represented by the 'we/they divide': 'Either you are with us, or you are with the terrorists', George W Bush declared in the aftermath of 9/11.[144] Derek Gregory argues that the War on Terror is epitomised in the 'split geographies of "us" and "them", "civilization" and "barbarism", "Good" and "Evil"',[145] and draws on Edward Said's argument that European colonial powers constructed a distorted, as well as racist, view of "others" with the aim of asserting their own position as 'enlightened and dominant'.[146]

The divide reflects the inner fight between allegedly legitimate governments and partisan-resistance movements.[147] At the same time, it cannot be confined within the legal (and geopolitical) precincts of the state. In legal geographic terms, the 'Nowhere' of the global war of terror is the ideo-logical response to

[142] Ward (2009), p. 3.

[143] Ward (2009), p. 13.

[144] See https://georgewbush-whitehouse.archives.gov/news/releases/2001/09/20010920-8.html (accessed 3 December 2021). See Dunn (2005), p. 13.

[145] Gregory (2004), p. 11.

[146] Stephens (2011), p. 255. See also Gregory (2004), pp. 19–24.

[147] Cumin (2009) and Fonseca (2017).

extraterritoriality; it is an unsafe logical space functional to carry out several undertakings outside, or across, borders to protect and pursue their own interests and exterminate the competing, and mortally dangerous, antagonistic logical space of terrorism.

In this lies a commonality between the ideo-logical spaces of domination, resistance, and terror; each tends to create a spatio-thought functional to its desired geographies. With twofold consequences. Firstly, the ideo-logical space of terror has the same essentialising and sanitising effects of our thinking along global lines (see Sects. 4 and 5).[148] Secondly, it normalises a new—and probably agreed—legal geographical lexicon. This was epitomised in Bush's Presidential Military Order of Nov. 13, 2001: Detention, Treatment, and Trial of Certain Non-Citizens in the War Against Terrorism, 66 Fed. Reg. 222 (2001), which 'contained no geographical limits for its campaign on terror'.[149] It created the *enemy combatant* in order to justify 'a war that had no fixed location'; it neglected the territorial and maritime war alike inasmuch as the 'floating war' against terror allowed for a preventive war in the light of the 'natural-law' justification of the defence of the safe (and clearly territorialised) normative spaces of the state.[150] Far from triggering 'global anarchy',[151] this logical space follows a logic of concatenation, which is patent in the construction of a space completely regulated by new basic legal tenets. To this extent, it 'legitimated a diverse range of extremely violent practices', such as 'the "black sites" of extraordinary rendition to detention camps', where the alleged connivence of Western governments made it possible to implement 'acts of torture.[152]

In the creation of spaces of concatenation, a limit cannot be trespassed, which makes the equation between them and our thinking along global lines untenable. As seen, in the Age of Discovery it was possible to suspend completely the Western legal derivatives beyond the amity lines. On the other side of them, European powers competed with each other without constraints; and the legal geopolitical scramble for the rest of the world was agreed on a common grammar and lexicon. By contrast, the ideo-logical space of terrorism and its logic of concatenation are limited by juxtaposition. If we are 'us', our mechanisms of legal securitisation cannot be the same as those adopted within the logical space of terrorism. We cannot create loopholes in our winning arguments and therefore deprive the enemy combatant of their life, liberty, or property, if it arbitrarily seems right to do so. Nor does our legal geopolitics endorse pieces of legislation limiting individual rights—also consenting states to strip citizens of their citizenship. Our strategic formalism traps our counter-terroristic response within the brackets of the rule of law. This accounts for the protection of the ideo-logical safe space through the law; in so doing, it preserves its

[148] Bahgat and Medina (2013), p. 47.

[149] Sciullo (2012), p. 330.

[150] Bradford (2004).

[151] Barber (2004).

[152] Stephens (2011), p. 255.

ideo-logical (if not ontological) opposition to the nihilism of terrorists. The law acts as a powerful, 'strategic'- but-traditional device. Being so strongly rooted in political communities, we cannot deny the due process of law even to those who hate us to death.

Our law is, albeit strategic, extremely formalistic. This was clearly stated by the U.S. Supreme Court in the 2008 decision in *Boumediene v. Bush*. As the war on terror is an admixture of law, language, and geography, fought in the nowhere of its logical space, we might be led to think that the territory of, say, Guantanamo is not part of U.S. land. This could mean that, with the 'Law of the Land' being synonymous with 'due process of law', the well-bounded order of constitutional liberties would not include federal territories, in general, and Guantanamo, in particular.[153] The Supreme Court did not adhere to such an ideo-logical conception. The whole of our Western liberal, as well as democratic, ideo-logical space is saturated with 'the constitutional privilege of habeas corpus, a privilege not to be withdrawn except in conformance with the Suspension Clause' set forth in Art. I, sect. 9, cl. 2 of the U.S. Constitution.[154] Even the creation of spaces of concatenation must follow the rules of juxtaposition. Although triggered by new processes of spatial production, our war against terror is still a variety of the traditional, and therefore formalistic, 'guerre en forme'.[155]

5.3.5 Fencing States: Walled Ideo-logical Spaces

In Sect. 1.1, I defined the nation state as the 'well bounded order of politics', as Richard K Ashley wrote in *Untying the Sovereign State* (1988). Ashley assumes that international politics is characterised by the '*absence of an orderer* ... enforcing rational global designs'.[156] If we understand him correctly, he assumes that this '*condition of anarchy*', which is the foundational truth of the international community, is boosted more by the fragmentation of national identities than by the lack of an international government. Such a want is a consequence of sovereignty, which gives rise to a competition regarding which nation state is better at arranging and embellishing national governance. Like in a contest, states compete in the politics of legal spaces, seeking to demonstrate that they are the best at turning theirs into meaningful places and 'culturally defined milieux within which economic, social and political decisions are made'.[157]

Within this contest, sovereignty acts as a narcissistic attribute, which strengthens the resilience of the 'territorial trap'. Our assumption goes hand in hand with

[153] See Mullen (1953), pp. 736–737.

[154] Boumediene V. Bush (Nos. 06-1195 And 06-1196) 476 F. 3d 981, *per* Kennedy J.

[155] Schmitt (2006), p. 141.

[156] Ashley (1988), p. 227.

[157] Johnston (1986), p. 452.

Taylor's state-as-a-container theory: the territorial trap reduces the porosity of borders and facilitates territorial demarcation, boundary restrictions, and the design of state national 'character'. At the same time, it reinforces the quality of the state as a 'power container'.[158] I termed such narcissism *hortus conclusus*, which points to the state as an enclosed garden developing its own politico-legal order. The process of enclosure gives consistency to the territorialised paradigm as the radical title of legal geography. Rooted, as it is, in the earth, this title accounts for the physical apprehension of the earth by humans; and the enclosure of the *erreichtes Land* constitutes an act of 'geo-graphing' (see Sect. 5). Indeed, 'the solid ground of the earth is delineated by fences, enclosures, boundaries, walls, houses, and other constructs', which make 'the orders and orientations of human social life become apparent' and 'forms of power and domination . . . visible'.[159] This is nothing else but the creation of a belong of public law.

Historically, this process of demarcation was undertaken by the English when gaining possession of the New World. The establishment of an English (then British) title to America followed this pattern; the colonisers constructed their right to the land by 'building houses and fences and planting gardens'. In so doing, they created a 'fixed place' that 'under English law created a virtually unassailable right to the land'.[160] The fence demarcated both the garden (i.e. the bounded politico-legal order) and the plantation, whose crops embedded the economic appropriation of the fruits of the earth according to the newly created legal title. Public finance concerns thus entered the state; investments, returns, and mortgaged societies soon ensued.

This also accounts for the very foundation of the ideo-logical space underpinning the territorial trap.

As Thorlac Turville-Petre states in *England the Nation* (1996), 'Defining a nation necessarily involves exclusion' of the others. This is exactly what the territorial trap is; fencing territories defines the idea of a nation 'in terms of its people (*gens*) and its territory (*provincia*)'. It is the 'graphic way': 'to circumscribe the nation is to map it'.[161] No wonder that the joint application of the territorialised paradigm and the territorial trap reinforces the narcissism of the Western state even in a distressing time like ours, where the very idea of borders is being undermined. Firstly, globalisation and its derivatives, such as free movement of goods, capital, and workers, favour border fluidity; processes of supranational integration do the same. The main example is cross-border cooperation within, for example, the EU, which makes member states' borders 'soft' and reduces their political-legal relevance.[162]

---

[158] See Giddens (1985). See also Taylor (1994), p. 151.

[159] Schmitt (2006), p. 42.

[160] Seed (1995), p. 17.

[161] Turville-Petre (1996), p. 1.

[162] Toniatti (2018), pp. 43–45.

In her *Globalization and the Postcolonial World* (1997), Ankie Hoogvelt formulates a similar argument.[163] As the latest phase of world capitalism, globalisation is no longer in an expansive stage, but rather in an 'imploding' one. During expansive stages, 'the core of the world system' engages with its periphery to maximise profits, whereas implosion entails 'an intensification of trade and capital linkages within the core of the capitalist system'.[164] In reviewing the book, Hongying Wang highlights that, 'While trade and investment linkages intensify among the countries in the core, the periphery is being selectively excluded'. Hoogvelt divides the ideo-logical space of capitalist globalisation into three circles. The inner circle comprises of the transnational elites benefitting from globalisation. The middle circle coincides with our spaces of acquisitiveness, where people engage 'in cut-throat competition to stay afloat' and enjoy the commodities recollected in tranquillity. The outermost circle is the periphery. Located outside the spaces of acquisitiveness, it comprised those that are outside globalisation, also bearing the direst socio-climatic consequences of capitalism-induced climate change.[165]

I shall consider the impact of climate change on the construction of spaces (see Sect. 8.7). Here, I shall focus on how our safer spaces react against the loopholes created by arguments (and spaces) of doubt and threat caused by migration.

To protect our comfort zones, we are fencing our safe places, which are now walled ideo-logical spaces. Again, the law is functional to the enforcement of spatial production. All around the world, this has compelled nation states to securitise their external borders. What is intriguing is that we are reinstating the *hortus conclusus* and the narcissism of the territorial trap; 'Physical walls and boundaries' indeed 'constitute the most rudimentary and geographically obvious form of [state] enclosure'.[166] Israel built a wall in the Occupied Palestinian Territory on grounds of national security reasons; India fenced its border with Pakistan; a U.S.-Mexico border wall is being erected. And litigation soon ensued.[167] The U.S. Supreme Court granted President Trump the petition for a writ of certiorari.[168] In the Court's reasoning, public-finance concerns are ranked as superior as regards solidarity because public funds are functional to the 'Plenary Power Doctrine', granting the federal government 'plenary powers to exclude noncitizens'.[169] The barrier may 'cause irreparable harm to the environment and to individuals'; however, if the federal government were 'unable to finalize the contracts [for the wall] then the

[163] Hoogvelt (1997).

[164] Hoogvelt (1997), p. 84.

[165] Hoogvelt (1997), p. 239.

[166] Jeffrey et al. (2012), p. 1249. See Brown (2010).

[167] See Delaney (2005), p. 143.

[168] On the U.S.-Mexico border policing see Boyce (2018), pp. 351–354. On Trump's implementation of global apartheid in the USA see Ceciliano-Navarro et al. (2021), p. 99.

[169] At least from the so-called 'Chinese exclusionary cases'. U.S Supreme Court *Chae Chan Ping v. United States*, 130 U.S.581(1889); *Fong Yue Ting v. United States*, 149 U.S. 698 (1893). See Chang (2021).

funds at issue will be returned to the Treasury'—and costs borne by the U.S. taxpayers.[170]

The same predicament is valid as regards the EU policies related to migration and refugeeism. Far from being inclusive, EU treaties and legislation are overpopulated with references to external borders. Schemes such as Frontex, for example, are considered 'the culmination of the process of securitization of the EU external borders', whereas the European Border Management (EBM) has 'multiple dimensions': it encompasses several methods of 'border checks and surveillance,' which makes it hard for migrants to access the EU.[171] Indeed, EU citizenship complements national citizenship, which means that third-country nationals residing within its borders are, in principle, excluded from it. It is as if it were 'necessary to draw a line between insiders and outsiders, between "us" and "them"'.[172] What is walled is not the border, but the societal ideo-logical space of the integration process; the EU cosmopolitanism and its *cité* establish 'a unitary basis for exclusion rather than a coherent set of criteria for inclusion'.[173] Evidently, the 'principle of solidarity and fair sharing of responsibility', which underpins the Area of Freedom, Security and Justice laid down by the Treaty on the Functioning of the EU, is a privilege reserved for EU citizens, whereas third-country nationals must merely be treated fairly when they try to access the EU.[174]

Although borders are now constantly blurred by geopolitical, economic, environmental, and migration-related variables, the prevailing conception still reflects the ideo-logical space of the territorial trap. Needless to say, 'The price of the supposedly unified world of capital is the brutal division of human existence into regions separated by police dogs, bureaucratic controls, naval patrols, barbed wire and expulsions'.[175]

5.3.6 *The* Cité *as an Ideo-logical Space*

The last variety of ideo-logical space reflects the dialectics between domination and resistance in urban environments. These places of commodification have become 'places of globalisation', interlocked within the transnational supply chain.[176] The holders of global capital that 'dominate modern urban investment' also move their

[170] U.S. Supreme Court, *Donald J. Trump, President of the United States, et al. v Sierra Club, et al.* on application for stay 588 U.S. (2019) 1, 26 July 2019.

[171] On the so-called 'Fortress Europe' see Delaney (2010), p. 153. See also de Witte (2022), p. 119.

[172] Lashyn (2021), p. 364.

[173] Bhabha (1998), p. 604.

[174] Article 80 TFEU. See Moreno-Lax (2017), pp. 31 and 34. See also Burridge et al. (2017), p. 242.

[175] Badiou (2008), p. 38.

[176] See Retaillé (2012). See also Ceciliano-Navarro et al. (2021), pp. 104–105.

'capital around the globe detached from place, rising above feelings for or attachments to the places in which they invest'.[177]

This global supply chain needs force labour to work. Capital has resort to new winning arguments to convince prospective workers to migrate to the places in which it is invested. Its ideo-logical space is made up of reassuring imagined geographies. This space is convincing, since it gives migrants the illusion of taking part in the global wealth distribution. Prospective workers are thus attracted by the pursuit of happiness and the perspective of improving their (and their beloved's) standards of life. Taking part in the global supply chain mobilises mass migration from the Global South towards the Global North with a 'promise of growth [that] has been the justification for inequality, justice, and exploitation,' with more wounds to heal 'in the near climate future'.[178]

When economic migrants legally (and sometimes illegally) cross the fences of our *hortus conclusi*, they enter a new ideo-logical space, into which (as well as into its supply chain) they become integrated. Richard Sennet reminds us that 'exclusion becomes more complicated when you need those whom you despise'.[179] In providing an ethics for the city, he continues, the ideo-logical space of the *cité* should be based on the 'common bond of humanity', as is laid forth in *Leviticus*: 'But the stranger that dwelleth with you shall be unto you as one born among you, and thou shalt love him as thyself; for ye were strangers in the land of Egypt'.[180] Our notion of integration into the *cité* is far from this; it is economical, not ethical. Their integration into our *cité* is functional to production and consumption of commodities. As consumers, shareholders, and entrepreneurs, we turn workers into something servient to the comfort zones of capitalist economies. Again, our motto revolves around the maximisation of financial returns: the quieter the space, the safer it is, and the higher the returns are.

What do these migrants expect from us? Integrated in our safe spaces economically, they must comply with the following rules: they must take part into the supply chain; like us, they must spend their stipends (their share in the global wealth) in the commodities they produce to sustain the economic cycle; finally, after having worked in our safer places, they must retreat from them. The last requirement, which is also known as the NIMBY (Not In My Back Yard) factor, completely undermines the precept laid down in *Leviticus*. Indeed, it reflects what has been termed as the 'Global Apartheid'.[181] In his *Reimagining Britain*, the Archbishop of Canterbury Justin Welby highlights how this resistance to the Levitical doctrine of inclusion makes communities (even parish communities) 'resistant to outsiders and reluctant to sponsor diversity if there is any possibility that it will diminish housing values or be perceived as a risk to the aesthetics of that community'. There is no

[177] Sennett (2019), p. 296.

[178] Wallace-Wells (2019), p. 166.

[179] Sennett (2019), p. 131. See also Farbotko et al. (2016).

[180] *Lev* 19:34 (KJV). We owe this reference to Sennett (2019), pp. 298–299.

[181] Ceciliano-Navarro et al. (2021).

'diversity, or mixing, no valuing of difference, no sight of the poor'; this sharply contrasts with the biblical (and Christian) mandate.[182] Ultimately, the spirit of the *cité* is that of gentrification, i.e. a course of action 'by which the bottom 70–75 per cent of an urban population become vulnerable to expulsion by the top quarter of the people in a city, either through raised rents or by poor homeowners being seduced into selling'.[183]

Gentrification recreates a safe space, whose community is homogenous, 'expensive', and 'exclusive'. Within it, the commodifying state of mind favours a change in our perception of spatiality. Even of our vertical spatiality. Take tower block apartments. Being traditionally 'connected . . . to social housing', they have undergone an incredible shift. Partly because building is processed within the global production of capital, partly because this gives gentrified people a 'privileged position of detached' housing that is alternative to suburbia, 'vertical living' has encouraged 'an "elite" takeover of the skies by the global rich'.[184] In Welby's words, 'the balance of property values and influence of location' are extremely 'materialistic' and point to the 'economic maximization' of investments in housing.[185] The presence of outsiders makes the space diverse and no longer exclusive, which may become less expensive if housing prices drop as a consequence of the *mixité*. This is unacceptable in the process of gentrification; the acolytes of the *cité* populate an ideo-logical mortgaged society based on acts of conveyance that are also secured on the mortgagors' property, i.e. the property of its inhabitants.

The spirit of the *cité* resides in the winning arguments of the socio-political hierarchies of domination. Displacement and spatial marginalisation triggers a 'really hostile environment for [allegedly] illegal immigrants',[186] i.e., a by-product of 'heightened border measures that restrict migration, measures that play well with anxious electorates and hawkish politicians'; in so doing, states disavow 'the West's complicity in the wider social, political and economic conditions that contribute to the migrations the West seeks to secure.'[187] Such ideo-logical space multiplies the iniquitous effects of the economic conveyance underpinning our politico-legal arrangements. In fact, the capital-state link generates sharp divides between within political communities and replicates 'colonial histories and racialised hierarchies of powers' between groups.[188]

Eaten up by an incredible zeal, we have unceasingly created gated communities. Suffice it here to consider the creation of urban ghettoes for the Jews throughout Europe in mediaeval and modern times. Like the migrants in a time of globalisation,

---

[182] Welby (2018), p. 145.

[183] Sennett (2019), p. 138.

[184] Hubbard (2021), p. 207.

[185] Welby (2018), p. 146.

[186] Goodfellow (2019), p. 2. The so-called 'hostile environment immigration' policy was coined by the then UK Home Secretary Theresa May. See Kirkup and Winnett (2015).

[187] Baldwin (2014), p. 521.

[188] Goodfellow (2019), p. 39.

they had their share in the global wealth by filling 'the niches of the informal economy'; they also had some rights 'only so long as [they] literally stayed 'where they belong', that is, they accepted their marginality.[189] Such separateness was also at the core of the U.S. policies of segregation; separate-but-equal access to public services (or racial redistricting[190]) on the basis of segregation did not imply the inferiority of African Americans (see Sect. 6.2).

Perhaps most striking of all, the same lexical item *apartheid* literally means separateness. Applied in South African constitutional law before its dismantlement in 1990–1994, it arranged its community after 'racial classification'. Writing in the 1980s, the constitutional lawyers Basson and Viljoen praised this policy and contrasted it with 'the doctrine of human rights as ... applied in the libertarian "Western' world", where it would have amounted 'to discrimination'.[191] The white, racially dominated South Africa provided for a separate-but-not-equal constitutional legal framework. From 1910 to 1948, the South African Dominion fostered social apartheid by progressively depriving the subordinate ethnic groups of their fundamental rights—even of their right to purchase land.[192] The apartheid system was strengthened after the Second World War—i.e. when the Afrikaner *Nasionale Party* won the 1948 general elections.[193] The national government passed several legislative acts and administrative measures, which established a legal system based on racial separateness and limitation of black people's rights. Segregationist policies eventually culminated in the 'territorial apartheid',[194] which led to the creation of *Bantustans*—i.e. quasi-independent Homelands for the blacks. Additional legislative provisions abolished the fundamental rights of Indians and coloureds.[195] No wonder, again, that the regime ignored (or minimised) 'numerous substantial Black settlements, particularly townships and squatter settlements'; to reduce the weight of the Black population, maps gave 'a misleading prominence' to 'small White-dominated towns, especially in rural areas'.[196]

The creation of divides between the sectionalities of the same political community may follow alternative patterns. Gentrification may be attained by expanding the part of the city palatable to the *cité*. And by ejecting the 'outsiders'.

Such an ejectment occurs when poor homeowners are seduced into selling their property. 'Since community of goods is impracticable,' as Thomas Wood put in

---

[189] Sennett (2019), pp. 130 and 132.

[190] Forest (2001).

[191] Basson and Viljoen (1988), p. 232.

[192] See among others, the Mines and Works Act (No 12 of 1911), the Native Land Law (No 27 of 1913), the Native (Urban Areas) Act (No 21 of 1923). The Natives Land Act (No 27 of 1916) 'was originally intended to be an interim measure to halt the acquisition of land by Africans outside "scheduled areas"': Bennett (1996), p. 81. See also Giliomee (2003).

[193] Lovell (1956), p. 324.

[194] Social apartheid was first recommended territorial in 1933. See Dubow (1992), pp. 211–212.

[195] See Nicolini (2015).

[196] Black (1997), pp. 58–59.

1704, 'in as much as some men are naturally more ambitious than others', the rich appear to be better at managing the levers of capital and get hold of the property in which they want to invest. Quite evidently, in coding capital, 'all men would not be equally industrious to the raising of the common stock', which means that 'we must suppose distinct property' and distinct ideo-logical spaces for the dwellers of the *cité* and the outsiders.[197]

Don Mitchell points to this when examining how the law annihilates the homeless through the law. Anti-homeless bylaws, indeed, 'both reflect and reinforce a highly exclusionary sense of modern citizenship'—and ultimately, of the *cité*. Towns would not probably attract capital if 'urban space . . . are chocked full of homeless people lying on sidewalks, sleeping in parks and panhandling' the gentrified 'every time they turn a corner'.[198] These laws usually outlaw 'common behaviors'; besides urinating and defecating, 'standing around, sitting, [and] sleeping' become prohibited 'in public property'. 'To the degree that laws can annihilate spaces for the homeless, they can annihilate the homeless themselves'.

With sour sarcasm, I may say that 'landed property'—which, in our mortgaged societies, is usually *upon condition*—'thus again becomes a prerequisite of effective citizenship'. Again, we have said; this interweaving of public and private property has been part of our narrative of exclusion and dispossession and still perpetuates the tragedy of ending public spaces, in general, and commons, in particular.[199] We have already mentioned the commons (see Sect. 5.3.1). Together with waste lands and forests, they had always been the major source of sustenance for commoners. To increase the production of commodities and maximise profits, these tracts of land were progressively enclosed. In England, enclosure became a further organising theme, embedding the idea of the *cite*, i.e. the domination of the 'up-standing bourgeoise' over the deviant poor. Indeed, it turned lands into capital, allowing the new owners 'to extract monetary values.'[200]

In enclosing waste and common fields, the state would have brought advantages to 'the poorest that dwell upon wastes.' These would have had 'a portion secured them to enclose about their cottages to raise herbs and roots, a cow and some corn for their better relief'.[201] The assumption proved to be simplistic in that not only did the enclosure of the commons displace the entire class of small proprietors, but it also swept away the public policy introduced by the Statute of Merton, according to which 'the lord of a manor may enclose so much of the waste as he pleases . . . provided he leaves common sufficient for such as are entitled thereto.'[202] 'In these utilitarian days', Anthony Trollope would have put it in the nineteenth century, public financial concerns turned the *Crown estates* into capital gains. In *Framley*

[197] Wood (1702), p. 63.

[198] Mitchell (1997). See also Blomley (2009), p. 584.

[199] Ellickson (2001), p. 20; Mitchell (1995, 2017). See also Hardin (1968).

[200] Pistor (2019), p. 29. See Turner (1984), p. 36.

[201] St. John (1792: 306).

[202] Bl. Comm. II, 34. See 20 Hen. III. c. 4. 29 Geo. II. c. 36, and 31 Geo. II. c. 41.

Parsonage, he vividly portraits how 'People still come from afar to see the oaks of Chaldicotes, and to hear their feet rustle among the thick autumn leaves;' but they would 'soon come no longer. The giants of past ages are to give way to wheat and turnips; a ruthless Chancellor of the Exchequer, disregarding old associations and rural beauty, requires money returns from the lands.'[203]

The ideo-logical space of the *cité* was able to infiltrate the public sphere because enclosure, as a public policy, was pursued by pieces of parliamentary legislation. As said, Parliament did not establish 'equality of goods and lands;' by contrast, free-holders were able to enforce their ideo-logical space and legally appropriate commons, wastes, and forests through pieces of legislation.

5.3.7 *Informality and Resistance*

The clearing of public space is only one of the examples whereby 'commonable lands' are turned 'into privately owned hands'. In making themselves attractive to capital, members of the *cité* have resort to enclosure, which 'has emerged in recent years as a key process of neoliberal globalisation'. Here, the trick is complemented by force. Jeffrey, McFarlane, and Vasudevan remind us that 'from Mumbai to São Paulo, New York to Glasgow, we are witnessing the entrenchment of urban inequality and the increasing fragmentation of the urban landscape'. In investing their money around the world, the holders of global capital have thus spread the spirit of the *cité* everywhere; gentrification (and its related processes: exclusion, ejectment, and annihilation) has become 'a global urban development strategy' and has accompanied the 'demolition of informal settlements' triggering 'mass slumming';[204] consequently, 'squatters are driven into ever larger, ever more crowded informal settlements'.[205]

The *cité* has reproduced the 'racially marginalised hierarchies' between the metropolis and its colonies within their urban areas. Migrants are located into cordoned-off 'periphractic spaces', which are often processes of 'spatial enclosure' along racialised lines.[206] This often constrains immigrants' participation in public life 'in terms of location and their limitation in terms of access – to power, to (the realization of) rights, and to goods and services.'[207]

In such an ideo-logical space, not only do migrants enter irredeemable conveyances with the capitalism-nation state nexus, but they also accept to bear the costs of such politico-legal marginalisation. It is undeniable that the capitalism-state nexus

[203] Trollope (2016), p. 26.

[204] Jeffrey et al. (2012), p. 1251; Verma (2002).

[205] Jeffrey et al. (2012), p. 1251.

[206] See Coleman (2018), p. 29. On the 'urban cultural geographies of race' see Nayak and Jeffrey (2011), pp. 182–186.

[207] Goldberg (2001), pp. 71 and 72.

arranges the political bonds of the community after 'structures of [economic] authority and [political] obedience.' Migrants—even climate-induced ones, who bear the costs of the politico-economic domination—'subject themselves' to this organising principle and accept its 'authority by their own consent and voluntary act.'[208] Within our wealthy communities, they accept 'these "unnatural" structures and strictures of authority, for 'they realize that without them . . . even their most basic rights will mean little.' [209]

The metaphor of the mortgaged society is incredibly fitting, even when it comes to describing the arguments of resistance to the winning arguments of the ideo-logical space. Within the private-law metaphor, if the deed so establishes, the mortgagor also has right of redemption, including an option to repurchase.[210] In our mortgaged societies, however, the 'terms and conditions' of the political obligation laid forth by the *cité* require unceasing acts of conveyancing—and this makes the political liens irredeemable. 'The current packages on offer', furthermore, 'could have the effect of removing or rendering illusory' this very basic right.[211]

When referring to resistance, I am not advocating the 'creation of spaces of collective rebellion.'[212] Quite the opposite; spaces of resistance are usually set at the margins of the ideo-logical spaces of the *cité*. Unlike those dominated by the forces of capital, the former are situated places. In enforcing gentrification, the global holders of capital do not have resort to situated legal practices; the recreation of safe spaces of globalisation is *cosmopolitan* in nature, and therefore has the 'capacity to move easily from place to place, culture to culture'[213] and transcends local customs and traditions. Even legal traditions. Its legal enforcement is indifferent to situated processes of legal spatial production, as well as to the role played by variables and practices (mainly social, legal, and geographical) within them. It is indifferent to our idea of 'space' *also* produced through situated legal practices,[214] where migrants 'learn to live the bitter-sweet lessons of displacement'.[215]

This also means that counter legal geographies of resistance also examine the practices that are unforeseen by the legal establishment of the space of the *cité*. Unlike these hyper-regulated spaces where exclusion, production, and annihilation are enforced by the law, spaces of resistance are also often made up of informal situated legal practices. We have already examined the notion of 'informal settlements', which usually populate the margins of the *cité* and are created in the interstices of formal law.[216] They usually generate a variety of 'illegal' spatial

[208] Witte (2007), p. 183.

[209] Witte (2007), p. 184.

[210] See e.g. LPA 1925, ss 5.1 and 205.1.xvi.

[211] Houghton and Livesey (2001), p. 167.

[212] On collective rebellion when confronting the state of flux see Silver (2018), pp. 161–168.

[213] Sennett (2019), p. 141.

[214] See e.g. Lefebvre [1974] (1991).

[215] Sennett (2019), p. 203. Bachelard.

[216] See, among others, Dovey and King (2011); Matthews Glenn and Bélanger (2003).

normativies. If, for example, public policies, bureaucratic requirements, and costs for formal housing are too elevated, individuals, groups, and families trigger new urban neighbourhoods. Caused, as they are, by the scarcity of land management policies, these spaces are informally established by the urban poor and squatters at the margins of the law. Albeit illegal, these are legally relevant notwithstanding the formal law, which is called to regularise them at a later time.

In these places there is plenty of room for resistance. As Blomley argues, challenging gentrification forces us to go beyond the *cité* and its private-property scheme 'to acknowledge the existence of counterposed property claims that are collective in scope'. Like the common law, the hyper-regulated spaces of the *cité* and their legally enforced policies of exclusion, production, and annihilation are countered by *stirring arguments*: Like equity stirs formalism, counter legal geographies insinuate the doubt that 'the poor also have a legitimate property interest in, and claim to, the site'. This equitable interest is a 'collective one ... and also a clearly localized one', pointing to 'the community' and its set of situated legal practices.[217]

The critical and subversive attitude of both comparative law and geography (see Sect. 2.1) sheds new light on these processes of spatial production. In particular, they insinuate into the ideo-logical spaces of domination and bridge them with their contexts and their situated legal practices, thus limiting their being 'only partially related to something called "society"'.[218] In so doing, they pluralise the debate in several ambits usually unexplored from a purely formalistic legal angle. Certainly, legal formalism 'commonly denies its work in the world'; at the same time, it is also productive, say, of several effects, which may be meaningful to the law. This is particularly apparent in the informal settlements mentioned above. Their paradigm is probably able to challenge the winning arguments and open our society to spaces of doubts, which do not aim to threaten the dwellers of the *cité*, but rather to turn societal engineering into effective legal change, thus radically transforming the conveyance of political obligations upon condition into a more inclusive political space.

References

Allen J (2018) The circulation of financial elites. In: Coleman M, Agnew J (eds) Handbook on the geographies of power. Edward Elgar, Cheltenham and Northampton, MA, pp 178–193

Ashley R (1988) Untying the sovereign state: a double reading of the anarchy problematique. Millennium 17:227–262

Athreya B (2011) White man's 'burden' and the new colonialism in West African cocoa production. Race/Ethn Multidiscip Glob Context 5:51–59

[217] Blomley (2008), p. 316.

[218] Blomley (2003), p. 17.

Attwood B (2020) Empire and the making of native title: sovereignty, property and indigenous people. CUP, Cambridge
Badiou A (2008) The communist hypothesis. New Left Rev 49:29–42
Bahgat K, Medina RM (2013) An overview of geographical perspectives and approaches in terrorism research. Perspect Terror 7:38–72
Baldwin A (2014) Pluralising climate change and migration: an argument in favour of open futures. Geogr Compass 8:516–528
Barber BR (2004) Fear's empire: war, terrorism, and democracy. W.W. Norton & Company, New York
Barkan J (2011) Law and the geographic analysis of economic globalization. Prog Hum Geogr 35: 589–607
Basson DA, Viljoen HP (1988) South African constitutional law. Juta & Co. Ltd, Cape Town et al., 1988
Bennett TW (1996) African land – a history of dispossession. In: Zimmermann R, Visser D (eds) Southern cross: civil law and common law in South Africa. Clarendon, Oxford, pp 65–94
Bennett TDC et al (2020) Legal imagination in troubled times. An introduction. In: Mullender R et al (eds) Law and imagination in troubled times: a legal and literary discourse. Routledge, Abingdon and New York, pp 1–13
Bhabha HK (1994) The location of culture. Routledge, London
Bhabha J (1998) 'Get back to where you once belonged': identity, citizenship, and exclusion in Europe. Hum Rights Q 20:592–627
Bhungalia L (2018) Governing banishment: settler colonialism, territory, and life in an economy of death. In: Coleman M, Agnew J (eds) Handbook on the geographies of power. Edward Elgar, Cheltenham/Northampton, MA, pp 313–331
Black J (1997) Maps and politics. The University of Chicago Press, Chicago
Blomley N (1989) Text and context: rethinking the law-space nexus. Prog Hum Geogr 13:512–534
Blomley N (2003) From 'what?' to 'so what?': law and geography in retrospect. In: Holder J, Harrison C (eds) Law and geography. OUP, Oxford, pp 17–33
Blomley (2007) Making private property: enclosure, common right and the work of hedges. Rural Hist 18:1–21
Blomley N (2008) Enclosure, common right and the property of the poor. Soc Leg Stud 17:311–331
Blomley N (2009) Homelessness, rights, and the delusions of property. Urban Geogr 30:577–590
Blomley N (2014) Disentangling law: the practice of bracketing. Annu Rev Law Soc Sci 10:133–148
Blomley N (2015) The ties that blind: making fee simple in the British Columbia treaty process. Trans Inst Brit Geogr 40:168–179
Boast RP (2013) Property rights and public law traditions in New Zealand. N Z J Public Int Law Special Issue: 21st Birthday of the New Zealand Bill of Rights Act 11:161–182
Boast RP (2017) Māori land and land tenure in New Zealand: 150 years of the Māori Land Court. Comp Law J Pac/Journal de Droit Comparé du Pacifique 23:97–133
Borrows J (2002) Recovering Canada: the resurgence of indigenous law. University of Toronto Press, Toronto
Boyce G (2018) Autonomy, human vulnerability and the volumetric composition of US border policing. In: Coleman M, Agnew J (eds) Handbook on the geographies of power. Edward Elgar, Cheltenham and Northampton, MA, pp 347–361
Bradford WC (2004) The duty to defend them: a natural law justification for the bush doctrine of preventive war. Notre Dame Law Rev 74:1365–1492
Braithwaite J (2005) Markets in vice, markets in virtue. OUP and Federation Press, New York and Sydney
Brink A (1976) English and the Afrikaans Writer. Engl Afr 3:34–46
Brink A (1988) Die eerste lewe van Adamastor: 'n novelle. Saayman & Weber, Kapstad. English edition: Brink A (1993) The first life of Adamastor. Secker & Warburg, London, 1993

Brink A (1996) Reinventing a continent (revisiting history in the literature of the New South Africa: a personal testimony). World Lit Today 70:17–23

Brown W (2010) Walled states, waning sovereignty. Zone Books, New York

Burridge A et al (2017) Polymorphic borders. Territ Polit Gov 5:239–251

Callon M (1998) An essay on framing and overflowing: economic externalities revisited by sociology. Sociol Rev 46:244–269

Calyx C et al (2020) Asserting land rights through technology and democratic expression. In: O'Donnell T et al (eds) Legal geography. Perspectives and methods. Routledge, Abingdon/New York, pp 37–57

Capra F, Mattei U (2015) The ecology of law: towards a legal system in tune with nature and community. Berrett-Koehler Publishers, Oakland

Carmody P (2018) Matrix governance and imperialism. In: Coleman M, Agnew J (eds) Handbook on the geographies of power. Edward Elgar, Cheltenham and Northampton, MA, pp 297–312

Ceciliano-Navarro Y et al (2021) Reflections on anti-immigration narratives and the establishment of global apartheid. In: Land M et al (eds) Beyond borders: the human rights of non-citizens at home and abroad. CUP, Cambridge, pp 94–110

Chang T (2021) The Chinese exclusion cases and policing in the fourth amendment–free zone. Stanford Law Rev Online 73:209–220

Clark G, Clark A (2002) Common rights to land in England, 1745-1839. J Econ Hist 61:1009–1036

Clarke A, Kohler P (2005) Property law. Commentary and materials. CUP, Cambridge

Coleman M (2018) Bodies. In: Coleman M, Agnew J (eds) Handbook on the geographies of power. Edward Elgar, Cheltenham and Northampton, MA, pp 27–33

Coleman M, Agnew J (2018) Introduction to the handbook on the geographies of power. In: Coleman M, Agnew J (eds) Handbook on the geographies of power. Edward Elgar, Cheltenham and Northampton, MA, pp 3–25

Comaroff JL (2001) Colonialism, culture, and the law: a foreword. Law Soc Inq 26:205–314

Correja JE (2021) All the land was stolen: investigating the aporia of justice through countertopographies of Indigenous land rights and settler colonialism across the Americas. In: Bartel R, Carter J (eds) Handbook on space, place and law. Edward Elgar, Cheltenham and Northampton, MA, pp 38–48

Crampton JW (2010) Mapping: a critical introduction to cartography and GIS. Blackwell, Oxford

Cumin D (2009) La théorie du partisan de Carl Schmitt. Stratégique 93-94-95-96:31–71

de Camões LV (1572) Os Lusíadas, book V. Gõnçalvez, Lisboa

de Witte F (2022) Here be dragons: legal geography and EU law. Eur Law Open 1:113–125

Delaney D (2003) Beyond the word: law as a thing of this world. In: Holder J, Harrison C (eds) Law and geography: current legal issues. OUP, Oxford, pp 67–83

Delaney D (2005) Territory. A short introduction. Blackwell, London

Delaney D (2010) The spatial, the legal and the pragmatics of world-making. Blackwell, Oxford

Dixon AD (2018) Reflections on the power in and the power financial markets. In: Coleman M, Agnew J (eds) Handbook on the geographies of power. Edward Elgar, Cheltenham and Northampton, MA, pp 123–136

Dodds K (2019) Geopolitics. A very short introduction. OUP, Oxford

Dovey K, King R (2011) Forms of informality: morphology and visibility of informal settlements. Informal Urban 37:11–29

du Toit SJ [1897] (1966) Hoe die Hollanders die Kaap ingeneem het. Human & Rousseau, Kaapstad

Dubow S (1992) Afrikaner nationalism, apartheid and the conceptualization of 'race'. J Afr Hist 33: 209–237

Dunn DH (2005) Bush, 11 September and the conflicting strategies of the 'war on terrorism'. Irish Stud Int Aff 16:11–33

Ellickson RC (2001) Controlling chronic misconduct in city spaces: of pandhanlders, skid rows, and public-space zoning. In: Blomley N et al (eds) The legal geographers reader. Blackwell, Oxford, pp 19–30

Farbotko C et al (2016) Climate migrants and new identities? The geopolitics of embracing or rejecting mobility. Soc Cult Geogr 17:533–552

Fish SE (1980) Is there a text in this class? The authority of interpretive communities. Harvard UP, Cambridge, MA

Fish SE (2017) Winning arguments: what works and doesn't work in politics, the bedroom, the courtroom, and the classroom. Harper Collins, New York

Fonseca D (2017) Parler du terrorisme. Une approche philosophique. Revista General de Derecho Público Comparado 21:1–29

Forest B (2001) Mapping democracy: racial identity and the quandary of political representation. Ann Assoc Am Geogr 91:143–166

Giddens A (1985) The nation-state and violence. Polity Press, Cambridge

Gilbert J (2007) Historical indigenous peoples' land claims: a comparative and international approach to the common law doctrine on indigenous title. Int Comp Law Q 56:583–611

Giliomee H (2003) The making of the Apartheid Plan, 1929-1948. J S Afr Stud 29:373–392

Goldberg DT (2001) 'Polluting the body politic': race and urban location. In: Blomley N, Delaney D, Ford RT (eds) The legal geographies reader: law, powers, and space. Blackwell, Oxford, pp 69–76

Goodfellow M (2019) Hostile environment. How immigrants became scapegoats. Verso, London

Gravells NP (2010) Land law, 4th edn. Sweet & Maxwell, London

Gray T (1751) An Elegy wrote in a Country Church Yard. R Dodsley, London

Gray S (1977) The myth of Adamastor in South African literature. Theoria: J Soc Polit Theory 48: 1–23

Gregory D (2004) The colonial present. Blackwell, Maiden, MA and Oxford, UK

Hardin G (1968) The tragedy of the commons. Science 162(3859):1243–1248

Harley JB (1988) Maps, knowledge, and power. In: Cosgrove D, Daniels S (eds) The iconography of landscape. CUP, New York, pp 277–312

Harley JB (1989) Deconstructing the map. Cartographica 26:1–20

Harley JB (1992) Rereading the maps of the Columbian encounter. Ann Assoc Am Geogr 82:522–553

Hashim AS (2001) The world according to Usama Bin Laden. Naval War Coll Rev 54:11–35

Hill C (2001) Puritanism and revolution: studies in interpretation of the English revolution of the 17th century. Pimlico, London

Hoogvelt A (1997) Globalization and the postcolonial world: the new political economy of development. Johns Hopkins UP, Baltimore, MD

Houghton J, Livesey L (2001) Mortgage conditions: old law for a new century? In: Cooke E (ed) Modern studies in property law, vol I, Property 2000. Hart, Oxford, pp 163–180

Hoyle R (1992) Introduction: aspects of the Crown's estate, c. 1558–1640. In: Hoyle R (ed) The estates of the English Crown 1550–1640. CUP, Cambridge, pp 1–57

Hubbard P (2021) Negotiating privacy in the 'vertical city': regulating the gentrification of the skies. In: Bartel R, Carter J (eds) Handbook on space, place and law. Edward Elgar, Cheltenham and Northampton, MA, pp 38–48

Hugues H (2007) Aesthetics of common law – domestic and international implications. La Law Rev 67:690–749

Husa J (2018) Advanced introduction to law and globalisation. Edward Elgar, Cheltenham and Northampton, MA

Jeffrey A, McFarlane C, Vasudevan A (2012) Rethinking enclosure: space, subjectivity and the commons. Antipode 44:1247–1267

Johnston RJ (1986) Four fixations and the quest for unity in geography. Trans Inst Br Geogr 11: 449–453

Kawharu IH (1977) Māori land tenure: studies of a changing institution. Clarendon Press, Oxford

Kedar A (2003) On the legal geography of ethnocratic settler states: notes towards a research agenda. In: Holder J, Harrison C (eds) Law and geography: current legal issues. OUP, Oxford, pp 401–441

Khorram S, Koch FH, van der Wiele CF, Nelson SAC (2012) Remote sensing. Springer, New York et al.

Kirkup J, Winnett R (2015) Theresa May Interview: 'We're Going to Give Illegal Migrants a Really Hostile Reception, (May 25, 2012), The Telegraph. <https://www.telegraph.co.uk/news/0/theresa-may-interview-going-give-illegal-migrants-really-hostile/>. Accessed 30 July 2020

Kusters et al (2013) Formalising participatory land-use planning. Experiences from Sanggau District, West Kalimantan, Indonesia. In: Both ENDS. World Agroforestry Centre. Available via Both ENDS. https://www.bothends.org/uploaded_files/document/203_07_13_PLUP_Background_document.pdf. Accessed 24 June 2021

Kwapena F (2021) Understanding the impact of customary land tenure and reform in Papua New Guinea. In: Bartel R, Carter J (eds) Handbook on space, place and law. Edward Elgar, Cheltenham and Northampton, MA, pp 99–108

Lashyn S (2021) The aporia of EU citizenship. Liverpool Law Rev 42:361–377

Lefebvre H (1974) La production de l'espace. Éditions Anthropos, Paris. English edition: Lefebvre H (1991) The production of space (trans: Nicholson-Smith D). Blackwell, Oxford

Lovell CR (1956) Afrikaner nationalism and apartheid. Am Hist Rev 61:308–330

Macintyre A (1985) After virtue: a study in moral theory, 2nd edn. Duckworth, London

Macpherson CB (2010) The political theory of possessive individualism. OUP, Oxford

Maitland FW (1936) In: Chaytor AH, Whittaler WJ (eds) The forms of action at common law. CUP, Cambridge

Mann G, Wainwright J (2018) Climate Leviathan: a political theory of our planetary future. Verso, London and New York

Matthews Glenn J, Bélanger V (2003) Informal law in informal settlements. In: Holder J, Harrison C (eds) Law and geography. OUP, Oxford, pp 281–304

McFarland Taylor S (2019) Ecopiety. Green media and the dilemma of environmental virtue. New York UP, New York

McHugh PG (2011) Aboriginal title. The modern jurisprudence of tribal land rights. OUP, Oxford

Mitchell D (1995) The end of public space? People's park, definitions of the public, and democracy. Ann Assoc Am Geogr 85:108–133

Mitchell D (1997) The annihilation of space by law: the roots and implications of anti-homeless laws in the United States. Antipode 29:303–335

Mitchell D (2003) The right to the city: social justice and the fight for public space. The Guildford Press, New York

Mitchell D (2017) People's Park again: on the end and ends of public spaces. Environ Plan A 49: 503–518

More T (1908) Utopia. In: Cayley A (ed) Memoirs of Sir Thomas More, vol II. Cadell & Davis, London

Moreno-Lax V (2017) Accessing asylum in Europe: extraterritorial border controls and refugee rights under EU law. OUP, Oxford

Muir Watt H (2012) Further terrains for subversive comparison: the field of global governance and the public/private divide. In: Monateri PG (ed) Methods of comparative law. Edward Elgar, Cheltenham and Northampton, MA, pp 270–288

Muir Watt H (2014) The relevance of private international law to the global governance debate. In: Muir Watt H (ed) Private international law and global governance. CUP, Cambridge, pp 1–19

Mullen JM (1953) The supreme law of the land. Va Law Rev 39:729–751

Mullender R (1997) Context, contingency and the law of negligence (or from Islands to Islands of Time). Bracton Law J 29:23–33

Mullender R (2018) Politeia's place in our practical life: Pierre Bourdieu on the modern state. Univ Tor Law J 68:694–718

Mullender R (2019) There is no such thing as a safe space. Mod Law Rev 82:549–576

Nayak A, Jeffrey A (2011) Geographical thought. An introduction to ideas in human geography. Routledge, Abingdon and New York

Nicolini M (2015) ‘n Droë Wit Seisoen in die Stormkaap: André Brink and the Fundamental Rights of the Afrikaners in Apartheid South Africa. In: Ward I (ed) Literature and human rights. The law, the language and the limitations of human rights discourse. de Gruyter, Berlin/Boston, pp 235–254
Nicolini M (2019) Praising the world “by geometrical terms”: legal metrics, science and indicators in Swift’s Voyage to Laputa. Pólemos 13:327–348
Nicolini M (2020) ‘Inequality of goods and lands’ in mortgaged democracies: paradigms and effects of global comparative law. Liverpool Law Rev 41:27–44
O’Tuathail G (1996) Critical geopolitics. The politics of writing global space. Minnesota UP, Minneapolis
Patton P (2000) The translation of indigenous land into property: the mere analogy of English jurisprudence. Parallax 6:25–38
Peluso NL (1995) Whose woods are these? Counter-mapping forest territories in Kalimantan, Indonesia. Antipode 27:383–406
Petzold J (2000) André Brink’s Magical History Tour: postmodern and postcolonial influences in ‘The First Life of Adamastor’. Engl Afr 27:45–58
Pistor K (2019) The code of capital. How the law creates wealth and inequality. Princeton UP, Princeton
Purves J (1909) Camoens and the Epic of Africa. The State, Cape Town
Relph E (1976) Place and placeness. Pion, London
Retaillé D (2012) Les Lieux de la Mondialisation. Le Cavalier Bleu, Paris
Rudden B (1994) Things as thing and things as wealth. Oxf J Leg Stud 14:81–97
Rudolph J (2013) Common law and enlightenment in England, 1689-1750. The Boydell Press, Suffolk
Ryan RR (2004) Postcolonial geographies. In: Duncan SJ et al (eds) A companion to cultural geography. Blackwell, Oxford, pp 469–484
Samson C (2016) Canada’s strategy of dispossession: aboriginal land and rights cessions in comprehensive land claims. Can J Law Soc/Revue Canadienne Droit et Société 31:87–110
Schmitt C (1950) Der Nomos Der Erde: Im Volkerrecht Des Jus Publicum Europaeum. Duncker & Humblot, Berlin. English edition: Schmitt C (2006). The Nomos of the Earth in the International Law of the Jus Public Europaeum (trans: Ulmen, G.L.). New York, Telos
Sciullo NJ (2011) The ghost in the global war on terror: critical perspectives and dangerous implications for national security and the law. Drexel Law Rev 3:561–582
Sciullo NJ (2012) On the language of (counter)terrorism and the legal geography of terror. Willamette Law Rev 48:317–342
Seed P (1995) Ceremonies of possession in Europe’s Conquest of the New World, 1492-1640. CUP, New York
Sennett R (2019) Building and dwelling. Ethics for the city. Penguin, London
Sharp JP (2014) Critical geopolitics. In: Cloke P, Crang P, Goodwin M (eds) Introducing human geographies. Routledge, London and New York, pp 530–541
Sidaway J (2000) Postcolonial geographies: an exploratory essay. Prog Hum Geogr 24:591–612
Siebert DT (2013) Mortality’s muse: the fine art of dying. University of Delaware Press, Newark
Silver D (2018) Everyday radicalism and the democratic imagination: dissensus, rebellion and Utopia. Polit Gov 6:161–168
Smith M (2015) Boko Haram: inside Nigeria’s unholy war. Bloomsbury, London
Stafford F (2013) Introduction. In: Wordsworth W, Coleridge ST [1798 and 1802] (2013) Lyrical Ballads 1798 and 1802. OUP, Oxford
Stephens AC (2011) Beyond imaginative geographies? Critique, co-optation, and imagination in the aftermath of the war on terror. Environ Plan D Soc Space 29:254–267
Sumner Maine H [1861] (1908) Ancient law: its connection with the early history of society, and its relation to modern ideas. John Murray, London
Taylor PJ (1994) The state as container: territoriality in the modern world-system. Prog Hum Geogr 18:151–162

Toniatti R (2018) How soft is and ought to be the law of interregional transborder cooperation? In: Kicker R et al (eds) Changing borders: legal and economic aspects of European enlargement. Lang, Bern et al, pp 43–45
Trollope A (2016) Framley parsonage. Penguin, London
Turner ME (1984) Enclosures in Britain 1750–1830. Macmillan, London
Turville-Petre T (1996) England the nation. Language, literature, and national identity 1290-1340. Clarendon Press, Oxford
van Wagner E (2021) Extracting Indigenous jurisdiction on private land: the duty to consult and Indigenous relations with place in Canadian law. In: Bartel R, Carter J (eds) Handbook on space, place and law. Edward Elgar, Cheltenham and Northampton, MA, pp 182–194
Veracini L (2014) Understanding colonialism and settler colonialism as distinct formations. Interventions 16:615–633
Verma GD (2002) Slumming India: a chronicle of slums and their saviours. Penguin, New Delhi
Wallace-Wells D (2019) The uninhabitable earth. A story of the future. Allen Lane, London
Walwyn W (1649) Tyranipocrit, discovered with his wiles, wherewith he vanquisheth. Rotterdam
Ward I (2009) Law, text, terror. CUP, Cambridge
Ward I (2021) The play of law in modern British theatre. Edinburgh UP, Edinburgh
Watson A (1988) Failures of the legal imagination. University of Pennsylvania Press, Philadelphia
Watt G (2009) Equity stirring. The story of justice beyond law. Hart, Oxford
Watts M (2014) In: Cloke P et al (eds) Introducing human geographies. Routledge, London and New York, pp 391–412
Welby J (2018) Reimagining Britain. Foundations for hope. Bloomsbury, London
Wilson S, Kornblut AE (2011) Obama: 'The world is safer . . . because of the death of Osama Bin Laden. The Washington Post, May 3
Witte J Jr (2007) The reformation of rights. Law, religion, and human rights in early modern Calvinism. CUP, Cambridge
Wolfe P (1999) Settler colonialism and the elimination of the native. Cassell, London and Washington
Wood T (1702) A new institute of the imperial or civil law. London
Wood D (1992) The power of maps. The Guildford Press, New York
Wood D (2010) Rethinking the power of maps. The Guildford Press, New York and London
Wordsworth W [1802] (2013) Preface. In: Wordsworth W, Coleridge ST [1798 and 1802] (2013) Lyrical Ballads 1798 and 1802. OUP, Oxford, pp 95–115
Wright R et al (2018) Spatial technologies of racialized knowing: on visuality, measurement, and the law. In: Coleman M, Agnew J (eds) Handbook on the geographies of power. Edward Elgar, Cheltenham and Northampton, MA, pp 64–80
Xifaras M (2016) The global turn in legal theory. Can J Law Jurisprud XXIX:215–243

Part III
Territorial Nomenclatures

Chapter 6
Legal Geography, Linguistics, and Borders

6.1 Absolute and Relative Belongs of Public Law

In the previous chapters I have explained how the territorialised paradigm is a 'disciplinary matrix' that assists us in identifying the operational rules of spatial production. Within such processes, the matrix was originally territorialised, turning the relation between human agencies and their territory into concrete *legal geographies* (see Sect. 2.6.2). As said, the matrix percolates processes of spatial production. Through the law of spatiality, the spatiality of law, and counter legal geographies, it undergoes unceasing regeneration; and spaces become 'places' (see Sects. 2.5, 3.1, 5.2.2, and 5.3).[1]

Part III of this book is dedicated to *Territorialised Legal Geographies.* The chapters in this Part apply this matrix to this type of legal spatialities. In so doing, they examine the lexicon of legal geography (see Sect. 6.3.1) and apply it to a set of territorialised paradigms, i.e. federal arrangements. Within these, the matrix is particularly patent (see Sect. 7.1). Territorialised logical spaces revolve around spatial reshuffling: physical apprehension, political appropriation, and creation of meaningful places. Once projected onto a given territory, the manufacturing of geographical (both natural and human) features is made functional to the imagined geographies of territorial communities.

These processes give relevance to what Maitland termed 'belongs of public law' (see Sect. 2.6.2). Communities project their desired politico-legal orders onto what they consider 'their' tract of the Earth's surface, and onto which they claim a title. The way groups *imagine themselves*,[2] conceive of the space-place-environment

[1] Delaney (2010), p. 59.

[2] The reference is Anderson's imagined communities: see Anderson (1991). See also Johnston (1991), pp. 69 et seq.

M. Nicolini, *Legal Geography*, Ius Gentium: Comparative Perspectives on Law and Justice 105, https://doi.org/10.1007/978-3-031-19410-8_6

nexus as regards 'their' tracts,[3] and display their spatial reshuffling through the law is nothing else but a belong of public law.

Like any other logical space, belongs are *absolute* and *relative* at the same time; and their being absolute or relative depends on the adopted geographical perspective.

From the angle of political geography, their absolute-ness is caused by Agnew's territorial trap; political authority makes the tract of land meaningful to itself to the exclusion of all other groups, communities, and political authorities. Even the flaws between territorial constitutional regimes and their operational rules are assessed against the absolute-ness of the territorial map (see Sect. 2.4.1). The conversion of imagined spaces into tangible territorial legal geographies is usually achieved by demarcating new boundaries and applying new operational regimes replacing the former ones. The matrix is always the same; it is the territorialised paradigm, whose application generates a new relationship between community and territory. In whatever way we label this relationship, the 'belong' takes (or mimics) the form of the nation state and performs similar functions.

Taylor's state-as-a-container theory puts the finishing touches on Agnew's territorial trap. In reducing the porosity of borders, the territorial trap facilitates territorial demarcation, boundary restrictions, and the design of the state national 'character'.[4] What is *absolute* is the container; boundaries can be altered, and territories secede or are occupied; but the territorialised paradigm remains. What changes is the content, i.e. the territory as a meaningful place; or, which is the same thing, the change in the meaning is a change in the idea of territoriality (see Sect. 2.4). Robert D Sack supplies a definition of territoriality that matches Taylor's theory, thus focusing on the relation between group and territory, whose porosity is reduced by the absolute-ness of demarcation. 'What I mean by territoriality', Sack states, is 'the attempt by an individual or group . . . to influence, affect, or control objects, people, and relationships . . . by delimiting and asserting control over a geographic area'.[5] In his words, 'Territoriality acts as a container or mold for the spatial properties of events'.[6] In such a definition the politico-geographical container already discloses hints of belongs of public law. Territory thus becomes the 'object to which other attributes are assigned'. As its meanings change, territoriality is a *relative* concept, which is reshuffled to fit into the space-place-environment nexus communities imagine in their own logical space. Evidently, 'place always involves an appropriation and transformation of space and nature', i.e. of both physical and human geography; and such a makeover is 'inseparable from the reproduction and transformation of society in time and space'.[7]

Logical space is *relative*, we must bear in mind, from the perspective of cultural geography, not from that of territorial communities. When asserting their exclusive

---

[3]Diener and Hagen (2012), p. 17.

[4]See Giddens (1985). See also Taylor (1994), p. 151.

[5]Sack (1983), p. 56. See also Sack (1986) and Johnston (1991), pp. 188–191.

[6]Sack (1983), p. 57.

[7]Pred (1984), p. 278.

relation with a given territory, human agencies usually turn their imagined legal geography into a concrete belong of public law. They code their spatial reshuffling through the law, demarcate their meaningful place, and make their well-bounded order of politics a new safe normative space. In other terms, they use what is absolute (i.e. the idea of the container) to turn what is relative (i.e. their imagined legal geography) into to an absolute relation between them and their territory. In so doing, they entrench it through the territorial trap and secure the 'material continuity both of the people who participate in that process and of any natural and humanly made objects employed in time-space specific practices'.[8]

The example of Cyprus is particularly fitting. The island is a place where a divide between two imagined legal geographies (i.e. the Greek- and Turkish-Cypriot ones) coexist alongside each other (see Sect. 2.4.1). Although this divide had always been a human geographical feature of the island, it became untenable when Greek-Cypriot movements attempted to introduce a new operational rule under the British rule: it is the *enosis*, i.e. the idea of politically uniting Cyprus with Greece.[9] Under the 1960 constitution, however, *enosis* was unconstitutional. As its Art. 185 stated, Cyprus territory was 'one and indivisible'.[10] As 'the indispensable foundation of the Cypriot constitution of 1960',[11] Greek Cypriots had to renounce it for the sake of the Island's unity. Though, a Greek-oriented coup took place on July 15th, 1974, and *enosis* with Greece was proclaimed. In other terms, Greek Cypriots used the 'Greek container' to turn their relative imagined legal geography into an absolute one, where their belongs counted as the belongs of Cyprus's bicommunal society. On July 20th, 1974, this led to the Turkish response; the north was invaded and occupied to 'protect' the Republic from unilateral *enosis* to Greece.[12]

The 1974 Turkish occupation of the northern part of the island split the state territory into two parts; as a consequence of this *taksim* (i.e. partition), the north fell under a new operational rule (the Turkish military authority), superseding the constitutional territorial regime. Again, the use of the state-as-a-container rule turned the relative (i.e. Turkish Cypriots' imagined legal geography) into an absolute belong of public law. The 1975 and 1983 Turkish Republics in Northern Cyprus created 'an indivisible whole with its territory and people'.[13] Northern Cyprus acquired a new meaning; the occupation led to the establishment of both an 'illegal' space in the form of the 1975 and 1983 Turkish Republics. It has also filled the

[8] Pred (1984), p. 279.

[9] See Brown (1967); Yilmaz (2010), pp. 39–40; Loizides and McGarry (2019), p. 45.

[10] According to Art. 181 of the 1960 Constitution, Cyprus accepted not 'to participate in any political or economic union with any State whatsoever'; likewise, 'any activity likely to promote, directly or indirectly, either union with any other State or partition of the Island' was prohibited. See Kedourie (2005), pp. 649–650.

[11] Kedourie (2005), pp. 649–650.

[12] Coufoudakis (1976), p. 469; Kaloudis (1999), pp. 7–8. See also Loizides and McGarry (2019), pp. 46–47.

[13] Arts. 2.1 and 11.2 Const. Turkish Republic of Northern Cyprus (1985).

territorial trap: a new frontier running along the 'Green Line' separates the illegally occupied north from the south.

6.2 Manufacturing Territorialised Legal Geographies

The manufacturing of territorialised legal geographies share several features with the law of spatiality, the spatiality of the law, and counter legal geographies.

The law of spatiality addresses how ideology normalises geographical features, which are subsequently coded through the law (see Sect. 3.3.3). From the perspective of cultural geography, the territorialised paradigm normalises the politico-legal discourse of a given community. It points to 'what takes place ceaselessly, what contributes to history in a specific context through the creation and utilization of a physical setting'.[14] Normalisation accounts for the physical apprehension of the land, whose natural geographical features become part of its cultural (and political) transformation into a meaningful place. This act of 'geo-graphing' (see Sect. 5.3.3) favours the enclosure of the *erreichendes Land*, i.e. the land of which communities want to get hold. The process applies the territorial trap and the state-as-a-container theory to create belongs of public law.

Processed to fit into the community's logical space, the spatiality of law projects the space-place-environment nexus onto territories. Not only do places (and their names) bear the meaning communities assign to them, but their significance is also embedded into their belongs. These reflect the culture-specific relation between the community and 'its' territory. The seal of the former is impressed into the grounds of the already *erreichtes Land* and subsequently entrenched into their culture-specific legal framework.

Embedded into legal devices making up the spatiality of the law, ideo-logical space becomes legally and culturally relevant. Communities try to enforce it through the law, thus triggering a critical legal geography. Spatial production becomes dialectical, depending upon the juxtaposition between domination and resistance. Through their belongs, human agencies impose their ideo-logical space of domination on other communities to replace former nomic settings with new ones conveying the cultural superiority of the colonisers. No room is left to native legal geographies.

I examined a variety of such spaces; periphractic, walled, and fenced ideo-logical spaces are all territorialised counter-legal geographies (see Sect. 5.3). Some of them are 'more' traditional. As Alexandre Kedar reminds us, counter-legal geographies usually take place in 'ethnocratic settler societies', i.e. political communities whose legal 'regime . . . facilitates the expansion of a dominant ethnic nation in an multi-ethnic territory'.[15] I shall consider what this entails in multi-ethnic federal

[14] Pred (1984), p. 279.

[15] Kedar (2003), p. 402. See also Kedar (2014), pp. 101–108.

arrangements (see Sect. 7.4.4). Suffice it here to examine the spatial production taking place under 'settler colonialism' (see Sect. 5.1.1). At the onset of European colonial empires, this was part of 'a strategy of ethnic migration and settlement' that aimed to 'alter the country's geography and structure' by arranging its politico-legal discourse along racial hierarchies. Due to their sanitising policies (see Sect. 4.2.2), Europeans refrained from 'mixing with indigenous populations and "inferior" groups', buried their legal geographies, impressed theirs into the fertile grounds on the colonies, and created societies 'based on a deeply ingrained patterns of segregation'.[16]

In *Empireland. How Imperialism Has Shaped Modern Britain* (2021), Sathnam Sanghera illustrates how racial lines were embedded in early twentieth-century India, 'with Europeans living, working and socializing separately from the people they colonised'.[17] He recounts what he terms an 'extraordinary story': 'an Indian servant dared to touch a memsahib [a British white lady], to warn her that she was about to step upon a krait [a venomous serpent]'. Although she acknowledged that, in so doing, the servant saved her life, the memsahib 'didn't like it all the same, and got rid of him soon after'.[18]

Settler colonialism is saturated with such narratives. Natives were marginalised in colonial Africa, Asia, and Latin America.[19] This meant disregarding both 'Indigenous peoples' belonging to a specific place' and their native titles, which acted as the repository—or, in Lee Godden's words, as the 'archive'—of their belongs.[20] In common-law jurisdictions, counter legal geographies of domination were even more complex, inasmuch as segregation often resulted—and still results—'in three major ethno-classes' that Kedar identifies as follows: 'founders, immigrants, and indigenous'.[21] In the U.S.A., for example, racial separateness avoided the prospect of indigenous peoples' resistance; their legal geographies were relegated within reservation areas. As 'Indians [were] not taxed' (they were not US citizens), they were irrelevant when apportioning representatives.[22] In the US formative period, immigrants were mainly African slaves. In apportioning representatives and direct taxes among states, they euphemistically counted as 'other persons'.[23] As a shameful constitutional practice, census had to add 'three fifths of' the blacks to 'the whole Number of free Persons', that is, the whites. After the Civil War Amendments and

---

[16] Kedar (2003), p. 403.

[17] Sanghera (2021), p. 38.

[18] Sanghera (2021), p. 39. The history is drawn from Wagner (2019).

[19] See Kedar (2014), pp. 103–108.

[20] Godden (2020), pp. 131 and 140 respectively.

[21] Kedar (2003), p. 403.

[22] See Art. I, s 2 cl 3 of the U.S. Constitution and XIV Amendment. Indians were eventually naturalised under the Indian Citizenship Act 1924. Reservations were spatially arranged by the General Allotment Act of 1887: see Delaney (2005), p. 142.

[23] See again Art. I, s 2 cl 3 of the U.S. Constitution. The text was altered by XIV Amendment.

till *Brown v. Board of Education of Topeka* (1954),[24] black people were segregated under an odd egalitarian doctrine, i.e. the 'separate-but-equal doctrine' that the Supreme Court favourably held in *Plessy v Ferguson.*[25] The ruling created a 'spatio-thought' for whites; and 'the legal regulation of social space [became] critical for the maintenance of American racial ideology'.[26] Segregation was not unconstitutional providing that everybody could have access to facilities on an equal footing, albeit along racial lines. The dictum in *Plessy* made 'the nomoscape of American slavery ... compatible with liberalism'.[27] At the same time, it provided the whites with a comfort zone, as well as with a 'safe normative space' of sanitation whose symbolism consisted in equal freedoms for everybody in distinct physical spaces.

Likewise, the white and racially dominated South Africa provided for a 'separate-but-not-equal' constitutional legal framework. The 1910, 1960, and 1983 Constitutions established a legal system based on racial separateness and limitation of black people's rights. No wonder that in the 1988 edition of their manual, Basson and Viljoen could boldly state that the whole of South Africa consisted in a series of interlocked ideo-logical spaces of domination: 'South Africa is in many respects a unique country and one of these is the fact that it differentiates on constitutional law terrain between people of different race groups to such an extent that one can talk of two constitutional law systems in one country: a systems for the Whites ... and a system of all the other race groups in the country'.[28]

The enforcement of apartheid legislation and policies sought to render 'black bodies undesirable to whites, and white bodies undesirable to blacks'.[29] The outcome was a legalistic and hierarchically organised legal system, within which whites and blacks were set at the top and at the lowest tiers of the caste-like system respectively. The separate-but-not-equal rule ended up with the transition to democracy (1994–1996). The result was 'the overriding importance of equality in the constitution itself and in the project of transformation envisaged by the [same] constitution.'[30]

When referring to ethnocratic states, Kedar focuses on Israel, whose legal geography of domination has been pursued through the enforcement of a new legal land regime.[31] In the aftermath of 1947, the progressive marginalisation of the Palestinian community was achieved by favouring the Jews through 'nationalization and Judaization of the land', a 'centralized control on it' performed by 'Jewish Institutions' (mainly the 'Jewish National Fund'), and a 'selective and unequal allocation of

[24] 347 U.S. 483 (1954).

[25] 163 U.S. 537 (1896).

[26] Forest (2001).

[27] Delaney (2010), p. 120.

[28] Basson and Viljoen (1988), p. 307.

[29] Sanders (2006), p. 102. See also Nicolini (2015).

[30] Grant (2006), p. 9.

[31] Kedar (2003), p. 422; Kedar (2011). For an overall examination of this new legal land regime see Holzman-Gazit (2007).

possessory rights' in favour of the settlers. By the 1960s, Kedar continues, 'approximately 93 per cent of the Israeli territory came into formal ownership and effective control of public and Jewish institutions aggregated together into *Israeli Lands*'.[32]

Perhaps most striking of all is what Irus Braverman tells us about the ecosphere of these contended territories. They are covered by trees acting as 'natural' markers of the spatio-thought of both Israelis and Palestinians. Two landscapes prevail 'along the mountainous strip that runs across the center of Israel/Palestine': 'conifer pine forests and deciduous olive groves'. They 'are not as random or natural as it first seems', though. Both landscapes reflect Palestine's critical legal geographies and the dialectics between domination and critique 'in both Zionist and Palestinian national narratives'. The pine embeds the legal geography of domination purported by Israel through a 'project of afforesting the Holy Land into a European-looking landscape'; olive trees symbolise 'Palestinian resistance'.[33]

Erasing and changing landscapes and trees are part of the dialectic between absolute-ness and relative-ness of territorialised legal geographies, inasmuch as they uphold 'the importance of *visibility* or *legibility* for establishing a legally recognized control over territory'.[34] This is made possible by Art. 78 of the 1858 Ottoman Land Code, which is still in force in the West Bank: 'every one who has possessed and cultivated *Miri* land [i.e. land held by public authorities] for ten years without dispute acquires a right by prescription . . ., and he shall be given a new title deed gratuitously'.[35] This article favoured nationalisation and Judaisation of lands formerly possessed by Palestinians in the West Bank.[36] Once seized, Palestinian lands became 'public'; Jews came and settled there, tilling it for the legally required time. The provision enforces legal geographies of domination and impresses a new meaning into these territories, creates new legal geographies, and encodes them in the landscape.[37] Palestinians critique and contest it. The dialectics goes on and on, since 'pines and olives not only inform each other's social and spatial identity but also interact and reify each other's strengths, acting like mirror images of each other in a tree carnival'.[38]

---

[32] Kedar (2003), p. 493.

[33] Braverman (2008), pp. 449–450.

[34] Braverman (2008), p. 452.

[35] The translation is drawn from Braverman (2008), p. 455. See also Delaney (2005), p. 124.

[36] On land law in Palestine under the British Mandate see Kedar (2011), pp. 936–939.

[37] On legal meanings implied in spatial production see, e.g., Delaney (2010), p. 5.

[38] Braverman (2009), p. 319.

6.3 The Linguistics of Legal Geography

6.3.1 Places: Meaning, Lexicon, and Grammar

The creation of meaningful spaces lies at the heart of spatial production. And the manufacturing of territorialised legal geographies is no exception to this rule. In Lefebvre's words, 'that space signifies is incontestable. But what it signifies is dos and don'ts, and that brings us back to power'.[39] But, as Delaney states, this also brings us back to the law;[40] like maps, it represents meaningful spaces. How this meaning is conveyed depends on the linguistic structure of the politico-legal discourse. Geo-graphing disseminates its legal meaning through vocabulary and phraseology; at the same time, it frames the interrelations between community and territory in terms of power.

The geographical construction of meaningful places entails a close interaction between legal geography and linguistics going beyond law and linguistics.[41] Here, I shall consider how and why words change meaning, and how geographical contexts impact on the use of legal language.[42] Linguistics studies how language is used to communicate ideas (that is, meaningful ideas) through the joint action of lexicon and grammar.

That legal geography requires its own lexicon is counterintuitive. To assert that *that* parcel of the world is its *belong* of public law, each community verbalises this idea through language. Spatial production has its own legal-geographical 'primitives' (see Sect. 3.2). Lexical items like placename policies, territorial identity, territorial constitution, borders, regional demarcation, and territorial alteration explain how communities imagine and arrange the space-place-environment nexus within their logical spaces. And they do it through this common lexicon. Not only do they account for how communities appropriate and transform physical and human geographies alike, but they also favour the transfer of meaning between different agencies (and their logical spaces).

The lexicon of legal geography makes us critically rethink the special relationship between law and geography, upholding the idea that it lies at the crossroads of several disciplines. This is apparent in bordering processes. Alongside human, physical, and imaginative geographies, their material scenario is supplemented, say, by linguistics, which discloses the dynamics of boundary-making.[43]

Linguistics is not confined to how meaning is expressed through grammar and lexicon. It also examines the processes of its transmission from one agency to another. Encoded in grammar and lexicon, meaning is conveyed through speech and writing; in legal geography, the act of geo-graphing is transmitted through maps

---

[39] Lefebvre (1991), p. 142.

[40] Delaney (2010), p. 4.

[41] See e.g. Mattila (2013) and Kurzon (1997), p. 121.

[42] See Kay and Allan (2015), pp. 70–91.

[43] For a theoretical and cross-disciplinary approach to bordering processes see Jones (2009).

and law. In particular, the law of spatiality makes physical (and human) geographical attributes meaningful through political-legal discourse (see Sect. 3.3.3). Maps do the same, since they 'allow the political community to render itself more place-like, thus bridging the phenomenological distance between the community's claimed territory and its subjects'.[44] The manufacturing of territorialised geographies lie in this exchange of meanings, which communities usually 'encode . . . in the landscape as facts on the ground'.[45] The meaning is communicated to the addressees through the politico-legal discourse, which acts as the grammar of legal geography.

Yet, the exchange does not occur in a vacuum. In territorialised geographies, the special relationship between law and geography deals with concrete processes of spatial production, where power, space, place, and the law are mutually constitutive. If we consider the two most important topics in legal geographical studies, namely the legal linguistics of places and the performativity of borders (see Sect. 6.4), we infer that these provide the linguistic evidence of spatial relations. Again, law and geography act first and foremost as performative practices that create normative spatialities, the meaning of which is pragmatically enriched by socio-geographical contexts.

6.3.2 *The Grammar of Legal Geography*

Performative practices lead us to a closer consideration of the grammar of legal geography. In embedding the meaning assigned to a territory, legal arrangements turn what is relative into the absolute community-territory relation. Their legal significance is normalised as a set of belongs of public law, which enforce the new meaning impressed onto the territory.

In linguistics terms, the law 'is not normally thought of just saying something' about that territory.[46] Quite the opposite; it has a perlocutionary attitude and possesses the rare ability to 'do [special] things with words'; that is, to 'produce certain consequential effects upon . . . the audience, or . . . other persons . . . with the design, intention, or purpose of producing them'.[47] As a series of speech acts, the law ascribes a legal meaning to things, whose performative practice has 'consequential effects' on the real world,[48] i.e. it assigns the real world a new meaning. Often, though, the meaning changes, or its significance is completely altered. Perhaps most remarkable of all is that such a perlocutionary force is able 'to impose compliance with' the knowledge of the world that a given community has legally established as

[44] Leslie (2016), p. 171.

[45] Blomley (2015), p. 168.

[46] Austin (1962), pp. 6–7. See also Conley and O'Barr (1988).

[47] Austin (1962), p. 101.

[48] Austin (1962), p. 102.

its own belong of public law.[49] Through legal discourse, the community makes its own knowledge the legal knowledge of reality. The perlocutionary force of legal speech acts prompts addressees to ascribe authority to the politico-legal structure, ultimately acquiring the status of the only (and legally) binding relation between that community and its territory.

The interrelations between legal speech acts and semantic knowledge of the world trigger concrete legal geographies. Their legal meaning is strictly related to contexts where the belongs are manufactured and situated. In normalising them, context acts as a 'background', i.e. a 'common knowledge between the conversational parties', which also helps us to decode what implicatures are hinted at in a specific legal geographical context.[50]

Yet, the process of making space a meaningful place is a wholly 'Historically Contingent Process'.[51] Delaney terms it 'nomoscapes' or 'nomic landscapes'; they 'are not simply occupied (as we might imagine figures in a landscape); they are *lived*'. Their legal meaning is 'socially constructed and ideologically informed'.[52] The performative force of the law rejects the possibility of its autonomy from the socio-geographical context in which it is uttered or performed. My point is this: if logical space is *relative* (and spatial production contingent), there might be cases in which flaws occur between the reality and what is made absolute through legal spatial reshuffling. Not only is communities' absolute relation with their territory challenged by secessions and annexations, but, as the case of Cyprus has demonstrated, new operational rules require new spatial manufacturing processes (see Sect. 6.3.1). Accordingly, communities must rearrange their belongs of public law if they want to take hold of their territory and exhibit attentiveness to the contingencies 'that may disrupt the order's operation' within their belongs of public law.[53] Tackling contingencies requires a constant exercise in legal geographical creativity to fit imagined legal geographies into the complexity of the real world.

Attentiveness to (human, political, and natural) geographical contingencies triggers the unceasing reshuffling of the place-space-environment nexus. The law turns the *relative* imagined legal geography into an *absolute* belong of public law, thus linking the 'reality to an imagined alternative', i.e. the real-word to the belong of public law.[54] The type of spatial manufacturing that the law provides accommodates the changes in geographical and human geographies alike to make them fit into the spatial reshuffling taking place in logical space.

This is possible because the law is an open texture (see Sect. 1.1). When conveying its prescriptive content, legal language becomes standardised to ensure

---

[49] See Didikin (2020).

[50] On pragmatic and semantic interpretation see Griffiths (2006), pp. 6–9; Marmor (2008), pp. 424–425.

[51] Pred (1984). See also Brighenti (2006), p. 70.

[52] Respectively: Delaney (2005), p. 1; Delaney (2010), p. 104.

[53] Mullender (2018), pp. 1020–1021.

[54] Watson (1988), p. 36.

'correct and unambiguous transmission of information', as well as 'maximum disambiguation'.[55] Every single provision may visualise and convey different legal meanings and contents. The features of the law may thus be pragmatically enriched by new factors, which impose a new sense on existing lexical items without the necessity of formally altering the law.[56] This gives expression to the view that, as the place-space-environment nexus changes over time, so too should logical space.

Ferdinand de Saussure's distinction between *langue* and *parole* explains such changes.[57] Whereas *parole* is the speech used by real people (the relative and contingent processes of spatial production), *langue* refers to 'the more abstract systems of linguistic signs'[58] encompassing the patterns of change and preserving the contingency of speech. Acting as the *langue*, the law adapts lexical items, mechanisms of affixation, phrases, and sentences to the new (and ongoing) contingent process of spatial production.

Take, for example, Art. 1 of the 1937 Constitution of Ireland. The Irish Nation has the right to 'develop its life, political, economic, and cultural, in accordance with its own genius and traditions'. This provision lays down 'a multi-layered representation of Ireland': 'outside the synthetic isolation of the law, [it] performs the symbolic role of representing the nation'.[59] The 'genius' is the shorthand of the unceasing regeneration of Irish people's relation with their Island. The spatial reshuffling has probably incorporated several counter-geographies of critique. I explained above (see Sect. 4.2.1) that St Patrick and Cáilte's joint survey of Irish placenames was a politico-legal response to English attempts to sever the relations between the Irish community and its territory. Reasserting Irish placenames conferred legal significance upon the physical geography of Ireland as *the* central aspect of its identity.

This also occurs in post-colonial environments, as East Timor. The open texture of the law has been pragmatically enriched by several factors. Firstly, the colonial factor. As Rebecca Strating puts it, 'the impact of 400 years of Portuguese colonialism' has been used to boost the 'political "imagining" of East Timor'.[60] 'Colonial borders', she continues, 'facilitated this emergence of an East Timorese identity (a name that reflects a position on the global map)'. Secondly, the Indonesian connection. Not only did the cartographical imaginative construction of East Timor provide 'the basis for its legal rights to self-determination', but it also favoured the formation of its distinctive national identity'. This was juxtaposed to Indonesia in the aftermath of its invasion on December 7th, 1975. Indonesians justified its annexation through the 'deliberate constructions' of a common insular past under Indonesian rule, as if the 'map of Indonesia' had extended over Timor

[55] Rissanen (2000), p. 120.
[56] Taavitsainen and Jucker (2015), pp. 5–7.
[57] See de Saussure (1914).
[58] Culperer and Kytö (2014), p. 4.
[59] Hanafin (2002), pp. 340, 341.
[60] Strating (2014), pp. 471, 473 and 490.

well 'before European imperialism'.[61] Thirdly, the Pacific way. Timorese people nurtured their own identity; it also 'legitimised claims of difference from Indonesia' and supplemented the East Timorese independence movement with a further 'ideological justification'. The politico-cultural differences between East Timor and Indonesia were accentuated by emphasising the 'Melanesian influences' in its culture, which indeed constituted a powerful 'political expediency of "thinking east" to the Pacific', heightening the gap between Timor and Indonesia's attempts to deliberately construct an allegedly common insular past.[62]

6.4 Legal Geographical Primitives

As a politico-legal discourse about communities and territories, legal geography, has its own lexicon (see Sect. 6.3.1). Likewise, territorialised spatial production has its own legal-geographical 'primitives'. In GIS, these are 'the smallest units of spatial information': points, lines, and polygons, on the one hand; pixels and voxels, on the other.[63] In legal geography, they verbalise the 'smallest units' whereby communities assert that *that* specific parcel of the Earth's surface is their own 'world', i.e. their belongs of public law. I shall consider place names (see Sect. 6.4.1), isoglosses and polygons (see Sect. 6.4.2), borders (see Sect. 6.5), and territorial identity (see Sect. 6.6). The processes of territorial demarcation and alteration will be examined in the subsequent chapter.

If primitives are useful in terms of spatiality of the law, the legal geographical ones provide a grid through which we examine spatial production. Their role is not confined to reproducing geographical information. Quite the opposite: they concur in manufacturing it. Contiguous polygons indeed seem to reproduce bordering processes, thus supplying a further argument in favour of the resilience of the 'territorial trap'. These 'building blocks' reinforce the ideological discourse underpinning the state-as-a-container theory fitting into the spatial reshuffling that is subsequently encoded as a series of belongs of public law.

In this ideological discourse about geography, what is remarkable is that legal geographical primitives are a shorthand for defining geographical artefacts (natural and geographical alike). As codes and symbols, placenames and borders, for example, stay in lieu of the reality they represent. As legal nomenclatures, they express political authority and belongs of public law. Like geographical primitives, they

[61] Strating (2014), pp. 471, 473 and 490. See also Anderson [1993] (2001).

[62] Strating (2014), p. 491 and Department of Foreign Affairs (1996), p. 13.

[63] See Borrough et al. (2015), p. 303. The former primitives are used in a 'vector data model', which records 'geographical data in the form of points, lines, and polygons (312). Pixels (2D) and voxels (3D) are used in a 'raster data model', which is 'a formal structure for recording geographical data in the form of tessellating units such as grid squares' (308).

distil real-word data: again, legal geographical primitives 'are never absolute' but are made absolute.[64]

Take, for example, the functional approach to boundaries. This is one of the criteria used to create 'units that are formed by boundary drawing to be viable units'. This means demarcating 'economically viable' units that are able 'either to produce for themselves what their members need for sustenance or to provide for them through trade in goods or services'. As if they were geographical primitives, the spatiality of law processes them as polygons and lines and identifies them as 'an optimal set of boundaries' irrespective of whether they match the belongs of public law.[65] This might be a normative (and abstract) statement, but it is useful as far as communities need to ideologically demarcate their belongs. Why do not use the functional approach (and its normative statement), if the viability criterion expands the scope of the land that the community actually controls? If the statement insists that the *erreichendes Land* is broader than that which signifies for the community, it follows that, due to this 'kind of nominalism', this 'feels perfectly free to draw' the line well beyond its human geographical border.[66]

6.4.1 The Linguistics of Places

The first legal geographical primitives are placenames and isoglosses. We may be tempted to draw an equation between them and geographical primitives, such as points and lines. To a certain extent, the equation seems plausible. From a cartographic perspective, placenames complement points on maps; the latter are usually displayed in italics or bold letters. A certain degree of scale is preserved: the bigger the font (and thicker the point), the more important the place; a variety of underlining techniques denote whether a place on maps is a capital city, a natural feature, or a cultural site. Lines are usually employed as geo-graphing markers of rivers, coasts, highways, and (evidently) borders. Like the belongs of public law, these signs on maps represent a relation between logical space and reality. '*Maps are about relationships*', Daniel Wood reminds us. 'What could be signified by any system of distributed dots, or branching lines, or nested lines? *Not much.*' Like legal geographical derivatives, indeed, they may undoubtedly acquire meaning: 'if juxtaposed with a sign system that we could recognize, or furnished with a nomenclature that allowed us to supply that system, they could' signify.[67]

Yet, the equation does not tell the whole truth. Once displayed on maps, legal geographical primitives represent the outcome of the spatial reshuffling that has taken place in logical space. We understand what the map signifies because its

[64] Espenshade and Morrison (1982), p. 73.

[65] Miller (2016), pp. 37–38.

[66] Wood (1992), p. 25.

[67] Wood (2010), pp. 103–104.

value-laden meaning has been normalised and conveyed through an agreed lexicon. Despite this, though, we do not know what lies beneath dots and lines. Are these ideo-logical spaces of domination? Have the demarcated areas previously been cleansed of specific ethnic groups by simply erasing their belongs of public law? This is the case of areas where genocides took place. In the aftermath of Bosnia and Herzegovina's independence (1991), for example, ethnic cleansing took place in the *Republika Srpska* from 1992 to 1995. This allowed 'Serbs to establish majority control over [their] territory and to hold a referendum of secession' from Bosnia and Herzegovina to join the Federal Republic of Yugoslavia. Bosniak Muslims were compelled to leave the country: 'from there to physically eliminating those who would not comply, there were only too few steps'.[68]

The example of Bosnia and Herzegovina testifies to the difficulties related to placenames. When it comes to territory, the idea of reducing legal geographical primitives to dots and lines is possible, even at the cost of human life. Though, the idea of precisely bordering spatial production is controversial: the limits between belongs of public law are often blurred, and the notion of 'sharp edges of homogenous territories' is replaced by a less 'cartographic vision of territory'.[69] This is not to impose the application of 'areal' order instead of 'firm frontiers', which are in turn bound up with the existence of 'assertive states.' It seems that, in logical space, the spatial reshuffling still meets this areal pattern; but, as it does not exactly know up to where it will be able to assert its 'demands for resources' and control over territories, it prefers to remain conversant in lines and figures in order to 'create [a] first line of defence'.[70]

As a second major point, dots and lines acquire meaning because they are complemented by placenames. These are undeniably the most relevant features when it comes to manufacturing territorialised legal geographies. The very act of naming is indeed powerful. Following in the footsteps of Paul Carter, Lauren Benton defines it 'as the activity that most clearly [blends] the ambition of geographic mastery with the production of a kind of knowledge inseparable from movement through space'.[71] As the descriptor of a given tract of land, it attains the whole congruence between logical space (and its spatial reshuffling), imagined geographies, and belongs of public law. In addition, it is the first and foremost connotation of the relation between territory and community. As a descriptor, it is then inherently related to the context where the belongs of public law are spatially situated. Once placenames are impressed into territory, the legal discourse about them makes communities turn the knowledge they have of their territory into its new legal meaning.

This also accounts why, as Michael F Goodchild states, the 'act of naming' is now rooted in official authority'; the 'renaming of geographic features [also] requires a

[68]Zahar (2019), p. 25.

[69]Delaney (2005), p. 64.

[70]Black (1997), p. 123.

[71]Benton (2010), p. 16. See also Carter (1987), pp. 31, 15.

lengthy process of review'. In the contemporary world, 'Geographic naming has been centralized and standardized'.[72] Since it embeds its territorial identity, the community usually 'self-identifies with the place name', which acts as the visual representation of the belongs of public law. This also explains why placenames indexes are in the hands of the political authority. Owing to their being part of legal geographical arrangements, their changes must comply with legal procedures. This is apparent in federal and regional countries, where the alteration of units' placenames is entrenched in the constitutions and must therefore follow a burdensome procedure to be altered (see Sect. 7.5).

Federal arrangements disclose that the equation between legal geographical primitives and geographical ones is not the whole story. The topological descriptor conveys only normalised placenames, which usually correspond to the outcomes of spatial reshuffling triggered by the dominant ideo-logical space within a given territory. Evidently, this means that, in the fight among territorialised ideo-logical spaces, some processes of legal spatial production remain contingent and are superseded by places meaningful to the winners. Within territorialised legal geographies, as is evident, 'Names are given intentionally, to impart a certain meaning . . . Yet, over time, people can fail to remember the original, specifically intended meaning and attribute other ones'. The placename 'becomes opaque and the original meaning is lost over time, the name comes to feel like a word'.[73] Let us consider some examples of English placenames. Remnants characterise, for example, the Celtic substrate in the geography of Anglo-Saxon England;[74] and Manchester, Worcester, or Doncaster (to name just a few) are remnants of the Roman rule; the descriptor is moulded into the soil since it was inherently related to the contexts where previous belongs of public law had been manufactured, i.e. the Roman *castra* (i.e. military camps) situated in the island. Several other toponyms are impressed into the English soil, thus revealing the presence of former belongs of public law. Let us consider, the nouns *by* and *thorp*. Whereas the latter derives from the Old English *ðorp*, the names of villages that contain the former show a different origin. *By* comes from the Old Norse *byr* (place where people dwell, hence village) and points to the Danelaw established in 878 AD in Northern England (see Sect. 6.4.2). These dwellings comprised of 'large groups in which they tend to be concentrated'; this points to 'the conditions of an age when the Danish settlers in England still felt themselves strangers in a hostile land'.[75]

The *Geography Fabulous* of Ireland supplies a further example. We have seen that, in *The Acallam na Senorach*, the tour undertaken by St Patrick and the noble

[72] Goodchild (2007), p. 121.

[73] Radding and Western (2010), pp. 395 and 396 for placenames that underwent a change in their connotation (e.g. Le Havre, Newcastle, and Buffalo). See also Sack (1986), p. 5.

[74] See Townend (2012), p. 80. On interactions between Old English and Old Norse in placenames see Baugh and Cable (2013), p. 94. On diatopic variation in English word geography see McIntosh (1973).

[75] Stenton (1942), p. 12.

warriors of the Irish mythical past constitute the literary and cultural reaction to the oppressive English colonial rule (see Sect. 4.2.1). This rule effectively threatened the Irish national territorial identity. Besides replacing Irish placenames, Elizabeth I attempted to force the reformed English liturgy upon the Catholic country; Cromwell repressed rebellions, forfeited Irish land, and subsequently parcelled it among English colonists (1641); William and Anne tried to suppress the Catholic religion. The Union with Great Britain (1800), the failures of the home rule (1886, 1893, and 1912–1913), the declaration of independence (1916), the *Dáil* Constitution (1919), the creation of the Irish free state largely dominated by the English (1921), and the 1937 Constitution reflect the constant rearrangement of the Irish territorial identity, as well as the endless mediation of the cultural and literary texts. The formation of Irish national identity asserts 'the unbreakable links between the Irish people and the land of Ireland'.[76] This is evident in the case of Geoffrey Keating's (or Seathrún Céitinn's) *Foras Feasa ar Éirinn* (*Foundation of Knowledge on Ireland*: c. 1634).[77]

The book was written in response to another process of spatial manufacturing triggered by the 'the wave of "New English" Protestant settlers who came to Ireland in the fifteenth and sixteenth centuries'. This led to a harsh confrontation between several imagined legal geographies; the Protestant ones, those of 'the "Old English" descendants of the Anglo-Norman invaders, such as Keating himself, and [of] the Gaelic, often called "mere," Irish populations'. In so doing, the *Fora Feasa* were also able to use, among other things, Ireland's placenames 'to counter British colonialism and assert an independent Irish national identity; it is also a reliable source text that can be cited in response to British claims about Ireland.[78]

That the topological descriptor tells only the normalised part of the whole history is also apparent in the case of the imagined legal geographies. I have already said that places may be contended, claimed, annexed, or occupied. In all these cases, an operational rule is created to challenge, at least at a politico-legal level, the belongs of public law (and the outcomes of spatial reshuffling) of the dominant ideo-logical space within a given territory. We have seen that Argentina's imagined geography still refers to the Falkland Islands (*Islas Malvinas*), South Georgia (*Georgias del Sur*), and the South Sandwich Islands (*Sandwich del Sur*). Likewise, 'Antarctandes' styles the production of Argentina's natural geography extending the Andes Mountains underneath the Drake Passage into Antarctica (see Sect. 4.3.2).

Ultimately, placenames disclose the presence of flaws between constitutional regimes and operational rules. In conveying the ideo-logical space of the dominant groups, they reveal how several legal geographies (such as those of the colonised) have been silenced. Indeed, power often changes 'names to gain control of a people', and 'a new government can choose names that aid in the establishment and legitimation of a state'.[79] The case of Bosnia and Herzegovina is complemented by the

[76] Connell (2011), p. 6. See also Cunningham (1986).

[77] Keating (1902).

[78] Connell (2011), p. 27.

[79] Radding and Western (2010), p. 402.

case of Uganda or the Berbers. Minority groups (religious and ethnic alike) have been silenced in all continents. In Italy, for example, the Fascist government entered into the 1939 South Tyrol Option Agreement with Nazi Germany to give the German- and Ladino-speaking minorities located in South Tyrol the right to move north of the Alps. These groups had the option either to relocate to Germany or to remain in Italy at the price, though, to be assimilated. The process of Italianisation was already in place; and all the German and Ladino toponyms were replaced with Italian ones. The same occurred, say, with the Italians in Croatia and Slovenia in the aftermath of WWII, with the Catalans, Basques, and Gallegos in Spain under Franco (1939–1975), in Eastern Europe and in Kurdistan under their authoritarian regimes.

There is, finally, a further hypothesis where placenames and identities overlap. We have already encountered it when examining imbrications, i.e. overlapping ethnicities that encroach upon each other on the same territory (see Sect. 2). This overlap is relevant as far territorial disputes are concerned: territory is, indeed, the major constitutive part of ethnic identity. Imbrications give rise to a paradox, which discloses how complex the relations between law and territory may be.[80] The stronger the territorial identity of the ethnicities is, the deeper these imbrications will be. Therefore, each ethnicity will claim an exclusive relation with the disputed territory, thus marginalising other ethnicities. This divide makes conflict resolution unfeasible. It follows that the accommodation of imbrications is likened to how human agencies reshuffle the concept of 'territorial identity', which rests on linguistic, religious, and ethnic cleavages, and presupposes a geographical interrelation between communities and territories. Genocides, occupations, secessions, and counter legal geographies of domination disentangle the puzzle by spatially reshuffling imbrications. Groups whose legal geographies have succumbed are silenced, buried, or swept away. And their placenames are rubbed out of the landscape.

In other cases, the belongs of public law rest on different policies. Rather than excluding other ethnicities, maps reveal that a dot (i.e. a place) is usually complemented by two or more placenames, each of which conveys the relation between several ethnic communities and territory they share. On the one hand, this policy upholds that it is possible to accommodate imbricated, and often extremely divided societies, through arrangements establishing multi- or bi-ethnic constitutional frameworks (see Sect. 7.4.4). Constitutional designs, in other terms, usually rest on power-sharing mechanisms, through which consociational participation and geographical cohesion are enhanced. On the other hand, this reveals that the belongs of public law may reflect imbrications in the multiplicity of placenames as an indissoluble feature of a given territory. Consequently, spatial production must take them into account. In accommodating territorial and ethnic divides, ethnic-based units are usually combined with linguistic policies, which acknowledge the co-officiality of several languages at the federal-national and subnational levels. Preserving imbrications through multilingual placenames maintains territorial and

[80] See Palermo (2016).

political integrity by holding together diversity and including different ethnic belongs of public law in a single legal spatiality.

6.4.2 *Getting Hold of Places. Lines, Polygons, and Isoglosses*

A further legal geographical primitive is the line which joins points on maps, such as the tops of the mountains or discrete military strongholds constituting a line of defence. They also represent natural features (such as rivers), artificial ones (such as railways and highways), and borders. There is, though, a stark difference between the line acting as a geographical primitive or as a legal geographical one. Whereas the former is usually a geometric straight line defined 'by at least two pairs of XY coordinates',[81] a line on a map may have a variety of shapes: straight, curved, or broken, as the case might be. Differences between types of lines are attenuated when they are employed to create polygons, which consist of 'a series of lines joined to create an areal unit'.[82]

The notion of polygon is useful in terms of spatiality of law. When it comes to creating (and cartographically representing) delimited areal units (such as lakes, ponds, and borders), it provides us with a geographical (and rational) explanation of how contingent processes of spatial production become absolute. It offers a mathematical argument in favour of the state-as-container theory, the linear borders of which account for the politico-legal construction of the territorial trap. At the same time, it is fitting to explain the cartographic representation of borders.

The polygon has two characteristics. Its first one makes it the template for all types of bordering processes, irrespective of their being based on natural, human, economic, and political features. In maps, the presence of polygons (which are abstract lines) allows states to select only those features that are functional to their geopolitical goals. Although 'borders began in nature', they 'were elaborated by humans'; natural borders are processed 'as the result of man's progress . . . [by] the removal of natural obstacles'.[83]

The same holds true for artificial boundaries, such as the spheric ones Europeans used for demarcating territories in America and Africa. The 1818 Oregon Treaty set the 49th parallel as the boundaries between Canada and America—and the second Oregon Treaty signed in 1846 extended it to the Strait of Georgia and the Pacific Ocean. As for Africa, I have noticed that 'astronomic' and 'mathematical lines' prevailed over 'geographical features',[84] disregarding human traits. Boundaries were both unfamiliar to African legal conceptions and incompatible with traditional societal organisation, which was built upon family settlements and non-territorial

[81] Borrough et al. (2015), p. 305.

[82] Borrough et al. (2015), p. 307.

[83] Respectively: Crampton (2010), p. 153; Dominian (1917), p. 327.

[84] Abraham (2007), p. 163 also quoted at Sect. 3.

arrangements. However, colonial policy disregarded borderless, communal arrangements: since they were divided among different states, communities were arbitrarily separated and subsequently merged with other groups to create political entities and territorial jurisdictions.[85] In the aftermath of WWII, decolonisation confirmed the 'territorial trap' into which European powers had led the world during the colonial era. Former colonial empires split into several nation states; groups started claiming the creation of territorial states based on ethnic grounds. Odd as it may seem, when the representatives of the newly independent African states met in Addis Ababa in 1963 to create the Organisation of African Unity (OAU), they immediately conformed to the status quo.[86]

The second characteristic points to how lines (and polygons) have some bearing on the framing of belongs of public law. Their relevance is patent as regards how cultural attributes are processed in spatial production. These signs on maps represent a relation between logical space and human geographical reality, which leads us to add a new primitive to our toolbox; it is the isoline, i.e. 'a line which joins points of equal value'.[87]

What counts as 'equal value' is dependent upon how signs, cultural attributes, and natural/human features spatially situated in a given territory are processed to become part of a meaningful space. In reshuffling spatial furniture, each logical space assigns a value to the reshuffled entities. This is apparent as regards borderland zones. These areal units are palatable to logical spaces because they may become places 'to be got hold of' in terms of both physical apprehension and political appropriation. Borderlands have been addressed in legalistic terms through the application of both the state-as-a-container theory and the territorial trap. These areas have been cordoned off as if they were in watertight compartments; their borders and adjoining districts have been turned into frontiers asserting the existence of a state. In other terms, the value of isolines coincides with that which logical space assigns to its imagined limit, whose border, as we have already said, primarily acts as the first line of defence of its meaningful place.

This first line of defence is usually related to the cultural (primarily linguistic) group identity. Territory is first and foremost the place where the language of the community is uttered. Language is a cultural marker of a place already made meaningful; it has been *erreicht* (i.e. got hold of) because, *there*, the language of the group has established its belong—at least in cultural and linguistic terms. Indeed, 'the limits of a population's language mean the limits of their place . . . at the same time . . . the limits of their place mean the limits of their language'.[88] No wonder, therefore, that the simplest definition of polygon is based on a qualitative element, that is, its being 'a homogeneous representation of a two-dimensional space'.[89] If we

[85] See Griffiths (1986); Black (1997), pp. 137–138.

[86] Art. 4.b of the 2000 Constitutive Act of the African Union does the same.

[87] Borrough et al. (2015), p. 304.

[88] Pred (1984), p. 285. See also Delaney (2005), p. 43; Barth (1998).

[89] Borrough et al. (2015), p. 33.

want to use the territorial-trap metaphor, the isoline gives consistency to the two-dimensional place where the language of a given group is actually spoken. It is the limit of the linguistic container, whose content is manufactured in terms of the cultural and linguistic belongs of a given community. It is a short step from language to nationalities and nation-states.

There are several relations between language and nationality as a means for drawing politico-legal boundaries. This is manifest in the linguistic studies of placenames, i.e. an intriguing research into the remnants of former communities and the merging of different identities (see Sect. 6.4.1). Isolines and polygons also saturate geographical linguistics, which draws boundaries, i.e. isoglosses, which may be used as preconditions upon which borders be demarcated. This practice is particularly studied in English linguistics; geographical linguistics draws isoglosses-as-boundaries in dialectology, which 'will not commonly coincide or bundle together with one another in such a way as to define a single firm and satisfactory dialect boundary'.[90]

Let us consider, for example, Scottish identity, in relation to which linguistic and geographical features overlap. Besides severing Scots from English, the geo-linguistic Scottish frontier is related to the formation of the Scottish nation. Firstly, the boundary drawn between England and devolved Scotland—which once 'stretched from the Humber to the Forth, but not further North'[91]—severs Scots (a variety derived from Old English) from English.[92] Secondly, the geo-linguistic frontier is related to the formation of the Scottish nation: 'the freeholders and whole community of the realm of Scotland were [thus] associated . . . with the struggle of independence' fostered by the Scots' folklore and literary works.[93] Thirdly, linguistic markers and variations draw geo-linguistic boundaries in the British Isles—among them, the 'rhoticity' (i.e. the pronunciation of /r/ after a vowel where it is present in the written word); a different pronunciation for 'wh-', which is indeed pronounced [hw] such as in 'when' [hwɛn]; the use of [u:] in /au/ words such as 'house', which is the sound before the onset of the Great Vowel Shift, which did not fully take place in Scotland.[94]

Scholars with a background in geographical linguistics have also completed linguistic atlases such as the *Linguistic Atlas of Late Mediaeval English*.[95] Linguistic variation[96] also draws boundaries in England and in the United States. There is a *Linguistic Atlas of England* and several atlases related to US word geography, which

[90] See Burrow and Turville-Petre (2005), p. 6. The pioneer of linguistic geography was Matteo Bartoli: see Bartoli (1925).

[91] Davies (1995), pp. 3, 8, 13 note 48.

[92] Smith (2005), pp. 9–10. See also Watt et al. (2014).

[93] Davies (1994), p. 5.

[94] Upton (2012), p. 386.

[95] McIntosh et al. (1986).

[96] See Beal (2011); Upton (2012), pp. 379–380.

'delimit the main speech areas ... by finding the boundaries of particular words and expressions' and therefore by detecting clear isoglosses.[97]

6.4.3 Overlapping Geographies of Law and Linguistics

Legal and linguistic geography often overlap. In a diachronic perspective, I consider the 878 Treaty of Wedmore between Guthrum, the Danish King, and Alfred, King of Wessex. In it, they asserted their respective 'belongs of public law' on a specific territory and established a closer connection between land, community, and law. It is the Danelaw, i.e. the 'territory ... subject to Danish law'[98] demarcated through a boundary running roughly from Chester to London, which outlined a legal-geographical relation between territory and community corresponding to the 'area to the north and east of the old Roman road known as Watling Street'.[99] The Danish 'belongs of public law' defined the legal relationship between territory and community, and comprised the single constitutive parts of the legal-linguistic geography of Danish rule: placename policies, linguistic borrowings, a legal system, and boundaries delimiting the area of the same Danelaw. The border drawn at Wedmore did not cancel the areal conception of territory. Indeed, it is still possible to detect several influences on placenames in those areas outside the Danelaw, but that were closer to the frontier, such as Northumberland and Durham. They 'lay outside the Danelaw, but they contain a number of names which indicate sporadic Scandinavian settlements of the ninth or early tenth century'.[100]

Legal and linguistic geography overlap in a synchronic perspective, too. Take, for example, Belgium, which experienced a progressive transformation into a federal state in the second half of the twentieth century. To accommodate the Flemish- and French-speaking communities, administrative linguistic borders were drawn in 1878, 1921, and 1932, thus creating three 'région linguistiques': the *région de langue néerlandaise*, the *région de langue française*, and the *région bilingue* in Brussels, where the communities coexisted alongside each other.[101] The cantons of Eupen, Malmédy, and Saint-Vith were separated from Germany and annexed to Belgium in 1919 as a war compensation. This meant creating a new linguistic area within the bilingual Belgium.[102] The current linguistic border was eventually set in

[97] See Orton et al. (1978), Kurath (1941), Kurath et al. (1939–1943), Kurath (1973) and Kurath and McDavid, Jr. (1961–1983).

[98] Baugh and Cable (2013), p. 89.

[99] Townend (2012), p. 81.

[100] Stenton (1942), p. 3.

[101] See the following Laws: 22 May 1878; 31 July 1921 (in force from 1 January 1922); 28 June 1932. See Rillaerts (2010).

[102] Art. 33 to 35 of the 1919 Treaty of Versailles. See the *Arrêté royal* (Order in Council) 4 October 1925 that laid forth the linguistic and administrative regimes of these three Cantons.

1962–1963 and subsequently entrenched into the constitution (1970).[103] This border already led to the creation of new unilingual communities (French, Flemish, and German) in 1970, and three regions in 1980 (eventually established in 1988: the Walloon Region, the Flemish Region, and the bilingual Region of Brussels). Communities now meet 'the demands of the Flemish movement for linguistic, cultural and educational autonomy', whereas regions were introduced at the request of the Walloon people 'to attain autonomy at the social and economic level'.[104] Perhaps most striking of all is that these two different kinds of constituent units reflect different logical spaces and processes of spatial reshufflings. No wonder, therefore, that the same tract of Belgian land may be regarded as the territory of a given Community but also of a region. As Flemish and Walloon logical spaces partly overlap, so do the territories of communities and regions.[105]

Linguistic and legal geography overlap in several federacies. As I shall consider them in the following chapter, a couple of examples will suffice here. The first is drawn from Europe. In Switzerland, such overlaps are entrenched into territorial demarcation. Cantons may indeed be monolingual (Italian is spoken in Tessin; French in Geneva, Vaud, Neuchâtel, and Jura; German is spoken is the remaining cantons), bilingual (both German and French is spoken in Berne, Fribourg, and Valais), and trilingual (as in the case of Graubünden). Thus, Swiss linguistic geography coincides with subnational borders, i.e. with the legal geography of cantons. Consequently, their linguistic identity is entrenched in the legal geography of the Swiss container as a constitutive feature of both federal and cantonal belongs of public law.

The second relates to Africa, where the Federation of Nigeria reflects an unceasing process of territorial rearrangements through which it has been possible to accommodate discrete (and often contrasting) ethnic groups living within its boundaries.[106] The process of rearrangement is due to several crosscutting cleavages. The Federation of Nigeria was created before independence (1946). Three regions were carved out of what was then a British colony; the 'colonial-inherited three-regional federation was formally blind to ethnicity but paid attention to the country's three main groups.'[107] From 1946 to 1960, 'many groups of people were arbitrarily sandwiched into territorial units that formed a geopolitical entity called the state',[108] and minority groups were forced to identify themselves with these three states. Consequently, minorities expressed fears regarding 'the regional

---

[103] See the *Lois Gilson* 8 November 1962 and 2 August 1963. See Rillaerts (2010), pp. 55–87.

[104] Peeters (2007), pp. 33 and 35 respectively.

[105] Peeters (2007), p. 31.

[106] Similarly, India has experienced a constant process of territorial adjustment. See Sect. 7.

[107] The three regions (North, West, East, and the Federal Territory of Lagos) reflected the three major groups (Yoruba, Hausa, and Igbo). See Elaigwu (2002); Ejobowah (2010), p. 255; Egwim (2020), pp. 19 et seq.

[108] Alapiki (2005), p. 50.

governments, which in each region were looked upon, with good reason, as synonymous with the majority ethnic group'.[109]

Ethnic conflict and fears of secession led to the creation of new states, and 'the ... segmentation of Nigeria's political structure [increased] from three regions at independence in 1960 to four regions in 1963, twelve states in 1967, nineteen states in 1976, twenty-one states in 1987, thirty states in 1991, and thirty-six states in 1996.[110] Territorial alteration also achieved the following additional targets: (a) no states could dominate or control the central government; and (b) each state should form a compact, geographical area. This is confirmed by territorial alteration processes; enshrined in Sects. 3 and 8(1)(b) of, respectively the 1979 and 1999 federal constitutions, the creation of a new state must be 'approved in a referendum by at least [a] two-thirds majority of the people of the area where the demand for creation of the State originated'.[111]

Similar overlaps characterise the belongs of public law in Cyprus. Before the 1974 Turkish invasion, there was no regional demarcation, as Greek and Turkish Cypriots were spread across the country and co-existed in the same areas. The Constitution established only consociational mechanisms to accommodate Cyprus's bicommunal character. The geographical and demographic reconfiguration of the island that followed the Turkish occupation of the north divided Cyprus into two separate ethnicity-oriented communities, which ceased to be dispersed, and became territorially concentrated along ethnic, religious, linguistic, and geographical lines.[112] This led to the creation of internal boundaries; power sharing, albeit necessary, was no longer sufficient. The 'illegal' spaces and the 'double' frontiers thus created (the *de jure* border and the *de facto* one running along the Green Line: see Sect. 2.4.1) undermined the mutual recognition of both ethnicities implicitly acknowledged by the 1960 Constitution.

6.5 From Primitives to Borders

Overlaps between geographical primitives and legal geographical ones may be extremely productive, especially in the case of polygons. We have seen that these supply a further explanation for the 'territorial trap', marking the boundaries within which communities (and states) exert 'territorial control', 'political authority', and sovereignty.[113] Borders are useful for outlining belongs of public law, turning abstract and nominal lines into real-world legal geographical borders.

---

[109] Alapiki (2005), p. 53.

[110] Alapiki (2005), p. 50. On the support military regimes granted to the progressive creation of states see Ejobowah (2010), p. 259.

[111] See Suberu (1991).

[112] Duchacek (1977), p. 13.

[113] Wood (2010), p. 36.

Polygons must not be confined within the spatial reshuffling taking place in logical space. Shifting from primitives to real-world features is essential as far as bordering processes are concerned. Territorialised legal geographies grasp real-world knowledge to be subsequently coded as belongs of public law. Fantasising about abstract lines, in other words, leads but to academic speculations. The real world has its own complications. It is indeed difficult to draw a clear divide between the law and the non-legal variables affecting bordering processes. We have already noticed that borders embed a relationship between a community and its territory. And this connection between land, community, and law highlights the manifold interactions between territory and institutionalised communities.

Mine is a cross-disciplinary project; and the study of borders is not immune from non-legal interferences. When critically rethinking the 'special relationship' between law and geography, such interferences corroborate the idea that the legal geography of borders lies at the crossroads of academic disciplines.[114] In his *Territory: A Short Introduction* (2005), David Delaney lists them: international relations, geopolitics, sociology, ethnology, and anthropology consider the connections between ethnic groups, boundary-making, and cultural spaces.[115] Cartography is also crucial when drawing borders, triggering the triad of 'bordering, ordering and othering'.[116] The lexical item 'border' then reflects the dominant territorialised paradigm. Its thorough application is patent as regards the sea; we have seen that Parts V and VI of UNCLOS treat them as landmasses, partitioning and delimiting the Earth's water surface and the seabed (see Sect. 3.5.2).

Maritime demarcation confirms that borders, as political artefacts, entail a dialogue between law, geography, and social sciences. The term borders points to a composite of several distinct, albeit interlocked, forms of connection: the physical, the imaginative, and the politico-legal. Focusing on territorial borders, the physical connection points to our legal geographical primitives: it is a 'line' separating 'one country from another', 'two jurisdictions or territories,' as well as the part lying 'along its boundary' or at 'the edge of a country or territory.' These definitions are shared by linguists and geographers, as the *Oxford English Dictionary* and the *Oxford Dictionary of Human Geography* uphold.[117]

The term is interchangeably used with 'frontier', 'margin', and 'boundary.' 'Limit' and 'boundary' refer to the line marking the border.[118] 'Margin' designates the outcome of bordering processes; border may have a broader meaning; it comprises either the line and the bordering areas, thus referring to 'any type of division';

---

[114] For an interesting cross-disciplinary approach to borders see Newman (2006).

[115] See Delaney (2005), pp. 39–49.

[116] Tripathi and Chaturvedi (2020), p. 173. See also Sect. 1.1.

[117] OED (Border); Castree et al. (2013), pp. 38–39.

[118] See Castree et al. (2013), pp. 39–40. See also Boucher (1963), pp. 789–818; Munge Sone (2017), p. 326.

or 'a line-on-the-ground political division' acting as a political border;[119] or it conveys the idea of othering, separation, and identity.[120]

Likewise, border populate imaginative legal geographies, where it is replaced by 'frontier,' i.e. the line and the lands lying on both sides.[121] The processes of othering may refer to political appropriations functional to the belongs of public law. Arcifinous coasts, the Great Wall of China, and Hadrian's Wall, for example, gave physical consistency to the imagined lines separating the empires from unoccupied areas (see Sect. 3.5.1). During the eighteenth century, the U.S. frontier stimulated the forging of the U.S. constitutional identity.[122] And the Scottish border designates the line, and its adjoining districts, suppressed by the Union with Scotland Act 1706 (in force from 1707) but reinstated by the *Scotland Act 1998*. From this it did not follow that Scots ceased to narrate (i.e. imagine) their own identity. The Scots deliberately preserved the ideal frontier with England. In the fourteenth century, John of Fordun's *Chronica Gentis Scotorum* (1363–1385) had already depicted 'an almost immemorial Scotland which stretched from the Tweed, or even a more southerly river, to the Orkneys.'[123] The renovation of Scottish constitutional identity now rests on several imaginative materials: Scots law, the National Church of Scotland (the Kirk), and Scots, which is one of the longest-attested varieties derived from Old English.

Finally, border is a synonym of borderland. Borderlands coincide with 'state borders.' For instance, in the parlance of the USA, the term points to states bordering on Canada.[124] Borderlands disclose the idea of othering, separation, and identity. Borderland regions usually belong to a state but are inhabited by people identifying with 'rival political communities', which means creating areal belongs of public law.[125] Take Northern Ireland, which the Irish Border severs from the Republic of Ireland. Its relations with the EU are now regulated by the EU (Withdrawal Agreement) Act 2020, whose Ireland/Northern Ireland Protocol prevents a backstop as a result of the UK leaving the EU.[126] The way in which it has been handled so far is being addressed in legalistic terms, as if borderland areas were in watertight compartments and their borders were lines and figures grounded in immutable frontiers. This imagined border does not exactly know where each country could assert its identity, but it is nonetheless functional to the process of two distinct, albeit interlocked, Irish and British legal spaces.

Borders and its synonyms share a common feature: all actively contribute to making contingent processes of spatial production absolute, turning imagined legal

[119] Jones (2009), p. 180.

[120] Newman (2006), p. 154.

[121] Castree et al. (2013), p. 166.

[122] Delledonne (2020).

[123] Davies (1995), p. 8.

[124] Blocher (2014), p. 243.

[125] Moore (2015), p. 123.

[126] de Mars et al. (2018) and de Mars and Murray (2020).

geographies into *the* only legally acceptable relation between communities and their territory. This is an outcome of legal geography as a performative practice. As said, this performativity takes the 'spatialities of law seriously'; belongs of public law cease to be 'a closed, formal, and acontextual system and see it instead as an assemblage of heterogeneous elements'.[127]

The performativity of territory thus extends over to borders. In this lies the most relevant difference between polygons and bounded political communities. Whereas the former merely demarcate areas in a way that is functional to the territorial trap, the latter concretely take part in the process of territorial identity to be subsequently enshrined into a belong of public law.[128]

Again, Cyprus is a case in point. The 1974 partition of the island turned two dispersed communities coexisting on the same territory into two territorially separated groups. This process led to a paradox; in disentangling the ethnic imbrication, the partition generated a territorial divide—and *new boundaries.* Cyprus now has different performative borders: expressing the legal geography of Cyprus, the *de jure* one makes the operational rule match the constitutional regime of the island. There is then a *de facto* border running along the 'Green Line', which is the visible expression of the flaw between a territorial constitutional regime, its ineffective operational rule, and the creation of two distinct, albeit interconnected, belongs of public law. However, the Green Line does not convey the idea, typical in international relations, according to which 'we are "sovereign" on this side of the line [the south] and they are sovereign on the other side'.[129] Nor has the line totally dissolved ethnic imbrications. Interferences, influences, and remnants of the original imbrication cross the Line—which is not in itself an obstacle to communication.[130]

Moreover, the border performs the function that was once staged by the imbricated territory: it shapes the territorial identity of both ethnicities. A legacy of the imbrication, it may be traced back to the attempts of both ethnicities to introduce a new operational rule capable of disregarding the constitutional regime. This also means that ethnic boundaries and the Cyprus Green Line convey a relational meaning: Greek and Turkish Cypriots define their territorial identity by mutually rejecting the interrelation between community and territory upon which the identity of the other ethnicity is established.

The Green Line thus matches a new belong of public law. It is the major manifestation of the operational rule that both ethnicities assert as the sole relation with Cyprus's territory. The Line thus makes power visible, and its spatial relation legally relevant. In addition, it physically draws distinct belongs of public law, which can be traced back to Turkish and Greek Cypriots' conceptions of their respective

[127] See Sect. 2. Quotations are drawn from Hogg (2002), p. 34.

[128] Diener and Hagen (2012), p. 8.

[129] Delaney (2005), p. 1.

[130] See Anastasiou (2002), p. 581.

relationships with Cyprus's territory.[131] Both conceptions played a central role in causing the flaw between the constitutional regime of Cyprus's territory and its operational rule. Greek Cypriots attempted to introduce *enosis* as the new operational rule; on the other, the Turkish occupation of the north represented the reaction to Greek-oriented *enosis*, as well as the achievement of *taksim*, i.e. the Turkish-Cypriot operational rule alternative to *enosis* and that called for the partition of the island.[132]

The Green Line triggered a transfiguration in the character of both ethnicities. Its performativity changed the claims, the belongs of public law, and the territorial identity of both ethnicities. Its performativity[133] is particularly evident if we focus on the so-called Buffer Zone, i.e. the inner part of the Green Line where 'the sovereign norm produced and enforced by the ... states [the 1960 Cyprus Republic, and the 1974 Republic of Cyprus and the 1983 Turkish Republic of Northern Cyprus] is suspended'.

The Green Line has generated three different 'belongs of public law'. There are the Greek- and Turkish-Cypriot ones, which are geographically demarcated by the northern and southern limits of the Green Line; there is the 'non-ethnically oriented' one, which lies within the Green Line and can override the paradox caused by the ethno-territorial divide.

6.6 Framing Territorial Identities

6.6.1 A Matter of Storytelling

In the previous sections, I examined how legal geographical primitives shape the spatiality of law and contribute to its production. We have seen how placenames, isoglosses, lines, and borders are used to make spaces meaningful. In so doing, their role in spatially fixing reality is pivotal; as legal nomenclatures, they convey the uniqueness of the relation between communities and their territories, i.e. a relation that political authorities encode as belongs of public law.

I have then pointed to the performativity of borders as a driving force that contributes to turning relative imagined legal geography into *the* relation between community and territory. Which means an absolute relation. In conveying the ideas of othering, separation, and identity, borders primarily act as first lines of defence of such meaningful places. Legal geographical primitives distil real-word data and territorialise legal geographies. On several occasions we examined how this occurred in Cyprus; in the aftermath of the 1974 Turkish intervention, the Green Line

---

[131] The two ethnicities 'understand Cyprus' place in history in radically different ways': Moulakis (2007), p. 532.

[132] See Kaloudis (1999), p. 6.

[133] See Green (2010).

triggered new meaningful places (see Sect. 6.5). Within the arcifinous coasts of Ireland, St Patrick and Caílte, toured the island to preserve Irish placenames. And the *Fora Feasa* did the same (see Sect. 6.4.1). Furthermore, the border between Indonesia and East Timor turned the post-colonial identity of the former into an absolute process of spatial production. This was made possible because, as Rebecca Strating observes, a 1914 Permanent Court of Arbitration award resolved a border dispute 'in favour of the Netherlands, and the final settlements in 1916 separated "Portuguese" (East) Timor from West Timor under international law'. These 'political borders' acquired a 'solidity' upon which 'contemporary understandings of East and West Timor' still stand.[134]

But: what do legal geographical primitives fix? What do borders protect as first lines of defence? What is fixed, I understand, is the creation of spatial, as well as legal, connections between a community and its territory; its relative (but turned-into-an-absolute) relation is nothing else that the concept of territorial identity, i.e. the politico-physical appropriation of a tract of land upon which the group's imagined community is projected. This self-identification as a territorial community is triggered by the reshuffling of several natural and human features; these may be economic, linguistic, religious, and ethnic.[135] Whatever the legal significance of these features may be, territorial identity presupposes a geographical interrelation. Hence, boundaries are their visible and concrete expression; together with their territorial units' denomination, they outline the territorial identity of the community, and this governs the formation of constitutional identity.

No wonder that human, political, and natural geographical contingencies trigger the unceasing reshuffling of the place-space-environment nexus. With its open texture, the law turns the *relative* imagined legal geographies into *absolute* belongs of public law, thus constantly accommodating changes in physical and human geographies within the reshuffling taking place in logical space. The texture of the law is pragmatically enriched by the contingencies of which territorial identity consists. Quite evidently, the law adapts to real-word primitives, as there is no need for formally altering it. The most intriguing example is, as said, Art. 1 of the Irish Constitution, where logical space, geographical factors, and their periodical reshuffling develop 'with its own genius and traditions'.

The open texture of the law turns territorial identity into what we lawyers term *constitutional (territorial) identity*. The constitution represents the higher law of the land, where the foundations of political communities are laid forth and its core values (their belongs of public law) are secured and preserved for future generations.[136] It is a composite of several materials, each of which has been craftily reshuffled to manufacture a specific territorialised legal geography. Impressing territorial placenames, isoglosses, and boundaries into territories is part of such a thick identity. Firstly, it asserts the legitimacy of placenames, for they are the linguistic evidence of

[134] Strating (2014), pp. 474–475.

[135] On the importance of these factors in comparative law see Grossfeld (1984), p. 1511.

[136] See Kumm (2012), p. 109. On the concept of constitutional identity see Jacobson (2006, 2010).

the spatial relations between territory and community. Secondly, the narrative of placenames testifies to the fact that territory and community are not separable. Thirdly, constitutional identity confers legal significance to the physical geography of a state as *the* central aspect of its identity, which turns physical geography into *legal geography*.

I have explained how such discrete materials are reshuffled in logical space to create belongs of public law. Besides physical features, constitutional (territorial) identity also processes materials stemming from the cultural tradition of the community. The reshuffle considers literary, historical, pseudo-historical, and antiquarian material; it may also regard histories related to the mythical past of the community, which makes the identity thicker. 'Old stories can retain a powerful hold', affecting 'the creation of law and its application' and mediating 'our perceptions of realities and truths'.[137] Stories are incorporated as values and traditions into constitutional provisions, from which interpreters will subsequently carve out constitutional identity and its legal significance.

This is what the Irish Constitution terms the 'genius' of the community. This is how constitution-makers establish a fine tuning with their societies. Constitution-makers purposely draw from the above-mentioned manufacturing tradition, and therefore uphold deliberate reengineering practices when they incorporate such materials into constitutional provisions. Reengineering 'involves the addition, amendment, or replacement of (at least) one cardinal element of a constitutional idea', which is central in identity-construction.[138] Constitution-framers consciously decide to use legal, literary, traditional, social, materials which preferably have their roots in an ancient past. In coding the territorial identity through the law, not only do they establish a first line of defence, but they also become storytellers of the constitutional identity. It is, thus, a matter of storytelling. Like poets, constitution-makers 'sing for [their] contemporaries' and recall to their 'minds well-known situations which could be conjured up by merely alluding to well-known events and personages'.[139]

6.6.2 Examples of Territorial Identities

The creation of constitutional territorial identity entails appropriating non-legal elements to match the imagined territorialised legal geography. Any geographical and cultural fact is, in sum, capable of been reengineered: the intent of constitution-makers acting as storytellers generates a process of spatial production that goes above and beyond time, real-world features, and the law. 'Mobilising the past' is a process of spatial production that corresponds to the 'manufacturing tradition'

[137] O'Donovan (1993), p. 90.

[138] See Stephenson (2013), p. 872. For more on these processes see Nicolini (2016).

[139] Williams (1926), p. 4.

typical in constitution-making processes. As Laurence Tribe has highlighted, 'the very identity of "the Constitution" – the body of textual and historical materials from which [fundamental constitutional] norms are to be extracted and by which their application is to be guided – is . . . a matter that cannot be objectively deduced or passively discerned in a viewpoint-free way'.[140]

The way identities are forged varies from place to place, taking the form which is functional to a given process of constitutional construction. Besides the case of Cyprus and East Timor, we have encountered a variety of processes whereby the relations between communities and territory are processed. In Ireland, the framers of the 1937 constitution acted as storytellers; the Irish Genius, as we said, went beyond time and law, and made it 'possible for the present to collect and even to converse with the venerable cultural identity of the past, an assertion that is the premise of the *Acallam*'.[141] This attempt aims to 'represent' a unified Ireland—that is, a nation in search of its wholeness, as Patrick Hanafin acutely points out.[142]

Further processes of spatial production are, for example, the belongs of public law generated, say, by the Danelaw, the Scottish and Northern Irish identities, colonial borders in Africa and Asia, and the landscapes opposing Israel and Palestine. We have also considered how the ethnic divide has favoured state demarcation in Nigeria. In the following chapter, I shall consider the impact of the place-space-environment nexus on federal arrangements.

The USA provides an additional example of how territorial identity may be forged. From a territorial perspective, it is an aggregative type of federalism; its territory originally coincided with that of the 13 colonies and was accrued by admitting new member states carved out of former federal territories (see Sect. 2.4.1; see also Sect. 7.3.3). Their admission may now be considered as completed in the continental United States. Its member states indeed occupy the entire middle reaches of the northern part of the Americas between Mexico and Canada, and there is only room left for the admission of new states that have an insular, i.e. a physically geographic demarcated, character, such as Puerto Rico.

The thick territorial identity of the USA stems from the reshuffling of human geographical attributes in logical space. Its territory has been treated as 'essentially neutral, that is, a blank slate to be filled in by whomever lives on the territory'. This territorial neutrality standardised the identity of member states as well.[143] The federal government shaped it and supervised the processes of territorial delimitation. The 'Congress [had] Power to dispose of and make all needful Rules and Regulations respecting the Territory [. . .] belonging to the United States'; states were carved out from those territories (*territorial clause*: Article IV, s. 3, cl. 2, of the U.S. Constitution). In addition, new 'States [might] be admitted by the Congress into

---

[140] Tribe (1983), p. 440. See also Strating (2014), p. 472; Coakley (2004), p. 531; Hobsbawm and Ranger (1983).

[141] Nagy (2001), p. 439.

[142] Hanafin (2002), p. 342.

[143] Kinkaid (2016), p. 565.

[the] Union' (Article IV, s. 3, cl. 1, of the Constitution), but admission implied the application of the criteria set forth in the *Northwestern Ordinance 1787*.[144] Finally, admission led to the conferral of statehood on the new states, and statehood implied the certification of state constitutions under the Republican Form of Government clause (or Guarantee clause: Article IV, s. 4, of the U.S. Constitution).[145]

We shall see that multinational federations do not rest on territorial neutrality, as their societies are divided along cultural, linguistic, and religious cleavages. Consequently, constitution-makers create storytelling by gathering fragments from the identities of the founding nations and accommodating them within the common framework of the territorial constitution (see Sect. 7.2.4). In Canada, both the English and the French participated as founding nations in the identity-building processes on equal footing. Quebec's distinctiveness was secured by the *Quebec Act 1791*, which demarcated the French-speaking provincial territory. Sect. 146 of the *British North America Act 1867* established different mechanisms for the admission of new provinces; at the same time, it maintained Imperial Parliament's responsibility for admitting new English-speaking provinces. This allowed Canada's English and French founding nations to live together in a multinational federation. The *Constitution Act 1982* introduced additional provisions to protect Quebec's identity: an amending formula for changing the extension of the existing provinces; the recognition of French minorities and aboriginal rights; an amending formula by unanimous consent in relation to the use of the English or French language. Quebec has an indirect veto on constitutional amendments and protects French linguistic minorities who reside outside its boundaries.[146]

The ethnic criterion was subsequently applied to carve the territory of Nunavut out of the Northwest Territories in 1999 and confer self-government on the Inuit people. It follows that the First Nations are also integrated into Canadian identity—their distinctiveness, however, was firstly legally stipulated by federal Parliament with the Indian Act 1951, and Sect. 35 of the Constitution Act 1982 subsequently recognised community-based aboriginal rights. Finally, the assemblage of ethnic identities is enriched by immigration, to which Canada has been devoting appropriate integration and multicultural policies enshrined in Sect. 27 of the Canadian Charter of Rights and Freedoms and in the Canadian Multiculturalism Act 1988. To sum up, assembled constitutional identities adapted as much as possible to the needs of the various cultural identities.

[144] See Duffey (1995), p. 942: 'Through the Ordinance, the states attempted . . . to create entities like themselves'.

[145] Lerche Jr. (1949).

[146] See Subsect. 42(e), Sect. 16–25, an Subsect. 41(c) respectively.

6.7 The Power of Naming: The USA and the EU

The way legal geographical primitives are legally processed to create meaningful spaces is particularly patent when it comes to territorial identity.

Take, for example, placenames. In purely legal terms, Michael Goodchild correctly reminds us that 'Modern naming in developed countries is closely regulated by a hierarchy of committees' that 'extends from the local to the national level'. He also adds that naming places—that is, making primitives meaningful to human agents—has undergone a process of normalisation: as it is part of how politico-legal discourse produces geographical derivatives, 'Geographic naming has been centralized and standardized, and assigns no role to' facts, individuals, events, or whatever political power deems insignificant, dubious, or even dangerous to the manufacturing of the belongs of public law, as well as to the legal coding of territorial identities.[147]

The 'renaming of geographic features' has also been highly standardised within federal arrangements. The 'lengthy process of review', naming, or denominational alteration must comply with burdensome procedures laid down in constitutional texts (see Sect. 7.5). Changing placenames is not impossible. Although they reveal belongs of public law, rRubbing them out of spatial production or complementing them with names in different languages must be normalised through constitutionally entrenched procedures, which fit them into the spatial reshuffling.

Yet, the history of the USA and the EU supranational integration account for processes where non-centralised and non-standardised geographical naming has been particularly productive. In 1507, for example, the cartographer Martin Waldseemüller was able to perform one of the major perlocutive acts in the history of Western world, whose consequential effects are still visible in terms of belongs of public law. Waldseemüller was able to draw 'an outline of a new continent and labeled it America'. Although he 'withdrew the name on a later map, the feminine form of Vespucci's first name stuck, and was eventually adopted as the authoritative name of not one but two continents.'[148]

Waldseemüller's process of naming has some bearing on the U.S. territorial identity, which counts as *constitutional identity*. We have already seen how it has been shaped during its history and evolution. Its formation reveals analogies and discrepancies with the creation of the EU's legal geography.

Like the USA, the EU may be considered an aggregative federalism, whose original demarcation coincides with the boundaries of its founding member states. Processes governing the gradual formation of the EU and the USA shed noticeable light on the legal geography of both types of federalism. With respect to the United States, the 'label' *America* designates the continent, and complements the United States' constitutional identity. Legal and linguistic geographies perfectly match. This is not the outcome of a standardised and centralised process of naming; rather, it is

[147] Goodchild (2007), p. 211.

[148] Goodchild (2007), p. 211.

the effect of a process of generalisation of the USA as the land, within the *Americas*, where individuals enjoy their freedoms. The admission of new states carved out of territories demonstrates this. The Portuguese equivalent for discovery (*descobrimento*) pointed to both logical space and spatial production; likewise, the production of new states allowed individuals to create meaningful spaces. In the newly produced spaces of the Western frontier, it was possible for individuals to enjoy fully their rights and freedoms.

By contrast, EU Treaties set a precise legal requirement for the admission of new states into the Union: its naming is centralised and standardised and assigns no role to facts or individuals. According to the lengthy process of admission, candidates must be *European states*. Despite the indication of a geographically oriented requirement for admission, the 'criterion of European-ness ... is best understood as a loose geographical' one:[149] the geo-linguistic concept of *Europe* does not have clear boundaries, and its application fades into the geographical continuity that characterises the Euro-Asiatic continent.

It follows that European linguistic geography does not match its legal geography, and this undermines EU 'territorial identity'. It could be argued that, when admitting prospective members, the Copenhagen criteria might complement the multifarious features upon which EU constitutional (and territorial) identity is built. These criteria are incorporated into Article 49 of the TEU, which, however, merely reflects a 'constitutional commitment to human rights, democracy and the rule of law' that is not specifically European.[150] The *abstract* commitment to human rights has a direct backlash on the *concrete* demarcation of the EU: new member states will be admitted provided that they respect human rights, democracy, and the rule of law. It follows that the European character of prospective members is irrelevant. This contradicts traditional legal geographical studies, where boundaries constitute the visible expression of clear territorial belongs (and divides) of public law. The absence of a clear boundary for a prospective EU territory makes its 'belongs of public law' even looser. A clear geographical interrelation between community and territory is lacking in the EU: as said, member states' territory merely defines the territorial scope of the Treaties.[151] Consequently, the formation of EU constitutional identity departs from that of traditional state-building processes.

There is another discrepancy between the legal geographies of the EU and the United States. The EU is a mere sum of national belongs of public law; indeed, member states' belongs are delineated by visible boundaries, concrete geographies, and historical placename politics—and national institutions reflect these belongs. Although the EU has traits in common with multinational federations, it only aims to demarcate the territorial scope of the Treaties by holding together the different EU *demoi* in a single quasi-federal structure.

[149] See Kumm (2012), p. 112.

[150] Kumm (2012), p. 109.

[151] Sees Art. 52 of the TEU and Art. 355 of the TFEU. See also Sect. 2.4.1.

This assumption reminds us of another discrepancy. Whereas the EU is a 'multinational', 'federal' polity, the USA is a homogeneous federation; during the process of spatial and state production, states' territorial neutrality was standardised: the settlers physically appropriating the frontiers gave 'life and meaning to a territory', but 'subsequent residents' may adapt 'the jurisdiction's institutions to changing times and their preferences'.[152] The open texture of the U.S. constitution turns the *relative* imagined legal geographies into *absolute* belongs of public law, constantly accommodating the contingent processes of spatial production within spatial reshufflings taking place in logical space.

The homogeneous nature of U.S. federalism can be traced back to the fact that constituent units were carved out of the federal territories and then admitted into the federation. The federal government thus shaped member states' territorial identity, boundaries, and placenames prior to their accession to the federation.[153] When admitting new constituent units, the U.S. federal government not only acknowledged their statehood but also their belongs of public law, which are, however, feebler and looser than those of the EU member states.[154] Hence, territorial identity of units is the outcome of a process through which U.S. territory and community have become inseparable, upholding the narrative of a strong constitutional identity and forging U.S. legal geography.

Admission to the EU generates the legal geography of the integration process, too: on the one hand, admission does not confer statehood on prospective member states, but bestows European-ness upon them. Thus, the EU generates *Europe* and shapes its own legal geography. Unlike U.S. legal geography, it will never be considered complete because the progressive admission of new 'European' states is a matter of politics, not of physical geography.

Hence, the EU does not have its own belongs of public law. Firstly, European-ness is the effect (and not the cause) of admission to the EU. Secondly, the EU lacks a definite territorial demarcation, and its territory is the mere sum of member states' territories and belongs of public law. Thirdly, as EU territory coincides with the territorial scope of the Treaties, there is no concrete and tangible process of self-identification between EU *demoi* and territory, and the EU does not conceive of its institutions as representative of this sum of different territories, peoples, and belongs, but of the individual member states. As far as European-ness is concerned, the EU confers legal and abstract significance onto member states' physical geography, which thus turns into an abstract EU legal geography.

By contrast, the USA possesses a concrete legal geography and a thick constitutional identity. Not only is this legal geography strictly connected to the formation

---

[152] Kincaid (2013), pp. 133–134.

[153] On state admission, see U.S. Supreme Court, *The American Insurance Company and The Ocean Insurance Company (New York) v. David Canter*, 26 U.S. (1 Pet.) 511 (1828).

[154] Apart from the South, which 'has long been the country's most distinctive region, so much so that the United States was, in important respects, a bicommunal federation from 1789 to about 1968': Kincaid (2013), p. 138.

and elaboration of national identity, but it also exhibits the above-mentioned legal and non-legal presuppositions linking thick constitutional identity to the 'social, cultural, as well as economic environment, within which [constitutional identity] effectively came into being'.[155] The U.S. narrative corresponds to what has been termed an identity shaped 'in the present from the connection with the struggles of the past and the ambitions for the future'.[156]

The U.S. storytelling succeeded in establishing its own thick constitutional identity because the USA possessed the legal and non-legal presuppositions allowing it to be establishment. These had a historical lineage that stretches back through the centuries, until the revolutionary 'big bang' caused by the Philadelphia Convention—but a national constitutional identity is also 'connected to the particular history, ambitions and current political practices of a particular community'.[157]

The USA guided the process of spatial production functional to the creation of shared belongs of public law, which reflects a peculiar relationship between its community, its received cultural traditions, and its territory, in relation to which identity is construed, i.e., a concrete legal geography. The EU *abstract legal geography*, by contrast, highlights the absence of EU belongs of public law: there has not been a shift from the aggregate of states' thick identities to the creation of common institutions capable of representing communities and territory, and this is one of the fallacies of the integration process. Although the EU tends to accommodate different European societies based on common values embedded in the Treaties, a constitutional identity comparable to that of its member states and of the USA is currently lacking.

The EU still must define what its European-ness means. When its legal and linguistic geographies match perfectly, that is, when Europe conceives of itself as a territorially, physically, and legally geographical demarcated polity, the criterion of European-ness will no longer be understood as a loose geographical tie. Europe will then be based on a *concrete legal geography*, which means that the time has come for a *thick constitutional identity*.

References

Abraham G (2007) Lines upon maps: Africa and the sanctity of African boundaries. Afr J Int Comp Law 15:61–84

Alapiki HE (2005) State creation in Nigeria: failed approaches to national integration and local autonomy. Afr Stud Rev 48:49–65

Anastasiou H (2002) Communication across conflicts lines: the case of ethnically divided Cyprus. J Peace Res 39:581–596

[155] Jacobson (2006), p. 384.

[156] Kumm (2012), p. 110.

[157] Kumm (2012), p. 110.

Anderson B (1991) Imagined communities. Reflections on the origin and spread of nationalism. Verso, London and New York
Anderson B (2001) Imagining East Timor. Lusotopie 2001:233–239
Austin JL (1962) How to do things with worlds. OUP, Oxford
Barth F (ed) (1998) [1969] Ethnic groups and boundaries: the social organization of cultural difference. Waveland Press, Long Grove
Bartoli M (1925) Introduzione alla neolinguistica. L.S. Olschki, Florence
Basson DA, Viljoen HP (1988) South African constitutional law. Juta & Co. Ltd, Cape Town et al
Baugh AC, Cable T (2013) A history of the English language, 6th edn. Routledge, London
Beal JC (2011) An introduction to regional Englishes: dialect variation in England. Edinburgh UP, Edinburgh
Benton L (2010) A search for sovereignty. Law and geography in European Empires, 1400–1900. CUP, Cambridge
Black J (1997) Maps and politics. The University of Chicago Press, Chicago
Blocher J (2014) Selling state borders. Univ Pa Law Rev 162:241–306
Blomley N (2015) The ties that blind: making fee simple in the British Columbia treaty process. Trans Inst Br Geogr 40:168–179
Borrough PA et al (2015) Principles of GIS, 3rd edn. OUP, Oxford
Boucher LJ (1963) The fixing of boundaries in international boundary rivers. Int Comp Law Q 12: 789–818
Braverman I (2008) 'The tree is the enemy soldier': a sociolegal making of war landscapes in the occupied West Bank. Law Soc Rev 42:449–482
Braverman I (2009) Planting the promised landscape: zionism, nature, and resistance in Israel/ Palestine. Nat Resour J 49:317–365
Brighenti A (2006) On territory as relationship and law as territory. Can J Law Soc 21:65–86
Brown N (1967) Cyprus: a study in unresolved conflict. World Today 23:396–405
Burrow JA, Turville-Petre T (2005) A book of middle English. Blackwell, Oxford
Carter P (1987) The road to Botany Bay: an essay in spatial history. Faber & Faber, London
Castree et al (2013) Oxford dictionary of human geography. OUP, Oxford
Coakley J (2004) Mobilizing the past: nationalist images of history. Natl Ethn Polit 10:531–560
Conley JM, O'Barr WM (1988) Just words: law, language, and power. University of Chicago Press, Chicago
Connell S (2011) Writing on the Land of Ireland: nationality, textuality, and geography in the Acallam na Senórach. Hortulus 7:1–27
Coufoudakis V (1976) United Nations peacekeeping and peace-making and the Cyprus question. West Polit Q 29:457–473
Crampton JW (2010) Mapping: a critical introduction to cartography and GIS. Blackwell, Oxford
Culperer J, Kytö M (2014) Early modern English dialogues: spoken interaction as writing. CUP, Cambridge
Cunningham B (1986) Seventeenth-century interpretations of the past: the case of Geoffrey Keating. Irish Hist Stud 25:116–128
Davies RR (1994) Presidential address: The Peoples of Britain and Ireland 1100-1400. I. Identities. Trans R Hist Soc 4:1–20
Davies RR (1995) Presidential address: The Peoples of Britain and Ireland 1100-1400. II. Names, boundaries and regnal solidarities. Trans R Hist Soc 5:1–20
de Mars S, Murray C (2020) With or without EU? The common travel area after Brexit. German Law J 21:815–837
de Mars S et al (2018) Bordering two unions: Northern Ireland and Brexit. Policy Press, Bristol
de Saussure F (1914) Course de linguistique générale. publié par Charles Bally and Albert Séhehaye avec la collaboration de Albert Riedlinger. Payot & Rivage, Paris
Delaney D (2005) Territory. A short introduction. Blackwell, Oxford
Delaney D (2010) The spatial, the legal and the pragmatics of world-making. Blackwell, Oxford

Delledonne G (2020) Depicting the end of the American Frontier: some thoughts on Larry McMurtry's Lonesome Dove Series. In: Mullender et al (eds) Law and imagination in troubled times: a legal and literary discourse. Routledge, Abingdon and New York, pp 195–205
Department of Foreign Affairs, Republic of Indonesia (1996) East Timor: building for the future: issues and perspectives. Jakarta
Didikin AB (2020) Speech acts and actions in legal language: conceptual analysis. Proc Inst State Law Russ Acad Sci 15:12–26
Diener AC, Hagen J (2012) Borders. A very short introduction. OUP, Oxford
Dominian L (1917) The frontiers of language and nationality in Europe. Henry Holton for the American Geographical Society, New York
Duchacek ID (1977) Antagonistic cooperation: territorial and ethnic communities. Publius 7:3–29
Duffey DP (1995) The northwest ordinance as a constitutional document. Columbia Law Rev 95: 929–968
Egwim AI (2020) The dynamics of federalism in Nigeria. Springer, Cham
Ejobowah JB (2010) Territorial pluralism: assessing the Ethnofederal variant in Nigeria. Reg Fed Stud 20:251–274
Elaigwu IJ (2002) Federalism in Nigeria's new democratic polity. Publius 32:73–95
Espenshade EB, Morrison JL (eds) (1982) Rand McNally Goode's World Atlas. Rand McNally, Chicago
Forest B (2001) Mapping democracy: racial identity and the quandary of political representation. Ann Assoc Am Geogr 91:143–166
Giddens A (1985) The nation-state and violence. Polity Press, Cambridge
Godden L (2020) Legal geography – place, time, law and method: the spatial and the archival in 'connection to country'. In: O'Donnell T et al (eds) Legal geography. Perspectives and methods. Routledge, Abingdon and New York, pp 130–148
Goodchild FM (2007) Citizens as sensors: the world of volunteered geography. GeoJournal 69: 211–221
Grant E (2006) Human rights, cultural diversity and customary law in South Africa. J Afr Law 50: 2–23
Green S (2010) Performing border in the Aegean. J Cult Econ 3:361–378
Griffiths I (1986) The scramble for Africa: inherited political boundaries. Geogr J 152:204–216
Griffiths P (2006) An introduction to English semantics and pragmatics. Edinburgh UP, Edinburgh
Grossfeld B (1984) Geography and law. Mich Law Rev 82:1510–1519
Hanafin P (2002) Constitutive fiction: postcolonial constitutionalism in Ireland. Penn State Int Law Rev 20:339–361
Hobsbawm E, Ranger T (eds) (1983) The invention of tradition. CUP, Cambridge
Hogg R (2002) Law's other spaces. Law/Text/Culture 6:29–38
Holzman-Gazit Y (2007) Land expropriation in Israel law, culture and society. Ashgate, Aldershot and Burlington, VT
Jacobson GJ (2006) Constitutional identity. Rev Polit 68:361–397
Jacobson GJ (2010) Constitutional identity. Harvard UP, Cambridge, MA
Johnston RJ (1991) A question of place. Exploring the practice of human geography. Blackwell, Oxford
Jones R (2009) Categories, borders and boundaries. Prog Hum Geogr 33:174–189
Kaloudis G (1999) Cyprus: the enduring conflict. Int J World Peace 16:3–18
Kay C, Allan K (2015) English historical semantics. Edinburgh UP, Edinburgh
Keating G (1902) Foras Feasa ar Éirinn, ed. and trans. David Comyn. The Irish Texts Society, London
Kedar A (2003) On the legal geography of ethnocratic settler states: notes towards a research agenda. In: Holder J, Harrison C (eds) Law and geography. OUP, Oxford, pp 401–441
Kedar A (2011) The legal transformation of ethnic geography: Israeli law and the Palestinian landholder 1948–1967. N Y Univ J Int Law Polit 33:923–1000

Kedar A (2014) Expanding legal geographies: a call for a critical comparative approach. In: Braverman I et al (eds) The expanding spaces of law: a timely legal geography. Stanford UP, Stanford, pp 95–119
Kedourie E (2005) The Cyprus problem and its solution. Middle East Stud 41:649–660
Kincaid J (2013) Territorial neutrality and coercive federalism in the United States. In: Mangiameli S (ed) Federalism, regionalism and territory. Giuffrè, Milan, pp 133–147
Kinkaid (2016) Territorial neutrality and coercive federalism in the United States. Swiss Polit Sci Rev 22:565–584
Kumm M (2012) The idea of thick constitutional patriotism and its implications for the role and structure of European legal history. In: Porsdam H, Elholm T (eds) Dialogues on justice. European perspectives on law and literature. Walter de Gruyter, Berlin, pp 108–137
Kurath H (1941) A word geography of the Eastern United States. University of Michigan Press, Ann Arbor
Kurath H, McDavid RI Jr (1961–1983) The pronunciation of English in the Atlantic states. University of Michigan Press, Ann Arbor. Reprint, University of Alabama Press
Kurath H et al (1939–1943) Linguistic Atlas of New England, 6 vols. Brown University for the American Council of Learned Societies, Providence
Kurath H et al (1973) Handbook of the linguistic geography of New England, 2nd edn. Brown University for the American Council of Learned Societies, Providence
Kurzon D (1997) 'Legal language': varieties, genres, registers, discourse. Int J Appl Linguist 7: 119–139
Lefebvre H (1974) La production de l'espace. Éditions Anthropos, Paris. English edition: Lefebvre H (1991) The production of space (trans: Nicholson-Smith D). Blackwell, Oxford
Lerche CO Jr (1949) The guarantee clause in constitutional law. West Polit Q 2:358–374
Leslie CA (2016) Territoriality, map-mindedness, and the politics of place. Theory Soc 45:169–201
Loizides N, McGarry J (2019) The 2002–2004 Annan Plan in Cyprus. An attempted UN-mediated constitutional transition. In: Anderson M, Choudry S (eds) Territory and power in constitutional transitions. OUP, Oxford, pp 43–59
Marmor A (2008) The pragmatics of legal language. Ratio Juris 21:423–452
Mattila HES (2013) Comparative legal linguistics: language of law, Latin and modern Linguas Francas, 2nd edn. Routledge, London and New York
McIntosh A (1973) Word geography in the lexicography of Mediaeval English. Ann N Y Acad Sci 211:55–66
McIntosh A et al (1986) A linguistic atlas of late mediaeval English, 4 vols. Aberdeen University Press, Aberdeen
Miller D (2016) Boundaries, democracy, and territory. Am J Jurisprud 61:33–49
Moore M (2015) A political theory of territory. OUP, Oxford
Moulakis A (2007) Power-sharing and its discontent: dysfunctional constitutional arrangements and the failure of the Annan Plan for a reunified Cyprus. Middle East Stud 43:531–556
Mullender R (2018) Transmuting the politico-legal lump: Brexit and Britain's constitutional order. Cardozo Law Rev 39:1020–1038
Munge Sone P (2017) Interstate border disputes in Africa: their resolution and implications for human rights and peace. Afr Secur Rev 26:325–339
Nagy JF (2001) Observations on the Ossianesque in Medieval Irish Literature and Modern Irish Folklore. J Am Folklore 114:436–446
Newman D (2006) The lines that continue to separate us: borders in our 'borderless' world. Prog Hum Geogr 30:143–161
Nicolini M (2015) 'n Droë Wit Seisoen in die Stormkaap: André Brink and the Fundamental Rights of the Afrikaners in Apartheid South Africa. In: Ward I (ed) Literature and human rights. de Gruyter, Berlin et al., pp 235–254
Nicolini M (2016) Narrators of fables or framers of the constitution? The Acallam na Senórach beyond time, place, and law. In: Carpi D, Leibhoff M (eds) Fables of the law. Fairy tales in a legal context. de Gruyter, Berlin, pp 137–168

O'Donovan K (1993) Marriage: a sacred or Profane Love machine? Fem Leg Stud 1:75–90
Orton H et al (1978) The linguistic atlas of England. Routledge, London
Palermo F (2016) Territory and the law of ownership: from misunderstanding to opportunity. In: Nicolini M, Palermo F, Milano E (eds) Law, territory and conflict resolution: law as a problem and law as a solution. Martinus Nijhoff Publishers, Leiden and Boston, pp 16–38
Peeters P (2007) Multinational federations: reflections on the Belgian federal state. In: Burgess M, Pinder J (eds) Multinational federations. Routledge, Abingdon and New York, pp 31–49
Pred A (1984) Place as historically contingent process: structuration and the time-geography of becoming places. Ann Assoc Am Geogr 74:279–297
Radding L, Western J (2010) What's in a name? Linguistics, geography, and toponyms. Geogr Rev 100:394–412
Rillaerts S (2010) La frontière linguistique, 1878-1963. Courrier hebdomadaire du CRISP 24:6–111
Rissanen M (2000) Standardisation and the language of early statutes. In: Wright L (ed) The development of standard English 1300-1800: theories, descriptions, conflicts. CUP, Cambridge, pp 117–130
Sack RD (1983) Human territoriality: a theory. Ann Assoc Am Geogr 73:55–74
Sack RD (1986) Human territoriality: its theory and history. CUP, Cambridge
Sanders M (2006) Undesirable publications. J.M. Coetzee on Censorship an Apartheid. Law Lit 18: 101–114
Sanghera S (2021) Empireland. How imperialism has shaped modern Britain. Penguin, London
Smith JJ (2005) Essentials of early English: an introduction to old, middle and early modern English, 2nd edn. Routledge, London
Stenton FM (1942) Presidential address: the historical bearing of place-name studies. The Danish Settlement of Eastern England. Trans R Hist Soc 24:1–24
Stephenson S (2013) Constitutional reengineering: dialogue's migration from Canada to Australia. Int J Const Law 11:870–897
Strating R (2014) Contested self-determination: Indonesia and East Timor's battle over borders, international law and ethnic identity. J Pac Hist 49:469–494
Suberu RT (1991) The struggle for new states in Nigeria, 1976-1990. Afr Aff 90:499–522
Taavitsainen I, Jucker AH (2015) Twenty years of historical pragmatics. J Hist Pragmat 16:1–24
Taylor PJ (1994) The state as container: territoriality in the modern world-system. Prog Hum Geogr 18:151–162
Townend MO (2012) Contacts and conflicts: Latin, Norse and French. In: Mugglestone L (ed) The Oxford history of English. OUP, Oxford, pp 75–105
Tribe LH (1983) A constitution we are amending: a self-defense of a restrained judicial role. Harv Law Rev 97:433–445
Tripathi D, Chaturvedi S (2020) South Asia: boundaries, borders and beyond. J Borderl Stud 35: 173–181
Upton C (2012) Modern regional English in the British Isles. In: Mugglestone L (ed) The Oxford history of English. OUP, Oxford, pp 379–414
Wagner K (2019) Amritsar 1919: an empire of fear and the making of a massacre. Yale UP, New Haven, CT
Watson A (1988) Failures of the legal imagination. University of Pennsylvania Press, Philadelphia
Watt D et al (2014) Language and identity in the Scottish/English border. In: Watt D, Llamas C (eds) Language, borders and identity. Edinburgh University Press, Edinburgh, pp 8–26
Williams RA (1926) The Finn Episode in Beowulf. An essay on interpretation. CUP, Cambridge
Wood D (1992) The power of maps. The Guildford Press, New York
Wood D (2010) Rethinking the power of maps. The Guildford Press, New York and London
Yilmaz EM (2010) Past hurts and relational problems in the Cyprus conflict. Int J World Peace 27: 35–62
Zahar M-J (2019) When the total is less than the sum of the parts. The lessons of Bosnia and Herzegovina. In: Anderson M, Choudry S (eds) Territory and power in constitutional transitions, 1st edn. OUP, Oxford, pp 22–42

Chapter 7
'Federal' Legal Geographies

7.1 Federal Arrangements

7.1.1 From Containers to Territories

When examining territorialised legal geographies, I focused on the ability of the 'disciplinary matrix' to produce legal geographical primitives. In this chapter, I shall apply them to a specific set of territorialised paradigms, that is federal arrangements.

Federal designs are described as a 'particular form of political organization', where at least two spheres of government coexist, namely the national and the subnational. Daniel J Elazar summarised this feature by using the formula 'self-rule plus shared rule'.[1] Extending over 'coordinate' and 'independent' spheres,[2] federations have 'two (or more) levels of government ... combining elements of shared rule (collaborative partnership) through a common government and regional self-rule (constituent units' autonomy)'.[3]

This definition perfectly fits the state-as-container approach (see Sects. 5.3.3 and 6.1), inasmuch as it complements the 'design' and 'character' of federations.[4] In adding a token of constitutional organisation to the container, federal attributes arrange the separation of powers on a territorial basis. Not only do constitutions distribute legislative, executive, and judicial powers horizontally (i.e. among the national branches of government), but they also allocate it vertically (i.e. among different tiers of government).[5]

---

[1] Elazar (1987), p. 12.

[2] Wheare (1963), p. 10.

[3] Watts (2008), p. 8; Watts (2013). See also Elazar (1994), p. xvi; Hueglin and Fenna (2015), pp. 31 et seq.

[4] Taylor (1994), p. 151.

[5] See Friedrich (1963b). On how horizontal separation and vertical division of powers interact see e.g. U.S. Supreme Court, *Printz v. United States*, 521 U.S. 898 (1997) at 2377.

M. Nicolini, *Legal Geography*, Ius Gentium: Comparative Perspectives on Law and Justice 105, https://doi.org/10.1007/978-3-031-19410-8_7

Although extensively used in federal studies, this definition is purely descriptive of the 'basic features' of federacies because it focuses on their 'forms', and levels of governments are addressed in terms of distribution of powers. It then treats federations as instrumentalities that states-as-containers use to govern, say, multinational societies, organise their territories, and match their needs. In sum, the definition points to the outcomes of spatial production without considering the context where manufacturing processes take place.

The fact that separation of powers is distributed territorially clearly points to the production of belongs of public law. In federal states, territory is the 'object to which other attributes are assigned';[6] though, the traditional definition of federacies does not indicate how this may be reshuffled to reflect the space-place-environment nexus that communities imagine in their logical space.

The focus on distribution of powers makes us lose sight of how legal geographical primitives contribute to manufacturing federal belongs. A broader notion of federal arrangements is required, which necessarily goes beyond the powers assigned to the different levels of government. In the following sections I shall reappraise federations as legal geographies operating at different scales (see Sect. 7.1.2). To this end, I shall introduce further legal geographical primitives, such as territorial constitution, regional demarcation, and alteration (see Sect. 7.2). These will also be probed in multinational federations, where the accommodation of diversity is based on units, and their boundaries are usually demarcated on an ethno-linguistic basis (see Sect. 7.3). The production of space will thus shed new light on the distribution of powers, which indeed reflects the demarcation of ethnic-oriented units created to meet the ethno-linguistic features of the groups they represent.

7.1.2 *Scalar Legal Geographies*

'On what basis should political units be formed, and what powers should they have? How widely or narrowly in space should their bounds be set?' With these questions, David Miller begins his essay on *Boundaries, Democracy, and Territory*, a true cross-disciplinary journey through a variety of territorialised legal geographies.[7]

The article is fitting for my purposes, because it examines the criteria 'that should inform political boundary-drawing'. Two reasons make the article *particularly* fitting. The first one relates to Miller's intuition that, in boundary-drawing, criteria are not limited to 'powers that political units should possess'. His predicament resonates with my going beyond the allotment of powers for defining federal arrangements. When discussing how to fill the state-as-container, lawyers, sociologists, political scientists, and even human geographers examine spatial production

[6] Sack (1983), p. 57.
[7] Miller (2016), p. 33.

by adopting a 'narrowed' notion of belongs of public law. Consequently, they study units where spatial reshuffling is understood in terms of competences, thus reifying 'regions as fixed containers rather than seeing them in relational terms' with the territorial production of space that makes them meaningful.[8] Spatial production becomes functional to creating responsibilities that act as instrumentalities for governing the ethnic, linguistic, and religious mashups in 'federal' belongs of public law.

Several reasons endorse a beyond-powers notion of federations. As Henri Lefebvre put it, 'going beyond' them entails 'a movement from *products* [the rule regulating the belongs] to *production*' of those rules.[9] This reminds us of the law as an open texture pragmatically enriched by its context, where natural-human geographies are constantly reshuffled. Lefebvre's predicament goes hand in hand with Miller's broadening the idea of boundary-drawing; powers cannot 'be divorced from their scope – the jurisdiction within which those powers will be exercised'.[10] Even in logical space.

The second reason that makes Miller's essay fitting relates to its scalar approach. It transcends international law and examines how spatial production works within states as containers. These processes, Miller continues, 'may choose how to define the regions that are to be granted devolved government, and they may redraw the borders of their counties or provinces'.[11] This chapter, like Miller's, broadens the notion of spatial production, addresses the territorial matrix from a scalar perspective, and focuses on the 'federal' belongs of public law, that is, the legal geography of subnational units. Their production takes place at a sub-national scale, since legal geography extends over all forms of 'territorial segmental autonomy'; among them, federal arrangements.[12]

Subnational units are a variety of territorialised legal geographies, to which the territorial matrix is applicable. Like states, they have their logical space and experience spatial reshufflings; physical apprehension, political appropriation, and the creation of meaningful places are replicated at their level. Through the law of spatiality, the spatiality of law, and counter legal geographies, the 'production of local jurisdictions and local cultures' pragmatically enriches territorial maps.[13] It is not important if 'Lines on a map may anticipate a jurisdiction'; what matters is that 'a jurisdiction itself consists of the practices that make the abstract space depicted on a map significant'.[14]

Furthermore, Miller assumes the inescapability of boundaries. Predicaments such as 'ideally there should be no boundaries at all'; or 'The world should form a single,

[8]See Agnew (2013), p. 11.

[9]Lefebvre [1974] (1991), p. 26.

[10]Miller (2016), p. 33.

[11]Miller (2016), p. 34.

[12]Lijphart (1979), p. 505.

[13]Ford (1999), pp. 845 and 855.

[14]Ford (1999), p. 856.

undivided political unit' are, in his opinion, hard to defend[15] even by the cosmopolitan global holders of capital. Like us, though, they are trapped by lines and polygons and accept the presence of subnational primitives: they also need 'smaller units under the global authority discharging many of the functions that states and their subdivisions now perform' (see also Sect. 8.3.1).[16]

7.2 Scalar Primitives

When applying the territorial 'matrix' to federal arrangements, we need further primitives. Indeed, 'we must constantly remind ourselves that jurisdiction [i.e. a subnational unit] is itself a set of practices, not a preexisting thing in which practices occur or to which practices relate'.[17] If federacies replicate processes of spatial production at a different scale, legal geographical primitives accommodate their discrete belongs of public law. Additional primitives are required since the interconnectedness between geographical features, territory, and federal arrangements goes beyond federalism as a mere distribution of powers. The production of scalar belongs takes indeed the form of units whose boundaries are demarcated territorially. It does not come as a surprise that geographers and lawyers alike extend the concept of territory to encompass lines 'separating jurisdictions or territories' of 'subnational administrative districts' within federal arrangements.[18]

Our first legal geographical primitive is the 'territorial constitution'. In broader terms, it points to the geographical character of federal constitutions, i.e. the governance through constituent units. So far, we may argue, nothing new under the sun. If we do not want to reduce it to the state-as-container approach, we must reappraise how it works within federations.

And to this legal geographical primitive I dedicate the following sections.

7.2.1 Synthetic Jurisdictions

Within a descriptive analysis of territory, the governance through constituent units usually refers to two types of units, i.e. 'synthetic' and 'organic' jurisdictions.

By 'synthetic jurisdictions', I point to units created 'by some institution in order to serve *its* purposes'. The state acts as the prototype of a container: synthetic jurisdictions do not 'define a prepolitical group, but are instead imposed on groups

[15] Miller (2016), p. 34.
[16] Miller (2016), p. 34.
[17] Ford (1999), p. 856.
[18] Castree et al. (2013), p. 38.

of people from "outside" or "above"'.[19] Jurisdictions like these identify neither 'subnational' processes of spatial production nor 'scalar' belongs of public law. Rather, they are instrumentalities of the state-as-container, which creates them 'for the sake of [its] convenience', i.e. for attaining its own desired ends. In unitary states, they manage the diversity of territory through a homogeneous 'culture of bureaucracy' and a nationally driven economic policy.[20] This occurs in France, where regions (not to mention departments) were created in 1982 by the national government. In 2014, a new territorial readjustment led to the creation of new regions: these were reduced from 22 to 13 by merging 'three former regions' (Aquitaine, Limousin, and Poitou-Charentes became Nouvelle-Aquitaine), and 'two former regions' (Nord-Pas-de-Calais and Picardie) into Hauts-de-France, while others, such as Île-de-France, did not change.[21] The process also characterised Eastern and Central Europe, where territorial divisions were demarcated to manage EU structural funds.[22] These are functional regions; a new 'design' and 'character' is impressed into their territory, thus filling the map with an externally driven imagined geography. This is also the case of Special Economic Zones (SEZs), that say, in China, India, Nigeria, and Russia are functional to attracting foreign investors and providing incentives for them. They are drawn into 'a defined geographical area' where 'separate customs and administrative procedures' take place, as well as other 'Unique benefits [including] operating under more liberal laws and economic policies compared with other parts of the country'.[23]

Synthetic jurisdictions are also part of the EU Cohesion Policy, overlapping with the 'diversity' of national areas to promote 'economic, social and territorial cohesion, and solidarity among Member States'.[24] Their being synthetic reflects the tenet of EU law, according to which member states' territories outline the scope of EU policies, upon which the EU projects its own priorities and objectives (see Sect. 2.4.1).[25] The 'Economic, Social, and Territorial Cohesion' policy pursues its own imagined geography with the aim to reduce 'disparities between the levels of development of the various regions and the adverse economic conditions of the least favoured regions or islands, including rural areas' (Art. 174.2 TFEU). To promote growth and cohesion, the EU demarcates its own 'regions', which become the recipients of EU actions and policies. EU spatial planning has thus triggered several regional areas, such as the 'Nomenclatures of territorial units for statistics' (NUTS), 'macro-regions', EU outermost regions, and cross-border areas within the

[19] Ford (1999), p. 860.

[20] Ford (1999), p. 860.

[21] Seys (2019a), pp. 55 and 58. See also Brennetot (2018); Perrin (2022), pp. 109–111.

[22] This occurred in Poland (1999) and the Czech Republic (1993). See Seys (2019a), p. 63; Domorenok (2013). See also Perrin (2022), pp. 34–35. Perrin and Seys (2019).

[23] Crane (2018), pp. 99–100.

[24] Art. 3 of TEU. On the structural differences between federalism, regionalism and EU multi-level governance see e.g. Panara (2015), pp. 166–168.

[25] On such a 'parasitic' approach to territory in EU law see Weinzierl (2021), pp. 652–653.

European Territorial Cooperation (ETC).[26] Each of them is servient to framing the EU regional policy and counts as the scope of the EU territory, where integration policies are enforced at a subnational scale. Usually, EU regions correspond 'to the territorial division of Member States in accordance with their own administrative arrangements'.[27] This, again, occurs for the sake of the EU's convenience, since these regions already possess that 'culture of bureaucracy' required to become instrumentalities for implementing EU measures.

7.2.2 *Functional Regions*

The concept of 'synthetic jurisdiction' resonates with the functional approach to boundary-drawing. Functionalism aims to carve out of territory 'viable units . . . able to perform the functions that states must discharge'.[28] To make containers more efficient, states may adopt a demarcation based on functional areas. In this event, economic factors are given prevalence over political, historical, and statistical ones. When redrawing regions for functional considerations, governments replace the traditional concept of territorial identity—which relates to a deep-seated 'belief' among people and captures how people define the region(s)—with a new one, which is mainly rooted in functional criteria (access to services, facilities, allocation of natural resources, distribution of financial proceeds, and so on).

However, it is methodologically incorrect to draw a biunivocal relation between synthetic regions and functional approach. Like other variables, functional criteria may be territorialised, spatially reshuffled, and processed together with human, physical, and cultural geographical attributes. The aim, it goes without saying, is to create a strong relation between community and territory. This occurred when processes of constitution-making took place, say, in Italy and South Africa. In Italy, the spatial reshuffling taking place in logical space led the Constituent Assembly to mix political, statistical, and historical criteria when demarcating regions. In South Africa, provincial boundaries were drawn in accordance with non-ethnic-oriented criteria to promote integrated development (see Sect. 7.3.4). Indeed, 'territorialised' functional criteria underwent a process of spatial production, whose outcome is usually the establishment of a scalar (i.e. subnational) belong of public law.[29]

Likewise, EU macro-regions and cross-border cooperation areas have the potential to go beyond the functional dimension of the EU, thus generating EU-related belongs of public law that transcend borders. Thomas Perrin, for instance, points to

[26] On such macro scalar areas see Gänzle (2019). For a critical approach to how regions may be used as instrumentalities *bonnes à tout faire* see Agnew (2013), pp. 10–12.

[27] See Bovis (2011), p. 81.

[28] Miller (2016), p. 38.

[29] Perrin (2019), p. 159; Perrin (2022), pp. 238 et seq.

the 'sense of commonness' between the discrete communities of the macro-regions. In this 'commonness', he explains, lies the 'driving force' conveying 'the will or the need to cooperate in a given macro-region'. Environmental or economic issues usually act as drivers broadening the scope of territorial cooperation projects and EU policies delivery. At this scalar level, functional territorial cooperation requires 'symbolical and intangible incarnation'; citizens and stakeholders must 'be able to project [their own] vision and/or a representation on a given institution or authority, and to relate it to their own perceived identity'.[30] Again, the 'third' (i.e. imaginative geography) acts as the interface between physical and human geographies in the production of legal spatialities (see Sect. 3.2).

7.2.3 Territorial Constitution

The territorialisation of functional-bureaucratic jurisdictions resonates with the second type of units mentioned above, i.e. 'organic' jurisdictions. These naturally grow out 'of circumstances ... defined socially rather than metrically, concretely rather than abstractly'. They may also be the outcomes of territorialised functional regions because their communities have come to share that sense of commonness that unites them. They combine 'both economic and cultural foundations'.[31]

In this interaction between communities and territories lies our new primitive: the territorial constitution. It is not a matter of design and character; it considers territorial governance a process of spatial production. Enforced 'in terms of autonomy, self-determination and cultural preservation', it is part of the constitutional-cum-territorial identity of a given federal arrangement.[32] Authority is thus distributed both *vertically* and *geographically*. Whereas the former denotes the allotment of powers between the different orders of government, the latter means that both units' constitutive elements (people, government, and territory) and boundaries are demarcated on a geographical basis. These elements count as units' belongs; at the same time, they complement the federal constitution identity (such as in the USA: see Sect. 6.6.2). In multi-ethnic federations, the territorial constitution also provides arrangements establishing multi- or bi-ethnic constitutional frameworks. Ethnic-based units enhance consociational participation and geographical cohesion, and complement power sharing, i.e. politico-legal arrangements that allow divided groups to share responsibilities in the governments of their respective federations.[33]

---

[30] Perrin (2019), pp. 162–163. See also Perrin (2022), pp. 206–213, and 235 for a broader examination of the idea of 'incarnation' of cultural identity in cross-border cooperation. See also Evrard (2021).

[31] Ford (1999), p. 859.

[32] Ford (1999), p. 859.

[33] On power sharing see Palermo (2016), p. 26.

Acting as a legal geographical primitive, the territorial constitution has several qualities. Firstly, it confirms that federalism is not only distribution of powers. It 'is a process rather than a design or pattern', Carl J Friedrich stated in 1963. With our parlance, it is a territorial production of space. No wonder that Friedrich termed these dynamics 'federalizing processes',[34] thus pointing to the federal idea as a process able to turn subnational spaces into meaningful places.

Secondly, the territorial constitution fills the gap in the traditional definition of federalism since it establishes an immediate relation between constitutions, subnational communities, and their territory. The constitutional relevance of territory is not limited to a matter of design. Quite the opposite: it broadens spatial production beyond the competences units should possess, opening them to the socio-geographical contexts that enrich them with new meanings.

Thirdly, the territorial constitution coordinates all other legal geographical primitives outlining subnational legal geographies. Certainly, 'lines on a map may anticipate a jurisdiction'; these are usually part of the federal constitution. Yet, the meaning they have been assigned in subnational logical spaces is not a line. It is a belong of public law. And it must be preserved as such. Its guarantee is represented by the same territorial constitution. Legal geographical primitives and scalar territorial identities are 'constantly being made real'. The territorial constitution protects units' belongs by entrenching political pluralism as the core feature, if not the constitutive element, of the federalising process. 'Without these practices' of spatial production at both federal and subnational level 'the lines would not "be real" – the lines don't preexist the practices'.[35]

Finally, the accommodation of federal-local belongs is facilitated by the open texture of federal constitutions, where scalar belongs are preserved for future generations. The territorial constitution does this by turning the *relative* subnational imagined legal geographies into *absolute* belongs of public law, accommodating them within the spatial reshuffling taking place in federal logical space. At the same time, the texture unceasingly adapts the changes that take place in subnational logical spaces; if these alter their belongs of public law, so does the federal logical space.

Attentiveness to subnational geographical contingencies triggers the unceasing reshuffling of the place-space-environment nexus and percolates federal spatial production. The territorial constitution thus manages all kinds of constitutional interactions (see Sect. 2.4.1). In terms of the law of spatiality, the legal geographical primitive processes different units' belongs irrespective of how federations have been created. Federalising processes of spatial production are pluralistic and incorporate subnational belongs irrespective of whether: (i) the federation integrates 'previously independent political communities' into a single system ('aggregative' federacies);[36] (ii) pre-existing units have been independent or enjoyed autonomy

[34] Friedrich (1963a), p. 238; Friedrich (1968), p. 24. See also Ford (1999), p. 855.
[35] Ford (1999), p. 856.
[36] Such as the USA, Switzerland, and Germany. See Aroney (2006), p. 282.

before their aggregation into the new federal structure;[37] (iii) 'a formerly unitary state has devolved governmental powers upon a number of regions within that state';[38] (iv) federacies have a 'quasi-aggregative' character, such as Argentina and Austria; (v) federalism was part of a more comprehensive strategy for fostering democracy within democratic transitions, such as in Nepal, Iraq, and Democratic Republic of Congo (DRC); finally (vi) asymmetrical and multi-national federacies are used to hold together different (ethnic, religious, linguistic) cleavages (see Sect. 7.3.5). The same holds true as far the unincorporated U.S. territories of Guam, the commonwealth of the Northern Mariana Islands (CNMI), American Samoa, the U.S. Virgin Islands, and Puerto Rico. These outline a quasi-federal (or, better, a confederal) constellation of federal territories; there, the U.S. Constitution is not entirely applied; as these territories are not part of the Federation, the constitutional principles extend over them under certain conditions and with noticeable limitations (absence of the jury trial, restriction to private property and citizenship), which were listed in 'a series of Supreme Court judgments known collectively as The Insular Cases'.[39]

As for the spatiality of law, the territorial constitution acknowledges the presence of manifold legal geographical systems within federal spatial production. Within federacies, several legal systems usually coexist alongside each other. Besides the presence of the mixed jurisdictions of Quebec, Louisiana, and South Africa,[40] they also display non-Western legal conceptions of the law: Hindu law (India and Nepal), Islamic law (Malaysia, Pakistan, Iraq, Comoros, and the UAE), Micronesian customary law (FSM and Palau), African law (DRC and Cameroon).[41] The South African Constitution also recognises tribal institutions, chieftaincy, and customary law, thus assigning jurisdiction thereover to national and provincial legislation.[42] The Constitution of Ethiopia does the same with tribal institutions and ethnic identities.[43] The Constitution of Malaysia grants a differentiated position to the *bumiputera* and *orang asli*, i.e., aboriginal peoples, to be subsequently implemented by federal and state legislation.[44]

Finally, federations undergo processes whereby counter legal geographies take place. In these cases, several flaws may occur between the legal regime and the

[37] Canada, Australia, Malaysia, St Kitts, and Pakistan were dominions and colonies enjoying autonomy and self-government within the former British Empire. The same holds true for Palau and the Federated States of Micronesia. See Nicolini (2022).

[38] Spain, India, Belgium, Italy, and South Africa. See Aroney (2006), p. 282.

[39] Jones (2021) and Duffy Ponsa-Kraus (2021–2022).

[40] In Canada, the USA, and South Africa respectively.

[41] See the following constitutions: Pakistan (Sects. 2 and 31); Iraq (Art. 2 and 92); UAE (sec. 7); Palau (Arts. V and VIII); the FSM (Art. V); DRC (Arts. 3 and 207), and Cameroon (Arts. 1.2 and 57).

[42] Sections 211 and 212 of the Constitution. See Himonga and Nhlapo (2015).

[43] See Tesfai Baraki (2013), pp. 233 et seq.; Fiseha (2019), p. 72.

[44] Articles 8, 153 and 160 of the Constitution. See Harding (2013), pp. 71–73 and 185; Bakar (2007), pp. 71 et seq.; Harding (2018), pp. 559, 568–573.

operational rule of the territorial constitution. Subnational logical spaces may experiment with a new spatial reshuffling aiming to turn new *relative* imagined legal geographies into *absolute* belongs of public law. If the federal territorial constitution is not able to accommodate such a reshuffling to its own logical space, subnational logical space may reframe the relation between community and territory as a new and reimagined alternative, such as a belong to be exercised through secession from the federation. Not only does the territorial constitution have to be flexible to hold together diversity, but it also has to prevent (and run counter to) the idea of secession. This was successfully achieved, say, in the case of the annexation of Crimea by Russia (2014), in Pakistan and Malaysia,[45] it was attempted (or is still wished-for) in several jurisdictions, such as Spain, the UK (Scotland), Canada (Quebec), Iraq, Comoros, the FSM, and St Kitts.[46] In Russia, the indigenous population represents the majority in only seven of the 21 republics (Chechnya, Chuvashia, Dagestan, Ingushetia, Kalmykia, North Ossetia-Alaniya, and Tuva), and claims for secession come from these territories.[47] In these cases, there is no dispute over demarcation, since it is perceived as an insufficient means for protecting territorial identity.

7.3 Regional Demarcation

7.3.1 Demarcating and Reshuffling

As a legal geographical primitive, the territorial constitution acts as the interface that, at the federal level, accommodates the varieties of imagined geographies, modes of spatial production, and belongs of public law that populate the federal legal space. It thus complements the rules of federalism and rearranges all other primitives (placenames, lines, isolines, borders, and territorial identity). It then accounts for how communities imagine (and rearrange) the space-place-environment nexus within their logical spaces. Finally, it makes it compatible with federal logical space.

The concept of territorial constitution gives rise to a paradox. Whilst securing units' territorial identity (and their belongs of public law), federal arrangements lay down procedures for altering their boundaries. Such belongs are, again, *absolute* and *relative* at the same time; and their being absolute or relative depends on how the

---

[45] With the separation of Bangladesh from Pakistan in 1971, and of Singapore from Malaysia in 1964. See Harding (2013), p. 43.

[46] On Scottish, Quebecois, and Spanish secession referendums see López-Basaguren and Escajedo San-Epifanio (2019) and Delledonne and Martinico (2019). On the 2017 referendum for secession in Kurdistan see Danilovich (2019). On the 1997 attempt of the island of Anjouan to secede from Comoros see Baker (2009). In the FSM, the states of Pohnpei and Chuuk make periodic callouts to secession. See Dziedic (2020), pp. 234–235. Nevis has been in constant 'flirtation with secession' since the nineteenth century. Ch. X, sec. 113 of the Constitution lays down a provision for its unilateral secession. See Griffiths (2020), pp. 306–307.

[47] See Ross (2007), p. 117.

perspective that the territorial constitution adopts for their examination. On the one hand, federal constitutions certainly consider units 'indestructible states' with absolute belongs.[48] At the same time, the necessity to constantly rearrange the production of space leads the territorial constitution to establish an enduring union of 'destructible' units. This means that it also lays forth mechanisms according to which such boundaries may be redrawn or altered to fit into the spatial reshuffling (with its new belongs of public law) that has already taken place in logical space. The dialectic between absolute-ness and relative-ness of scalar spatial production is even more contentious in multinational federations, where ethnic-based states are deemed to secure a perpetual compact between the 'indestructible' different territorial groups. Despite this, the open texture of federal and regional constitutions clearly reflects the rationale of all legal geographical primitives, which give expression to the view that, as the place-space-environment nexus changes over time, so too should the reshuffling that takes place in logical space.

Among the legal geographical primitives that allow territorial constitution to reshuffle the territorial matrix in logical space there is regional demarcation and territorial alteration. Both legal taxonomies are worked out by comparative scholars when devising law-based analyses of federal systems. Demarcation presides the division of a country's internal territory into two or more territorial constituent units, whereas alteration entails either the total or the partial reconfiguration of a country's territorial constitution. Both assume the creation of 'regionally defined constituent units to respond to the demands of a territorially concentrated population'.[49]

7.3.2 *Demarcating Subnational Logical Spaces*

The nature of regional demarcation is a matter of scholarly disagreement. From a geographical perspective, delimitation of geographical areas is considered 'arbitrary',[50] because, as Rogerson argues, 'events outside the study area can affect those inside of it'. When demarcating, he suggests a 'buffer zone' be created 'to include features that affect analysis within the primary area of interest.'[51] Arbitrariness means nothing else than demarcation is a political choice to be normalised. In legal geographical terms, I have already come across this thorny issue when examining how the performativity of territory and its related spatial practices extend over borders.[52] This, I added, is the most relevant difference between polygons and

[48] Huber (1967), p. 256.

[49] Anderson (2008), pp. 72–73. See also Ramutsindela and Simon (1999) and Nicolini (2019, 2022).

[50] Seys (2019a), p. 52.

[51] Rogerson (2019), p. 19.

[52] On the symbolism and meanings attached to spatial borders see Ojalammi and Blomley (2015), p. 52.

bounded political communities: the former demarcate areas in a way that is functional, for what concerns us here, to creating synthetic regions, whereas the latter assume that borders-as-primitives are constitutive of territorial identity and part of their belongs of public law (see Sect. 6.5). That is why the territorial constitution takes borders seriously. If it fails to accommodate the legal geographies of its several scalar communities, it may fall prey to the creation of flaws and territorial operational rules that might lead to the disruption of the federal order.

This also explains the crucial role regional demarcation played in several federacies. In younger federal and devolutionary states, constituent assemblies were responsible for both the creation of sub-national units and the drawing of their boundaries. To avoid territorial flaws, units' borders were drawn up by the same constituent assemblies, which carved them out of the map of former unitary states or colonies. Legal geographical research on regional demarcation is therefore indebted to the debates that took place in the Constituent Assemblies of, say, Italy, India, and South Africa. They encompass all the most relevant issues in regional demarcation: the choice of federal-regional design; the role played by demarcation in governing diversity; recourse to economic, ethnic, linguistic, and religious factors to draw sub-national boundaries.[53]

7.3.3 *Demarcation Criteria*

Boundary-drawing processes reflect human/physical geographical variables, as well as ethnic, political, economic, and religious ones. The territorial constitution has to find an equipoise between them, particularly when ethnicities are imbricated on a given territory, thus considering *that* territory the repository of their own identity. In this event, borders perform the function once staged by the (imbricated) territory, i.e. they disentangle the imbrication and shape distinct, albeit territorially related, ethnic identities.

Several criteria are applied to give a response to these issues. The traditional divide is between historical and functional criteria. The former advocates for historical continuity in boundary-making. It does not seek 'a historical break in the delimitation of regions . . . boundaries are old and have not been called into question, though they may have undergone gradual modification' over time. To the extent they undergo minor modifications, they represent a legacy of the past.[54]

Such a demarcation may be found, for example, in insular federacies, such as the Comoros, the FSM, and Palau. Processed as physical geographical features, their coasts become arcifinous borders and generate the 'island model' as the prototype of

[53] For Italy see Pedrazza Gorlero (1979) and Nicolini (2019). For India see Bhattacharyya (2007a); Bhattacharyya (2019), p. 85; Bhattacharyya (2021), pp. 61–76. For South Africa see Muthien and Khosa (1995).

[54] Seys (2019a), p. 54.

a natural bounded territory (see Sect. 3.5.2). Historical criteria also characterise the legal geography of aggregative federations, where demarcation overlaps with the boundaries of pre-existing units, which have come together to create a new federation with its territorial constitution. The U.S. territory originally coincided with that of the 13 colonies; minor readjustments led to the creation of Maine, carved out of Massachusetts in 1820. Likewise, in Australia, the former colonies of New South Wales, Tasmania, Victoria, South Australia, Western Australia, and Queensland became the Commonwealth of Australia (1900). In Canada, the Imperial Parliament passed the confederative scheme settled at the Quebec Conference (1864) as the British North America Act 1867, thus uniting the former colonies of Canada, New Brunswick, and Nova Scotia into a single dominion. The Act divided Canada into two provinces, Ontario and Quebec, thus restoring the territorial identity of the French-speaking province that had been suppressed by the Union Act 1840. Likewise, in Malaysia, the Federation of Malaya Independence Act 1957 demarcated the national territory by considering the historical territorial identity of the previously existing Malay States. The admission of Sabah and Sarawak under the Federation of Malaya Agreement and the Malaysia Act 1963 reflected this historical criterion. State boundaries coincided with those of the former colonies and protectorates, which had been historically created on the basis of predominant ethnic groups.[55] In Argentina, ten provinces signed the San Nicolás de Los Arroyos Agreement in 1852, which meant renovating the former 1824 federation on a compactual basis.

The second criterion is a variant of the historical one and points to ethnic boundaries, which 'circumscribe pre-existing homeland'.[56] It reflects the point of view of each territorial community: the territorial constitution should reflect the congruence between their logical space, 'the land that they currently occupy, [and] the land that historically has been theirs'.[57] Due to the presence of concurrent claims to the same territory, its application may lead to conflicting views as regards the scope and meaning of the federal territorial constitution. Belgium and Nigeria reshuffled their scalar belongs of public law, thus recreating ethnic territorial units for their respective constituent communities (see Sect. 6.6.2). Bosnia Herzegovina is another case in point, whose 'territorial restructuring agreed at Dayton' (1995) was aimed at preventing further ethnic cleansing. The Federation is made up of two units, the *Republika Srpska* (dominantly Serb) and the Federation of Bosnia and Herzegovina (where Muslims and Croats coexist). The territorial constitution is shaped to reflect the 'ethno-territorial nature' of such arrangements.[58] Groups have the power to reshuffle the areas where they are dominant without triggering ethnic segregation. This was reflected, for example, in the management of the Brčko enclave within the *Republika Srpska*, i.e. an imbricated entity now jointly hold 'in condominium' by both units. We may say that the territorial constitution enshrines a

[55] See Bhattacharyya (2021), pp. 56–61; Harding (2013), pp. 20 et seq.; Harding (2017), p. 262.

[56] Miller (2016), p. 44.

[57] Miller (2016), p. 44.

[58] Zahar (2019), p. 30.

multi-ethnic geography which prevails over the subnational belongs agreed at Dayton. The Constitutional Court was able to avoid segmentation and homogenisation in its landmark decision on 'constituent peoples'.[59] Although they are dominated by ethnicities, units cannot discriminate against any member of the constituent people, particularly ethnic minorities; nor can they assign privileges to majorities.[60] According to the Court, 'Even if constituent peoples are . . . in a majority or minority position in the Entities, the express recognition of Bosniacs, Croats and Serbs as constituent peoples . . . can only have the meaning that none of them is constitutionally recognised as a majority'. To put it differently, the 'constitutional principle of collective equality of constituent peoples' prevails over homogenisation and 'segregation based on territorial separation'.[61]

Far from thoroughly applying the ethnic criterion, the prevalence of the territorial constitution over subnational belongs accounts for the application of a further criterion, which I term the political one. Its application aims to make the 'boundary proposal' accepted as the 'legitimate one'. It is not a matter of breaking with history; nor is it related to ethnic issues. Its main purpose is to give the territorial constitution a strong political legitimisation. In Austria, the federal arrangement was adopted to accommodate the newly created *Länder* within the framework devised by the Constitutional Act of 18 December 1918. The fears of secession induced the assembly and the *Länder* to sign a compact and therefore establish a highly centralised federal framework.[62]

Processes like these are often 'works in progress'. In the USA and Canada, for instance, the political criterion complemented the historical one when it came to admitting new member states into the federations. Both accrued their territory by welcoming new states carved out of former federal territories.[63] In the USA, the federal government shaped territorial identity, boundaries, and denomination of the states prior to their accession to the federation. As admission meant conferring statehood to them and certifying their constitutions (Article IV, s. 4, of the U.-S. Constitution), their creation reflects synthetic processes upon subnational production.[64] As for Canada, provinces carved out of Rupert's Land and the North-Western Territory required the prior admission of these territories to Canada, and the creation of the provinces out of them. Prince Edward Island and British Columbia were admitted by an imperial Order in Council on addresses from the Houses of the

[59] Const. Court of Bosnia and Herzegovina, U-5/98 (Partial Decision Part 3). 1 July 2000.

[60] Zahar (2019), p. 34.

[61] Const. Court of Bosnia and Herzegovina, U-5/98 (Partial Decision Part 3), para 60.

[62] Kelsen (1920).

[63] Art. IV Sect. 3 cl 1 U.S. Constitution. For Canada, see sect. 146 of the British North America Act 1867; Newfoundland Act 1949.

[64] See Sect. 7 and Ford (1999), pp. 890–891.

Parliament of Canada and from the Houses of the respective legislatures. Newfoundland was admitted in 1949 by an imperial statute, which followed two referenda.[65]

As for the functional criterion, I have already noted that it creates 'viable units', i.e. 'economically viable, with the means either to produce for themselves what their members need for sustenance or to provide for them through trade in goods or services', without 'monopolizing the natural resources in a particular region, or creating pollution that engulfed their neighbors'.[66] The aim, it goes without saying, is to establish a strong relation between community and territory in terms of efficiency, public services, and goods delivery, as well as better territorial governance from above. François-Olivier Seys reminds us of three demarcation models for creating functional regions. The first one is 'delimitation by aggregation'; smaller units are merged into bigger ones, which are usually 'unconnected to former, historical territorial units'. The 1790 creation of departments in France followed this pattern: 'the capital of the department had to be reachable from every municipality inside it within a one-day horse ride'. In any case, aggregation is unilaterally decided from the centre, without any consultation of the communities concerned.[67] Reys points to a second model, i.e. the centre-periphery model without any prior aggregation, which is used 'in countries which have only recently created regions'. These are 'organised around large cities and extend their boundaries to encompass their catchment areas', as in the Polish Voivodeships (*Województwo*) created in 1999 as useful instrumentalities for managing prospective EU structural funds.[68]

7.3.4 Mixing Criteria Politically

I have already referred to South Africa as an example of how reshufflings led to functional (i.e. economic) criteria. After the dismantlement of the apartheid, the Transkei, Bophuthatswana, Ciskei, and Venda *Bantunstans* were reincorporated into the country. However, the creation of a federal-regional state was a matter of fierce disagreement. The African National Congress (ANC) advocated a strong unitary state, whereas the National Party (NP) and the Inkatha Freedom Party (IFP) supported the establishment of a strictly federal system, as well as the devolution of substantial powers to sub-national units. The ANC initially rejected the federal solution. A compromise was reached in favour of a feeble form of regionalism. On May 28th, 1993, the Negotiating Council of the Multi-Party Negotiating Process created the 'Commission on the Demarcation/Delimitation of States, Provinces and

[65] Sect. 146 could not be used because it 'required the admission be requested by the Legislature of the province seeking admission, and in 1949 Newfoundland lacked a Legislature': Hogg (2011), pp. 2–16.

[66] Miller (2016), p. 38.

[67] Miller (2016), p. 55.

[68] Seys (2019a), p. 59; Seys (2019b), pp. 6–8.

Regions' (CDDR), i.e. an independent technical body which was 'acceptable to all political parties and represented a balanced political, race and gender profile'.[69] The CDDR was entrusted with the task of redrawing South Africa by taking into account the delimitation criteria proposed by the political parties. The Commission reduced the criteria into four categories: economic features, geographical coherence (the boundaries of the development regions and the magisterial and district boundaries created under apartheid rule), sociocultural issues, and institutional capacity.[70] On the basis of such mixed criteria, it demarcated provincial boundaries. The outcome was the creation of nine provinces then embedded in the 1996 Constitution.[71]

As said, this process is a vivid example of how functional criteria undergo territorialisation. It also accounts for how the territorial constitution masters a variety of factors when matching its political imagined geography. If arbitrariness means that demarcation is always a political choice (see Sect. 7.3.2), in legal geographical terms mixing criteria politically may be a successful way for accommodating communities, territories, and their spatial practices, also giving consistency to all legal geographical primitives used to outline subnational legal geographies.

Another extremely relevant case study is Italy. Legal scholars and geographers have been endlessly debating the criteria on which Italy's territorial constitution, quasi-federalism, and top-down regionalism rest. The debate is still open.[72]

The transcripts of the Constituent Assembly reveal that several criteria were considered. In 1946, Gaspare Ambrosini, then President of the standing committee which drafted the constitution, acknowledged the existence of such a variety. Firstly, there was the historical criterion: 'In Italy regions are clearly demarcated upon a geographical and traditional basis'. At the same time, Ambrosini accepted the functional criterion: Italy needed 'regions that prove to be vital; therefore, it may be not proper to rigidly apply the historical criterion when demarcating regional territories'. The sole application of the historical criterion would have led to the creation of an elevated number of 'non-vital' (i.e. too small) regions. Finally, he introduced the political one: 'regions should be merged or changed in their historical dimension when it appears right to do so'.

Functional, historical, and political criteria generated different proposals for demarcation.[73] The plenary of the Constituent Assembly eventually approved a proposal for demarcation based on a mixed criterion. The historical and political ones were adapted to the map of Italy set forth in 1912 for statistical purposes. The historical-cum-statistical (and political) demarcation led to the list of regions now enshrined in Art. 131 of the Constitution.

---

[69] See Muthien and Khosa (1995).

[70] See De Visser et al. (2013), pp. 277–279.

[71] Sect. 103(2) and Schedule 1A of the 1996 Constitution.

[72] See e.g., Pedrazza Gorlero (1979), Pedrazza Gorlero and Nicolini (2010), Desideri (2013) and Nicolini (2019).

[73] Camera dei Deputati (1970–1971), p. 3602; Pedrazza Gorlero (1979), pp. 76–91.

We must also remember that the Italian territorial constitution comprises of 20 regions. Fifteen are called ordinary regions, since their belongs of public law are directly enshrined in the constitutional text. Five regions are denominated 'autonomous' regions, and their belongs are contained in separate basic laws adopted by an act at the constitutional level.[74] Several criteria were employed for their demarcation. These were mainly ethnic and geographical: Trentino-Alto Adige/*Südtirol*, Friuli-Venezia Giulia, and Aosta Valley/*Vallée d'Aoste* are border regions with a huge presence of linguistic minorities, whereas Sicily and Sardinia are islands. The historical and geographical criteria totally overlap when it comes to Sicily and Sardinia, whose insularity facilitated the demarcation process.[75] The legal geography of Trentino-Alto Adige/*Südtirol* is particularly asymmetrical since the provinces of Trento and Bolzano/*Bozen* have distinct belongs (and legislative powers).

As regards the Trentino-Alto Adige/*Südtirol* and Valle d'Aosta/*Vallée d'Aoste* regions, their basic laws clearly apply several criteria. Firstly, Art. 116.1 of the Constitution (as amended in 2001) enshrines the linguistic one since it entrenches the bilingual denomination of both autonomous regions. Secondly, minorities—which constitute the majority within their respective territories—are accommodated by being given a strong voice in these regions. The consequences of the geographical accommodation are illustrated in the Trentino-Alto Adige/*Südtirol* regions, where three communities coexist: the German-speaking majority, and the Italian- and the Ladino-speaking minorities. The interweaving of the linguistic criterion and power sharing triggers a close connection between demarcation and territorial identity and accommodates diversity. The 'ethnic' criterion thus supplements the 'historical' and 'geographical' ones.

The Trentino-Alto Adige/*Südtirol* region is also the outcome of a process of spatial production that dates back to the Austro-Italian Agreement signed in Paris on September 5th, 1946 and annexed to the Treaties of Peace. Its Art. 1 states that: 'German-speaking inhabitants of the Bolzano Province and of the neighbouring bilingual townships of the Trento Province will be assured complete equality of rights with the Italian-speaking inhabitants, within the framework of special provisions to safeguard the ethnical character and the cultural and economic development of the German-speaking element.' Furthermore, its Art. 2 grants 'The populations of the above-mentioned zones . . . the exercise of autonomous legislative and executive regional power.'

Likewise, the Constituent Assembly adopted linguistic, geographical, and historical criteria when demarcating the Aosta Valley/*Vallée d'Aoste* region. The regional basic law entrenches bilingualism (French and Italian are the regional official languages: Art. 38); furthermore, it lists the municipalities that are comprised within the regional boundaries (Art. 1.2). These are the French-speaking municipalities that had historically been part of the former Aosta province, to which the 1944–1946

[74] These are: Trentino-Alto Adige/*Südtirol*, Friuli-Venezia Giulia, Aosta Valley/*Vallée d'Aoste*, Sicily, and Sardinia. See Art. 116.1 Const.

[75] Art. 1 of the basic law for Sicily; Art. 1 of the basic law for Sardinia.

provisional government had granted administrative autonomy before the entry into force of the Constitution.

The demarcation of the Friuli-Venezia Giulia region was extremely contentious. This region comprises two distinct territories: the Friuli and the Venezia Giulia regions. On the one side, Friuli was part of the Venetian Republic (until 1797), of the Kingdom of Lombardy-Venetia (1815–1866), and of the Veneto statistic compartment. On the other side, the Venezia Giulia region was historically part of the Julian March and was a trilingual territory, where Italians, Croatians, and Slovenians coexisted. Together with Istria and Zara, the region was transferred to Italy after World War I. In the aftermath of World War II, the majority of the Julian March (i.e. Istria, Zara, and Carnaro) was ceded to Yugoslavia as war compensation, except for the provinces of Gorizia and Trieste, which were assigned to Italy.[76]

The Constituent Assembly added finishing touches upon the legal geography of the special regions. The two distinct territories of Friuli and Venezia Giulia were merged into a single region, and the Austro-Italian Agreement was extended to include the Italian-speaking province of Trento in the Trentino-Alto Adige/*Südtirol* special region; the 'neighbouring bilingual townships of the Trento Province' mentioned in the Agreement were aggregated to the Bolzano/*Bozen* province.

The proceedings of the Italian constituent assembly reveal that the territorial constitution did not merely readjust political entities to match the needs of decentralised governance. When adopting the regional territorial constitution, the nascent Italian republic deliberately mixed the criteria to strengthen the transition from fascism to democracy.

In 'mixing criteria politically', Italy's delimitation was a reaction to the processes of political and administrative centralisation Italy had experienced under monarchic rule and the fascist regime. After the annexation to Piedmont of the former independent Italian pre-unitary states, fear of secessions led to the rejection of a bill which the then Prime Minister, Marco Minghetti, had introduced into the Houses of Parliament in 1862 to establish administrative regions. Parliament eventually imposed legislative unification and administrative centralisation throughout the Italian Peninsula (law No. 2248 of 1865); and this had long-term consequences for the territorial constitution, which culminated in a huge concentration of power in the hands of the national government under fascist rule.

7.3.5 *Multi-National Territorial Constitutions*

The territorial constitution accommodates a variety of subnational logical spaces, spatial reshufflings, and belongs of public law. Its major task is therefore the holding together of the overlapping nomoscapes within multinational federations, where

[76]Fascism precluded the establishment of a multi-ethnic region, which was supposed to be established under the royal decree No. 1319 of 1921 and the royal decree law No. 1353 of 1922.

different identities and interests encroach upon the same territory. It also aims to govern national societies divided both ethnically and economically. In paving the way to multi-level-territorial identities, it has to find an equipoise between them, particularly when ethnicities imbricated on a given territory may lead them to consider *that* territory the repository of their own identity. The process of federal spatial production is thus unceasingly *à la recherche* of disentangling such imbrications and shaping distinct, albeit territorially related, ethnic identities.

For that purpose, scholars usually draw a distinction between societies governed under federal-regional constitutional arrangements. Some federations are said to have homogeneous societies, such as the USA, Germany, Austria, and Australia. This does not mean a lack of diversity. Certainly, we find 'significant black and Hispanic minorities', Indian communities, and a Southern 'distinctive' culture in the USA (see Sect. 6.7). Whatever their relevance, 'in no state do [these factors] constitute a majority'.[77] Likewise, Australia is a homogeneous federation, since 'the configuration of its constituent units, or states, is not coincident with marked cultural diversity'.[78] Germany and Austria are linguistically homogeneous; in Germany the main cleavage is probably drawn on political and geographical divides—those opposing the former West Germany to former Socialist East Germany. What I understand is that, in these countries, regional variations in belongs of public law are limited to the reshuffling of social, economic, and political factors.

The second type of federal societies are characterised by diversity. These multi-ethnic (or multi-national) societies are often 'divided' because segments of the federal society are differentiated upon linguistic, religious, and ethnic factors. The territorial constitution usually acknowledges such diversity. Not only does it enshrine discrete subnational belongs of public law, but their variety is also reflected in the federal framework. In adding a token of constitutional organisation to the container, federal constitutions infuse the distribution of powers on a territorial basis. Asymmetric (that is, multi-national) societies should possess differentiated (that is, asymmetric) competences reflecting their distinct processes of spatial production. And, it goes without saying, such a token may be a powerful and strategic mechanism for the governance of divided societies to avoid secession.

In drawing an equation between the qualities of their societies (homogenous-divided) and the descriptive analysis of their design (symmetric-asymmetric), scholars fall prey to federal arrangements as instrumentalities of states-as-containers. A legal comparative survey demonstrates that, as regards competences units should possess, a symmetric design also corresponds to federations built around societies that are divided on the ground of cultural, linguistic, and religious cleavages. Take, for example, the Constitutions of Switzerland and Canada, which apply symmetry to the distribution of powers. The Constitution of India now does the same. The sole asymmetry was related to the state of Jammu and Kashmir enshrined in Sect. 370 of the Constitution. This precedent for adopting asymmetric features was eventually

[77] Watts (2008), p. 29.

[78] Saunders (2015), p. 22.

repealed in 2019, when the Union Parliament created the Indian Union Territories of 'Jammu and Kashmir' and 'Ladakh' carved out of the former State of Jammu and Kashmir.[79]

Several other federal countries have tried to standardise diversity by imposing the characters of the dominant groups. We have mentioned the case of Canada, Belgium and Nigeria, whose territorial constitutions reflect unceasing processes of territorial rearrangement through which it has been possible to accommodate discrete (and often contrasting) groups. In Canada, the *British North America Act 1867* made the coexistence of English and French founding nations possible. Despite its attempts to secede in 1980 and 1995, Quebec's distinctiveness is still accommodated within the multinational federation. In addition, the ethnic criterion was applied to carve the territory of Nunavut out of the Northwest Territories in 1999 and confer self-government on the Inuit people. In Belgium and Nigeria policies of homogenisation were implemented. In Belgium, the nineteenth-century national logical space favoured the adoption of a French-related territorial constitution. In Nigeria, the Federation was created in 1946—that is, before it gained its independence in 1960—by creating three regions out of the former British colony. This arrangement favoured the three predominant ethnic groups and was fiercely resisted by the minorities living within the three ethnic-dominated regions.

Homogenisation rarely succeeds. It is often counter-productive. In Belgium and Nigeria counter legal geographies took place. Subnational spatial reshufflings started challenging it, trying to make their imagined legal geographies independent belongs of public law. To avoid this, both federal territorial constitutions were reframed on an ethno-linguistic basis; and their logical space was able to accommodate societies ethnically and economically divided, thus paving the way to multi-level-territorial identities (see Sect. 6.4.3).

To this end, the territorial constitution establishes an immediate relation between constitutions, subnational communities, and their territory. In multi-national federations, for example, territorially concentrated groups are accommodated by creating units that reflect their inherent pluralism. Besides Italy, where some special regions are established to protect linguistic minorities (see Sect. 7.3.4), this also occurs in Spain, where the national Constitution safeguards the so-called *hechos diferenciales*, i.e. linguistic and legal factors that make subnational communities extremely diverse and require ad hoc territorial units be created, such as Catalonia, the Basque country, Galicia, and Navarra.[80] In Switzerland, territorial diversity is the prerequisite of the federal constitution. Its logical space is a cross-cutting area where two dominant religious denominations (Protestant and Catholic) and four linguistic groups

---

[79] See Schedules I and II of the Constitution of India, as amended by the Jammu and Kashmir Reorganisation Act, 2019.

[80] See Aja (2014).

coexist.[81] The federal territorial constitution accommodates the reshufflings by allowing cantons to decide on State-Church relations[82] and their official languages. Cantons must respect the traditional territorial distribution of languages and take into account indigenous linguistic minorities. I may say that the territorial constitution enshrines a multi-linguistic geography which prevails over the subnational belongs of public law also supporting plurilingual cantons in the fulfilment of their duties.

The number of countries, which accommodate ethno-linguistic and religious cleavages, has increased at a steady rate in the past decades by resorting to federal arrangements. Throughout the book, we have examined a vast array of cases: Belgium, Canada, Bosnia and Herzegovina, South Africa, and Malaysia. In Malaysia, the 1963 Constitution adopted a two-tier federal system: 'Malaysia is . . . a Federation of three subjects (Malaya, Sabah, and Sarawak), one of which (Malaya) is itself a Federation'.[83]

Other cases in point are Russia, Ethiopia, and India. In Russia, the collapse of the Soviet Union led to the 1993 federal constitution, which retained the rationale of ethno-federalism. Due to the extremely fragmented Russian society, the demarcation is based on the principles of ethnicity and territory. With some exceptions, the 'dominant' ethnic groups represent a minority of the population in their respective ethnic-based constituent units.[84] Likewise, the Partition between Pakistan and India in 1947 led the nascent Union to manage the variety of ethno-linguistic cleavages populating the Subcontinent through demarcation. Although the Indian Union was created on the basis of a federal framework, the Constitutional Assembly did not use the term 'federation'. This was a result of the elevated ethnic fragmentation of the former *British Raj*: the term was perceived as a 'recipe for disintegration'.[85] Despite this, ethnic diversity was accommodated through territorial demarcation. Indeed, the Constituent Assembly admitted the former *Raj* provinces and territories, as well as numerous princely states. This gave rise to three categories of constituent units—provinces, princely states, union territories—that were then embedded in the First Schedule to the Constitution. Thus, Indian asymmetry goes far beyond the distribution of powers, and concerns several features, such as territorial identity.

In the aftermath of the military rule in 1991, Ethiopia also experienced 'ethnic pluralism . . . as an organizing principle to establish a federal state, creating primarily ethnic-based territorial units'.[86] Despite the ongoing civil war, the 1994 Constitution still entrenches ethnic pluralism in Art. 39; and Art. 46.2 organises units' boundaries

---

[81] There are four national languages (German, French, Italian, and Romansh: Art. 4) and three official ones (German, French and Italian; Romansh is an official language when communicating with persons who speak Romansh: Art. 70).

[82] Art. 72 of the Constitution; the federal level accommodates the subnational belongs by adopting measures to preserve public peace between the members of different religious communities.

[83] Harding (2013), p. 134. See also Sect. 78–80 and the Ninth Schedule to the Constitution.

[84] Ross (2007), p. 109.

[85] See Bhattacharyya (2007a, b), p. 58; Bhattacharyya (2019), p. 85.

[86] Habtu (2004), p. 92.

along ethnic lines. In Tigray, Afar, Amhara, Oromia, and Somali, the majority of inhabitants belong to a predominant ethnic group. Several ethnic groups coexist in the southern regional state of Benshagul-Gumz, and in Gambela, where a 'mix of factors like political, economic, settlement pattern, similarity of culture and language have been taken into consideration' for territorial demarcation.[87]

As Russia, India, and Ethiopia demonstrate, it is often difficult to assign each community a state of its own. Units usually comprise a majority corresponding to the demarcating ethnic criterion, at the same time including one or more minority groups. This usually occurs when multi-national societies are 'dispersed' throughout the country. This implies that regional processes of demarcation and spatial production are based on crosscutting constituent units—that is, several factors (ethnic, economic, and social) and numerous communities intertwine on the same territory, as in South Africa: although societies are divided on ethnic and linguistic lines within provincial boundaries, demarcation tends to create compact socio-economic units. In Malaysia, ethnic groups are more concentrated in some states than in others, and minorities are generally spread fairly evenly across the country—and the federal government accommodates diversity by promoting integration programs.[88]

Scholars pointing to the equation between societies (homogenous-divided) and design (symmetry-asymmetry in the distribution of powers) lose sight of the role of legal geographical primitives in manufacturing federal belongs of public law. Symmetry is a legal fiction (*fictio juris*), which conceals the constitutional relevance of territory and opens the production of spaces to the socio-geographical contexts that enrich them with new meanings. Relating the territorial constitution to the distribution of powers has indeed a noticeable consequence: it turns the process of spatial production and primitives into size, population, wealth, and economic growth. Hence, a narrowed notion of territorial constitution merely focuses on *legal* equality, disregarding the contexts where belongs of public law are manufactured.

Slyly kept in the background, these are treated as *de facto* federal features. Such a formalistic approach towards territorial governance may lead to underestimating the territorial nature of the federal constitution. This may also lead to dismiss the relevance of overlapping nomoscapes within multinational federations. Far from finding an equipoise between them, it may improve territorial imbrications leading them to consider *that* territory as the repository of their own identity. The process of federal spatial production, as said above, is unceasingly *à la recherche* of disentangling such imbrications and shaping distinct, albeit territorially related, belongs of public law.

---

[87] Tesfai Baraki (2013), p. 235; Fiseha (2019), pp. 73–74.

[88] Bakar (2007), pp. 74 et seq.; Harding (2013), pp. 135 et seq.

7.4 Territorial Alteration

7.4.1 Appropriating Spatial Reshufflings

The last legal geographical primitive is territorial alteration. Throughout the chapter, I have highlighted how legal geographical primitives are constantly being made real through scalar territorial identities. Furthermore, the territorial constitution entrenches the outcomes of such processes, by accommodating subnational belongs within federal arrangements. As is evident, the territorial constitution presupposes the division of national territory into constituent units, as if units (and their belongs) were 'indestructible' subnational containers.

I have also said that it secures their territorial identity; and that, at the same time, it lays down the procedures for altering their borders. I have already explained this apparent paradox (see Sect. 7.3.1). Units may be regarded as 'indestructible states' because the territorial constitution turns *relative* subnational legal geographies into *absolute* belongs. Yet, the relation between contingent-relative and absolute production of subnational spaces is more complex. We know that processes of spatial production are contingent; community may indeed forge 'its own – *appropriated* – space'[89] again and again, unceasingly adapting the repository, which is their territory, to new meanings and spatial reshufflings that unceasingly take place in their logical spaces. What is absolute today may become relative tomorrow. Consequently, the territorial constitution must turn new reshufflings into renewed *absolute* belongs of public law, constantly accommodating them into the federal logical space.

In most cases, subnational reshufflings affect neither units-as-containers nor their borders. The container remains. What changes is the content, i.e. the subnational territory as a meaningful place. It may happen, though, that reshufflings require a change in borders. In this event, what changes here is the scope of the relation between communities and territories; or, which is the same thing, the change in the territorial meaning is a change in the same idea of subnational territoriality. As territory is a constitutive feature of constituent units, its alteration, creation, and dissolution—as well as any other form of alteration of their boundaries and territories—require the consent of the units affected. This assumption was upheld, among others, by the Federal Constitutional Court of Germany.[90] Although the German territorial constitution entrenches the division of the Federation into *Länder* as its basic feature,[91] the territories of *Länder* may be altered by applying the procedures laid forth in the same Basic Law.[92] The Court addressed the 'destructible character' of *Länder* by defining Germany a *Labilen Bundestaat*, i.e. a territorially feeble Federation. The alteration of the territorial constitution is thus admissible,

[89] Lefebvre [1974] (1991), p. 31.

[90] See ruling 23 October 1951 – 2 BvG 1/51, BVerfGE 1, 14 (*Südweststaat-Urteil*).

[91] See Art. 79.3 of the German Basic Law.

[92] See Art. 29.

providing that it reflects changes in subnational legal geographies that adapt their territorial identity to new meanings previously assigned to territory in logical space. Again, territorial readjustments among two or more *Länder* require that the communities of the units concerned approve the proposed reshuffle and territorial readjustment.

Territorial alteration triggers a change in how all other legal primitives are located within the federal logical space. Such territorial changes indeed may alter the collocation of lines, points, and polygons, thus affecting the belongs of public law of other units. All these primitives lie in the territorial constitution as if they were on a map and are usually a constituent part of it. It thus coordinates all legal geographical primitives and find an equipoise between changes affecting subnational logical spaces. If federacies want to preserve the equipoise, they must exhibit attentiveness to subnational contingencies and reshufflings that may disrupt the operation of the federal order. Tackling contingencies requires a constant exercise in legal geographical creativity aimed at accommodating imagined legal geographies into our complex world.

No wonder, therefore, that mechanisms of territorial alteration are enshrined in federal constitutions and require the consent of affected units.[93] This consent may also be expressed by subnational legislatures or through a territorial referendum, which counts as the will of the units.[94] With the exception of India (see Sect. 7.4.2), alterations cannot be unilaterally approved by national parliament. The territorial constitution commits the decision on the proposed reshuffling to the same territorial community, which thus ascertains whether the proposed alteration reflects a new belong of public law. If the population accepts the proposed reshuffling, it decides its own 'self-identification' as a new subnational community, and therefore sets a new relation between itself and its new territory.

7.4.2 *Altering Logical Spaces: Procedures*

Procedures for altering units' territories vary from system to system. Nevertheless, a legal comparative perspective reveals that there are at least three different formulas. Firstly, the territorial constitution may call for constitutional amendments to accommodate the proposed subnational reshufflings with the belongs of the other units. Whatever the case may be, alterations cannot be unilaterally approved by national parliament. In the United States, for example, the creation of new states 'within the Jurisdiction of any other State', the erection of new ones 'by the Junction of two or

[93] See e.g., the USA (Art. IV Sect. 3 of the U.S Constitution), Malaysia (Sect. 2.a), India (Sect. 2), South Africa (Sect. 74.3), and Canada (Sect. 3 of the Constitution Act of Canada).

[94] See Art. 29 of the German Basic Law, Art. 132 of the Italian Constitution; Art. 53 of the Swiss Constitution, Sect. 124 of the Australian Constitution. The constitutions of Australia (Sect. 123) and Switzerland (Art. 53.4) contain ad hoc provisions for altering limits of States.

more States, or Parts of States', requires the 'Consent of the Legislatures of the States concerned as well as of the Congress'.[95] Likewise, in Canada and Australia, territorial alteration requires constitutional amendments.[96] In Russia, the number of subnational units and their constitutional legal status may be altered by adopting a constitutional law, i.e. a piece of legislation intermediate between the constitution and federal laws. A less rigid form of amendment is foreseen if changes are made to their denominations. The President of the Federation replaces the old denomination with the new one—as approved by the subnational unit that has exclusive jurisdiction thereover—by issuing a presidential decree.[97]

The second formula for territorial alteration requires a constitutional amendment provided that the proposed alteration affects the very existence and the number of constituent units, which are listed in the constitution as in Switzerland, Austria, and Italy.[98] By contrast, territorial alteration does not call for an amendment when the very existence or the comprehensive number of constituent units is not affected. For example, minor territorial adjustments point to the separation of a part of a subnational unit and its aggregation to another one. According to Art. 53.3 of the Swiss Constitution, this may occur when the alteration is approved both by the population of the Cantons affected and federal parliament. Article 132.2 of the Italian Constitution allows provinces and municipalities to be separated from one region and aggregated to another one. Requests for territorial regrouping must be submitted to referendum. Once the population of the provinces and municipalities concerned has approved the regrouping, Parliament can pass the act for territorial readjustment. Either the population (Switzerland and Italy) or subnational legislatures (as in Austria) must consent to alteration.[99]

The third model is similar to the first one, inasmuch as it also requires a constitutional amendment. According to Sect. 74(3)(ii) of the South African Constitution, constitutional amendments altering provincial boundaries have to be approved both by two-thirds of the members of the National Assembly and by the National Council of Provinces with the supporting vote of at least six provinces. However, territorial alterations concerning 'a specific province or provinces' must

[95] See Art. IV.3.1 of the Constitution.

[96] For Canada, see Sections 3 of the Constitution Act, 1871, and 42 of the Constitution Act, 1982. Sect. 43 of the Constitution Act 1982 requires the consent of subnational legislatures. In Australia, the consent (of federal and state parliaments) is complemented by the will of 'the majority of the electors of the State' affected (Sections 123 and 128 of the Commonwealth of Australia Constitution Act 1900). Their will is not required when a 'new State [is to] be formed by separation of territory from a State . . . [or] by the union of two or more States or parts of States' (Sect. 124).

[97] Units are listed in Art. 65 of the Constitution, whereas territorial alteration is laid forth in Art. 137.1.

[98] See Art. 53 of the Swiss Constitution; Art. 132.1 of the Italian Constitution (requiring a constitutional act to create new or merge existing Regions and the consent of the electors of the regions affected). In Austria, Art. 2.2 of the Constitution (as amended in 2008) states that territorial alteration implies an amendment to both federal and provincial constitutions. Se Bußjäger (2009).

[99] In any case, consent is not necessary for the mere alteration of regional boundaries in Switzerland (Art. 53.4) and Austria (Art. 3.4).

comply with an additional requirement: territorial alteration cannot be passed by 'the National Council of Provinces . . . unless it has been approved by the legislature or legislatures of the province or provinces concerned' (see Sect. 74(8)). Likewise, Sect. 239.4 of the Constitution of Pakistan prescribes that constitutional amendments which may alter the limits of a Province must also be approved 'by the Provincial Assembly of that Province by the votes of not less than two-thirds of its total membership'. In sum, the federal legislature may alter several provincial boundaries, propose minor territorial readjustments, and create (or merge) provinces by *having recourse to a single constitutional amendment process.* Thus, a general territorial regrouping is admissible by approving just one constitutional amendment act. This occurred in South Africa with the Constitution Twelfth Amendment Act of 2005 (see Sect. 7.4.3). The act modified Sect. 103 of the Constitution (which sets the number and denominations of Provinces), provided for abundant minor alterations, and introduced Schedule 1A, thereby establishing a new criterion for territorial delimitation, which is now based on the municipal demarcation map.

There is, then, the case of India, which sets an important exception to the rule of constitutional amendment. I have already said that its territorial constitution is an admixture of several belongs of public law. After the 1947 partition with Pakistan, 11 ethnically heterogeneous provinces, and 561 princely states represented the 'geographical' legacies of the British Raj, which were spatially reshuffled by the 1950 Constitution so as to create 27 states and Union Territories 'of different types, powers, and status'.[100] After independence, the reorganisation of Indian legal geography required a comprehensive alteration of state boundaries. To this extent, the Union government appointed the States' Reorganisation Commission in 1953, and then passed the State Reorganisation Act in 1956. This led to a continuous process of state creation and territorial alteration.[101]

The lists of states enshrined in the First Schedule to the Constitution have been augmented continuously from the 1950s onwards. In this respect, the Constitution sets down a process for territorial alteration that is flexible and capable of serving different purposes. Pursuant to Sect. 3 and 4 of the Constitution, national parliament may create new states by separating a territory to form a new state, by merging two or more states or parts thereof, by increasing the area of a state, or by diminishing or otherwise altering boundaries by passing a federal law. Such a bill is referred to the legislatures of the affected states, and parliament is not bound by the view of the state concerned. The territorial constitution thus gives 'Union Parliament virtually unilateral power to' reshuffling spatially the belongs of public law of states and territories.[102] This is confirmed by the case of Jammu and Kashmir, whose asymmetric status based on ethnic and religious grounds was altered by the Jammu and Kashmir Reorganisation Act, 2019. This piece of federal legislation, as said, created the

[100] See Bhattacharyya (2019), p. 85.

[101] See Singh (2013); Bhattacharyya (2019), pp. 86–88.

[102] See Bhattacharyya (2019), p. 86.

Union Territories of 'Jammu and Kashmir' and 'Ladakh' carved out of the former State of Jammu and Kashmir in 2019.

7.4.3 Coherence Test

When it comes to territorial alteration, the change in subnational belongs of public law also implies altering the same criteria upon which the original demarcation was drawn. I have already referred to the different criteria which the territorial constitution may adopt when making the original demarcation and its legal geographical primitives (see Sects. 7.3.4 and 7.3.5).

It does not come as a surprise that territorial alteration gives rise to relevant questions. This is what I term the 'coherence test': should the criteria enabling territorial alteration be consistent with those applied to the original demarcation? The answer to the question depends on whether the territorial constitution exhibits a neutral attitude towards the belongs of constituent units. The territorial constitution may consent to alteration without imposing any criteria. If this is the case, 'territorial neutrality' (see Sect. 6.6.2) will only prescribe that territorial readjustments be achieved in accordance with the procedures enshrined in the constitution.

In most cases, constitutions favour spatial reshufflings generating organic jurisdictions; whether territorialised or not, they accept those territorial alterations that prove to be consistent with the economic, political, ethnic, religious, and linguistic criteria originally used for demarcation and that are still compatible with those underpinning the belongs of public law of other units. This is patent when it comes to detecting the criteria that can avoid ethnic conflicts, say, in multi-ethnic federations, where the original demarcation is drawn along ethnically oriented lines. In these cases, divided societies assume that there must be consistency between the criteria adopted for the original demarcation and those employed for subsequent territorial alterations. This avoids undermining the whole legal geography and the existence of the federation, thus accommodating the belongs of public law of groups through the creation of new ethnic-based units.

The practice of territorial alteration has also been consistent with the ethnic, linguistic, and religious criteria governing the original demarcation. We have already seen the cases of Nigeria, and Belgium, where the numerous and often contrasting groups living within its boundaries have been accommodated along ethno-linguistic belongs of public law (see Sect. 6.4.3). Switzerland did the same. The Canton of Jura was carved out of the Canton of Bern along linguistic and religious lines. Indeed, the Catholic, French-speaking inhabitants of Jura never identified with the Protestant, German-dominated Canton of Bern and, in 1979, a popular vote led to the creation of a new canton.[103] However, a small French-speaking community remained within the boundaries of Bern. After 1979 Jura continuously tried to incorporate both the

[103] Belser and Setz (2013), p. 12.

French-minority and the corresponding territory under Bern's jurisdiction. In 2013, another referendum was held to allow the French-minority to join the Canton of Jura. The initiative failed, however, as the French-minority did not accept the territorial readjustment proposed by Jura.[104]

More recently, the Canton of Fribourg had its boundaries altered. The municipality of Clavaleyres, which had historically been part of the Canton of Bern, was merged with the city of Murten located in the Canton of Bern. The communities of Murten and Claveleyres consented to the merger in 2020, thus accepting the proposed reshuffling. In so doing, the community of Claveleyeres accepted that its belongs of public law coincided with those of Murten, also setting a new relation between itself, its territory, and Fribourg. In 2021, an intercantonal agreement between Bern and Fribourg ratified this new territorial identification, thus paving the way to the territorial alteration that took place on January 1st, 2022. What makes this case intriguing is that both communities are linguistically homogenous. Having complied with the coherence text, the alteration gave consistency to the geographical criterion, inasmuch as Claveleyeres was a German-speaking exclave in the canton of Fribourg.

Likewise, in Italy, territorial alterations were consistent with the original demarcation criteria. The most relevant change in the territorial constitution was the creation of the Molise region, which was separated from the Abruzzi region in 1963. Far from altering it, the creation of this region was consistent with the territorial constitution. Firstly, the Constituent Assembly had already approved the creation of the Molise region, and its establishment had been merely delayed. Secondly, its creation still reflected 'specific historic and cultural elements': these had been sacrificed, and 'the creation of Molise then was a sort of "reparatory action" that in any case followed in the wake of "historic" regionalism'.[105] There have also been minor variations. In 2009, several municipalities were separated from the Marche region and aggregated to the Emilia-Romagna region: a referendum was held, and the population of the municipalities assented to the alteration because they self-identified with the proposed reshuffling. In 2017 alteration concerned the Friuli-Venezia Giulia region, i.e. a special region. The change in its territorial identity reflects the ethnic criterion, which was also used for its demarcation. In 2017 Parliament separated the bilingual municipality (Italian and German) of Sappada/*Plödn* from the Veneto region and aggregated it to the Friuli-Venezia Giulia region by passing a piece of ordinary legislation.

Likewise, in India, the ethnic criterion has been constantly applied. This was particularly evident in in the territorial management of the North-East. This was indeed 'distinct from the rest of India in terms of history, demography, and political

[104] See Ruiz Vieytez (2014), p. 101.

[105] Desideri (2013), pp. 191–192. Parliament did not resort to the procedures envisaged in Art. 132 of the Constitution, but applied those enshrined in Art. XI of the Transitory and final provisions of the Constitution.

institutions'.[106] Originally split between Assam and the princely states of Manipur and Tripura, it experienced the creation of the 'Christian-dominated states' of Nagaland (1963), Meghalaya (1971), and Mizoram (1987). In the secular Indian logical space, the criterion was not religious, but ethno-tribal: 'the legitimacy basis of their creation ... was ... tribal ethnicity', thus concealing the presence of the religious one.[107]

In South Africa, claims for territorial alteration proceeded from the so-called affected areas. These were 14 areas, whose incorporation within a province was supposed to be extremely contentious. The Demarcation Commission had suggested such areas be entitled to a petition for a referendum. Indeed, Sect. 124 of the 1993 Interim Constitution endorsed recourse to a referendum as a 'mechanism for subsequent appeal and the resolution of disputed new boundaries'.[108] Political parties preferred a political solution rather than a legal one. Consequently, territorial alteration process was not activated. Several geographical disputes arose, and the affected area of Bushbuckridge—aiming to separate from Limpopo and to join Mpumalanga—filed a case before the High Court in Pretoria.[109] In order to resolve such a dispute, national parliament passed the Constitution Twelfth Amendment Act of 2005, which disregarded both territorial identity and affected areas' requests. As a result, local communities challenged the validity of the Amendment Act before the Constitutional Court, which held that the amendment was unconstitutional. Although alterations were consistent with the procedure set forth in Sect. 74(8) of the 1996 Constitution, national parliament did not meet the levels of public participation required by the constitution while altering provincial boundaries.[110] Nonetheless, such alterations match the coherence test. In the absence of ethnic criteria, the sole criterion governing alteration is the territorial identity of the people concerned.

In sum, territorial alteration exhibits a relatively high degree of 'coherence' with the original demarcation. In Switzerland, India, Belgium, South Africa, Italy, and Nigeria, the practice of territorial alteration has proven to be consistent with the ethnic, linguistic, and religious criteria governing the original territorial demarcation. As for multi-national federations, in particular, the purpose of holding together divided and conflicting groups seems to set a substantive limitation to territorial alteration. The readjustment of territorial complexity should, therefore, be consistent with the rationale of the territorial constitution in multinational federacies.

---

[106] Bhattacharyya (2019), p. 90.

[107] Bhattacharyya (2007a, b), pp. 54–55.

[108] Ramutsindela and Simon (1999), p. 482. The *affected areas* were listed in the second part of Schedule 1 to the 1993 Interim Constitution.

[109] High Court of South Africa, *Judgment in the High Court of South Africa, in the matter between Bushbuckridge border committee, Michael Mangisi Mnisi and the Government of Northern Province, the Government of Mpumalanga, the Government of the Republic of South Africa, the African National Congress*, Case No. 15607/97.

[110] See Constitutional Court of South Africa, *Matatiele Municipality and Others v. President of the Republic of South Africa and Others* CCT 477/07 [2006] ZACC 12 ('Matatiele 2').

7.4.4 Functional Reshufflings

Territorial alteration also gives rise to a further question: is it possible that economic factors affect the coherence test and gain relevance over criteria that are based on political, historical, and statistical factors? The question at stake is particularly fitting in terms of legal geographical primitives, particularly in multi-national federations. If the territorial constitution could exhibit a neutral attitude towards economic factors, then it might promote a paradigm shift in how its own federal space has been construed. When proposing to redraw regions on functional considerations, federal governments may endorse spatial reshufflings able to replace the way that subnational communities conceive of their relationship with their territory with a new one imposed from above. Territories would probably lose their ability to be repositories of deep-seated 'beliefs', thus capturing how people define their regions and belongs of public law.

Functional criteria are particularly appealing. Access to services, facilitation, allocation of natural resources and distribution of financial proceeds are indeed 'winning arguments' that may convince communities to adopt the ideo-logical space of globalisation made up of reassuring imagined geographies.

We have already seen how ethnic-based criteria presiding territorial demarcation and alteration may play an even less relevant role in shaping territorial identity. This is evident in South Africa, where the original demarcation favoured the creation of crosscutting communities. Within provincial boundaries, ethnic societies are still divided on ethnic and linguistic lines, but they tend to become compact in a socioeconomic perspective. The same thing has occurred in Malaysia. Although some groups are more concentrated in some states than others, ethnicities are spread across the country,[111] and the federal government accommodates diversity by promoting integration programs. In India, geographical reorganisation along linguistic lines did not grant any privileged protection to linguistic communities and facilitated the creation of states based on crosscutting cleavages and identity.

Economic factors also reflect the safe ideo-logical space of globalisation, thus giving us the illusion of taking part in the global distribution of wealth by improving our (and our communities') standards of life. As is evident, ideo-logical spaces favour a change in our perception of spatiality: we have seen that commercial comfort zones are functional to safe spaces of consumption (see Sect. 5.3.1); and we shall also see how these zones are the outcomes of a specific production of space, which aims to reshuffle the border-less logical space of commerce and trade to maximise financial returns (see Sect. 8.5). At the same time, economic factors are also directly implicated in subnational spatial reshufflings. Just to mention an example, in 2012 the Swiss federal government delivered the *Raumkonzept Schweiz* (Switzerland Space Concept): the homogenisation of federal society, the globalised economic system, increased mobility throughout the country, and intra-federal immigration toward metropolitan areas have induced the federal government to

[111] Harding (2013), p. 135.

deliver 'strategies for sustainable spatial planning' based on geographical and functional areas.[112]

The idea of spatially reshuffling Swiss logical space confirms that the intersection of traditional and socioeconomic factors represents the most noticeable challenge to an ethnically construed territorial identity. Such an intersection adds additional layers of complexity to the governance of divided societies, which, through constituent units, implies a reassessment of the criteria presiding internal political demarcation and new coherence tests in territorial alteration.

Natural resources or economic and financial areas may lie in a territory that is deemed to be a constitutive element of two or more conflicting ethnic groups. I have said that the territorial constitution has to find an equipoise between them, particularly when ethnicities imbricated on a given territory may lead them to consider *that* economic territory the repository of their own identity. In this event, borders perform the function once staged by the (imbricated) territory, i.e. they disentangle the imbrication and shape distinct, albeit territorially related, identities.

However, when several territorial identities and a single economic (and territorial) interest overlap, demarcation are perceived as insufficient means for governing a crosscutting, divided society. Indeed, societies would not be divided on the grounds of ethnic and linguistic cleavages, but on the basis of economic interests. To this extent, multinational federations must cope with new cleavages, as well as with new criteria for the creation of sub-national units not merely ethnically oriented. Indeed, it would be very bizarre if the governance of divided societies were attained on ethnic grounds while, at the same time, federations were divided on the basis of economic factors, which are neutral with respect to ethnicity and language.

We do not know how these processes will impact on the legal production of subnational spaces. What is clear is that, as I said above, functional criteria are variables like the others, which may be territorialised, spatially reshuffled, and processed together with human, physical, and cultural geographical attributes alike. 'Territorialised' functional criteria may thus undergo a process of spatial production, whose outcome is usually the establishment of a scalar (i.e. subnational) belong of public law.

The aim, it goes without saying, is to create a strong relation between community and territory. Thomas Perrin is right when delivering his idea of the 'sense of commonness' between the discrete communities of macro-regions. The pattern is the same; like in macro-regions, we have overlapping economic and ethnic interests; we have distinct territorial identities but one territory to manage under a new common framework. What we need to find is the 'commonness' that may act as the 'driving force' conveying 'the will or the need to cooperate' in a given region.[113] Again, we need that functional-but-territorially-located criteria act as the 'third' (i.e. imaginative) geography, that is, as the interface between physical, human, and now economic geographies in the production of legal spatialities (see Sect. 3.2).

[112] See Belser and Setz (2013), pp. 6–8.

[113] Perrin (2019), pp. 162–163. See also Evrard (2021).

References

Agnew JA (2013) Arguing with regions. Reg Stud 47:6–17

Aja E (2014) Estado autonómico y reforma federal. Alianza, Madrid

Anderson G (2008) Federalism: an introduction. OUP, Oxford

Aroney N (2006) Representation and amendment in federal constitutions. Am J Comp Law 54:277–336

Bakar I (2007) Multinational federation: the case of Malaysia. In: Burgess M, Pinder J (eds) Multinational federations. Routledge, Abingdon and New York, pp 68–85

Baker B (2009) Comoros: the search for viability. Civ Wars 11:215–233

Belser EM, Setz R (2013) The territorial structure of federal Switzerland revised: about the difficulty of being small. In: Mangiameli S (ed) Federalism, regionalism and territory. Giuffrè, Milan, pp 1–30

Bhattacharyya H (2007a) Federalism and competing nations in India. In: Burgess M, Pinder J (eds) Multinational federations. Routledge, London and New York, pp 50–67

Bhattacharyya H (2007b) Federalism and competing nations in India. In: Burgess M, Pinder J (eds) Multinational federations. Routledge, Abingdon and New York, pp 50–67

Bhattacharyya H (2019) States reorganization and accommodation of ethno-territorial cleavages in India. In: Anderson M, Choudry S (eds) Territory and power in constitutional transitions. OUP, Oxford, pp 81–98

Bhattacharyya H (2021) Federalism in Asia. India, Pakistan, Malaysia, Nepal, and Myanmar, 2nd edn. Routledge, Abingdon and New York

Bovis C (2011) The role and function of structural and cohesion funds and the interaction of the EU regional policy with the internal market policies. In: Panara C, De Becker A (eds) The role of the regions in EU governance. Springer, Berlin and Heidelberg, pp 81–109

Brennetot A (2018) A step further towards a neoliberal regionalism: creating larger regions in contemporary France. Eur Urban Reg Stud 25:171–186

Bußjäger P (2009) Grenzänderung und Bundesstaat. Eine verfassungsrechtliche Analyse der Neuregelung der Art 2 und 3 B-VG mit der Novelle BGBl I 2/2008. Zeitschrift für öffentliches Recht 64:115–131

Camera dei Deputati (1970–1971) La Costituzione della Repubblica nei Lavori Preparatori della Assemblea Costituente, Segretariato Generale, Rome, vol I–VIII

Castree et al (2013) Oxford dictionary of human geography. OUP, Oxford

Crane B et al (2018) China's special economic zones: an analysis of policy to reduce regional disparities. Reg Stud Reg Sci 5:98–107

Danilovich A (ed) (2019) Federalism, secession, and international recognition regime. Iraqi Kurdistan. Routledge, Abingdon and New York

De Visser J et al (2013) The role of ethnicity in the demarcation of internal boundaries in South Africa and Ethiopia. In: Mangiameli S (ed) Federalism, regionalism and territory. Giuffrè, Milan, pp 267–289

Delledonne G, Martinico G (eds) (2019) The Canadian contribution to a comparative law of secession. Palgrave Macmillan, Cham

Desideri C (2013) Regionalism and territorial politics in Italy. In: Mangiameli S (ed) Federalism, regionalism and territory. Giuffrè, Milan, pp 149–196

Domorenok E (2013) EU regional policy and the process of institution-building in Central and Eastern European countries. In: Palermo F, Parolari S (eds) Regional dynamics in Central and Eastern Europe new approaches to decentralization. Martinus Nijhoff, Leiden and Boston, pp 35–58

Duffy Ponsa-Kraus C (2022) The insular cases run amok: against constitutional exceptionalism in the territories. Yale Law J 131:2449–2541

Dziedic A (2020) Micronesia. In: Griffiths A et al (eds) The forum of federations handbook of federal countries. Springer, Cham, pp 227–236
Elazar DJ (1987) Exploring federalism. University of Alabama Press, Tuscaloosa
Elazar DJ (1994) Federal systems of the world, 2nd edn. Longman Group, Harlow
Evrard E (2021) Reading European borderlands under the perspective of legal geography and spatial justice. Eur Plan Stud. https://doi.org/10.1080/09654313.2021.1928044
Fiseha A (2019) Emergence and transformation of territorially based cleavages and constitutional responses in Ethiopia. In: Anderson M, Choudry S (eds) Territory and power in constitutional transitions. OUP, Oxford, pp 60–80
Ford RT (1999) Law's territory (a history of jurisdiction). Mich Law Rev 97:843–930
Friedrich CJ (1963a) New dimensions of federalism. In: Proceedings of the American Society of International Law at Its Annual Meeting (1921-1969), vol 57, pp 238–240
Friedrich CJ (1963b) Separation of powers. International encyclopaedia of social sciences, vol 13. MacMillan, New York, pp 663–666
Friedrich CJ (1968) Trends of federalism in theory and practice. Praeger, New York
Gänzle S (2019) Experimenting governance and EU macro-regional strategies: new dynamics in European regional and territorial cooperation? In: Abels G, Battke J (eds) Regional governance in the EU. Regions and the future of Europe. Edward Elgar, Cheltenham and Northampton, MA, pp 140–158
Griffiths A (2020) St. Kitts and Nevis. In: Griffiths A et al (eds) The forum of federations handbook of federal countries. Springer, Cham, pp 301–314
Habtu A (2004) Ethnic pluralism as an organizing principle of the Ethiopian federation. Dialect Anthropol 28:91–123
Harding A (2013) The constitution of Malaysia. A contextual analysis. Hart, Oxford and Portland, OR
Harding A (2017) Devolution of powers in Sarawak: a dynamic process of redesigning territorial governance in a federal system. Asian J Comp Law 12:257–279
Harding A (2018) 'A measure of autonomy': federalism as protection for Malaysia's indigenous peoples. Fed Law Rev 46:557–574
Himonga C, Nhlapo T (eds) (2015) African customary law in South Africa. Post-apartheid and living law perspectives. OUP South Africa, Cape Town
Hogg PW (2011) Constitutional law of Canada. Carswell, Toronto
Huber H (1967) Die Gleichheit der Gliedstaaten im Bundesstaat. Österrechische Zeitschrift für öffentliches Recht 17:247–261
Hueglin TO, Fenna A (2015) Comparative federalism: a systematic inquiry. Broadview Press, Peterborough
Jones CPA (2021) The islands that ate the constitution. Liverpool Law Rev 42:51–69
Kelsen H (1920) Die Verfassung deutsches Österreichs. Die Offentliche Recht der Gegenwart 11:232–274
Lefebvre H (1974) La production de l'espace. Éditions Anthropos, Paris. English edition: Lefebvre H (1991) The production of space (trans: Nicholson-Smith D). Blackwell, Oxford
Lijphart A (1979) Consociation and federation: conceptual and empirical links. Can J Polit Sci/Revue canadienne de science politique 12:499–515
López-Basaguren A, Escajedo San-Epifanio L (eds) (2019) Claims for secession and federalism. Springer, Cham
Miller D (2016) Boundaries, democracy, and territory. Am J Jurisprud 61:33–49
Muthien YG, Khosa MM (1995) The Kingdom, the Volkstaat and the New South Africa: drawing South Africa's new regional boundaries. J South Afr Stud 21:303–322
Nicolini M (2019) Reforming the territorial constitution in Italy: some reflections on durability and change. In: Abels G, Battke J (eds) Regional governance in the EU. Regions and the future of Europe. Edward Elgar, Cheltenham and Northampton, MA, pp 106–123

Nicolini M (2022) Borders. In: Max Planck Encyclopaedia of Comparative Constitutional Law [MPECCOL]. OUP, Oxford

Ojalammi S, Blomley N (2015) Dancing with wolves: making legal territory in a more-than-human world. Geoforum 62:51–60

Palermo (2016) Territory and the law of ownership: from misunderstanding to opportunity. In: Nicolini M et al (eds) Law, territory and conflict resolution. Law as a problem and law as a solution. Martinus Nijhoff, Leiden and Boston, pp 16–38

Panara C (2015) The sub-national dimension of the EU. A legal study of multilevel governance. Springer, Cham

Pedrazza Gorlero M (1979) Le Variazioni Territoriali delle Regioni. Contributo allo Studio dell'art. 132 della Costituzione, I, Regioni Storiche e Regionalismo Politico nelle scelte dell'Assemblea Costituente. Cedam, Padua

Pedrazza Gorlero M, Nicolini M (2010) Il principio costituzionale del consent nella decisione delle variazioni territoriali degli enti intermedi. Diritto Pubblico Comparato ed Europeo 12:1070–1106

Perrin T (2019) The cultural dimension of macro-regions: a prospective reflection. In: Abels G, Battke J (eds) Regional governance in the EU. Regions and the future of Europe. Edward Elgar, Cheltenham and Northampton, MA, pp 152–175

Perrin T (2022) La region au XXIe siècle. Peter Lang, Bruxelles

Perrin T, Seys FO (2019) Editorial : La région, vous dîtes ? Le kaléidoscope régional de l'Union européenne », *Belgeo* [En ligne], 2, http://journals.openedition.org/belgeo/34552; https://doi.org/10.4000/belgeo.34552

Ramutsindela MF, Simon D (1999) The politics of territory and place in post-apartheid South Africa: the disputed area of Bushbuckridge. J South Afr Stud 25:479–481

Rogerson PA (2019) Statistical methods for geography: a student's guide, 4th edn. SAGE, London et al

Ross C (2007) Russia's multinational federation: from constitutional to contract federalism and the 'war of laws and sovereignties'. In: Burgess M, Pinder J (eds) Multinational federations. Routledge, London and New York, pp 108–122

Ruiz Vieytez EJ (2014) ¿Derecho a decidir en clave interna? El conflicto del Jura y el referéndum de 2013. Revista d'Estudis Autonòmics i Federals 20:96–131

Sack RD (1983) Human territoriality: a theory. Ann Assoc Am Geogr 73:55–74

Saunders C (2015) The interdependence of federalism and democracy in Australia. In: Palermo F, Alber E (eds) Federalism as decision-making. Changes in structures, procedures and policies. Brill-Nijhoff, Leiden-Boston, pp 20–39

Seys F-O (2019a) Neo-regionalism in Europe: the process of delimiting regional boundaries. In: Abels G, Battke J (eds) Regional governance in the EU. Regions and the future of Europe. Edward Elgar, Cheltenham and Northampton, MA, pp 52–70

Seys F-O (2019b) Des régions territoriales aux régions fonctionnelles, une proposition de classification des régions de l'Union Européenne, *Belgeo*, 2 http://journals.openedition.org/belgeo/33263; https://doi.org/10.4000/belgeo.33263

Singh AK (2013) Region and state formation in India. In: Mangiameli S (ed) Federalism, regionalism, and territory. Giuffrè, Milan, pp 291–309

Taylor PJ (1994) The state as container: territoriality in the modern world-system. Prog Hum Geogr 18:151–162

Tesfai Baraki G (2013) The making of states and local governments in the multi-ethnic federation of Ethiopia. In: Mangiameli S (ed) Federalism, regionalism and territory. Giuffrè, Milan, pp 229–266

Watts RL (2008) Comparing federal systems, 3rd edn. McGill-Queen's University Press, Montreal and Kingston

Watts RL (2013) Typologies of federalism. In: Loughlin J et al (eds) Routledge handbook of regionalism and federalism. Routledge, London and New York, pp 19–33
Weinzierl J (2021) Territoriality beyond the state: the EU's territorial claims and the search for their legitimacy. German Law J 22:650–672
Wheare KC (1963) Federal government, 4th edn. OUP, London
Zahar M-J (2019) When the total is less than the sum of the parts. The lessons of Bosnia and Herzegovina. In: Anderson M, Choudry S (eds) Territory and power in constitutional transitions, 1st edn. OUP, Oxford, pp 22–42

Part IV
De-territorialised Legal Geographies

Chapter 8
Un-Bounded Legal Spaces

8.1 Beyond the Territorial Matrix

8.1.1 A Matrix with 'Proprietary Traits'

At the onset of my cross-disciplinary journey through law and geography, I stated that spatial production follows a territorialised paradigm. In acting as a 'disciplinary matrix', it represents a cultural competence, a faculty of mind, a capacity for comprehension and synthesising world knowledge. The way we humans have traditionally made spaces meaningful has indeed several hints of 'territoriality', which I have exemplified by perusing the feudal adage *Nulle terre sans seigneur, nul seigneur sans titre* (see Sect. 2.6.2). The matrix possesses the ability to outline the relation between a community and its territory in a variety of ways, also evolving over time and adapting incrementally to the most diverse processes of spatial production.

The paradigm is far from being purely historical. Quite the opposite: it conveys the way we produce spaces by physically, politically, or imaginatively apprehending territories. In sum, it renders the relation between communities and territories spatially visible in terms of belongs of public law. In this manner, the special relationship between law and geography turns into legal geography (see Sect. 6.1).

Paraphrasing Kuhn's conception of scientific evolution,[1] there is a broad consensus around the matrix, which provides legal scholars and geographers with the ability to synthesise world knowledge. In addition, the matrix is productive in scientific research since it helps us to identify the operational rules governing spatial production. Finally, it allows us to grasp the complex relations between law, geography, power, and territory.

[1] Kuhn (1970), pp. 170–173. For a thoughtful examination of his conception of scientific (r)-evolution see Bird (2018).

M. Nicolini, *Legal Geography*, Ius Gentium: Comparative Perspectives on Law and Justice 105, https://doi.org/10.1007/978-3-031-19410-8_8

In the previous chapters I probed the matrix by applying it to several processes of spatial production. In Parts I and II, I examined how spatial normativities are manufactured, also explaining how the matrix turns spaces into places, i.e. archives and repositories of situated legal practices. Maitland had already hinted at this possibility. In *Township and Borough* (see Sect. 2.6.2), he argued that 'almost all the land in the world "belongs" to "communities" . . . The story of the map of Europe is upon the surface a story of inheritance, conveyance, dower and marriage portion'.[2] In this excerpt, Maitland pointed to spatial production as physical apprehension and political appropriation, through which communities impressed their space-place-environment nexus into what they consider 'their' own tracts of land. Spatial reshuffling thus generated their belongs of public law. Similar patterns percolate spatial production; the law of spatiality, the spatiality of law, and counter legal geographies are all applications of the territorial matrix, where politico-economic actors and 'ideologies' compete against each other to normalise their own representation of the world. The competition may also oppose domination and resistance, such as in counter legal geographies (see Sect. 5.2.1). It does not come as a surprise that this 'juxtaposition' resonates again with Maitland's belongs of public law; and that, as he acutely stressed, 'all political power exhibits proprietary traits, and every ownership of land is actually or potentially a right of governing and doing justice'.[3]

8.1.2 De-Territorialising Spaces?

So far, my reflections have revolved around the 'valorisation of place and community' as a device that forms 'a bridge across the methodological divides of physical and human geographies'.[4] With the exceptions of legal hydrology (Sect. 3.5.2), outer space (Sect. 3.4.2), British 'integrated geography' of land and sea (Sect. 4.3), and ideo-logical spaces of doubt and terror (Sect. 5.3.4), I have prevalently focused on territorialised legal geographical competitions and their insatiable appetite for blank spaces (see also Sect. 4.2.1). Part III of the book examined them by demonstrating how the 'disciplinary matrix' percolates spatial production. The territorialised paradigm was productive for devising legal geographical primitives, i.e. the simplest meaningful units that can help us to handle, store, and reproduce legal geographical knowledge (see Sects. 6.4 and 7.2). The matrix has evolved over time, adapting real-world knowledge to new contexts, variables, and contingencies. It coarsely corresponds to Kuhn's 'normal' phases of science with its 'cumulative picture of scientific progress'.[5]

---

[2]Maitland [1897] (1964), p. 29.

[3]Maitland [1897] (1964), p. 29. See also Cohen (1927), p. 9.

[4]Godden (2020), p. 134.

[5]Kuhn [1962] (1970), p. 92.

Located in Part IV (*De-Territorialised Legal Geographies*), this chapter engages with further processes of spatial production (the Anthropocene, globalisation, digital spaces, and climate-change related ones). Apparently, they depart from the territorial matrix. What I shall do is to ascertain whether they may still reflect the territorialised paradigm and Kuhn's 'normal' phases of science. Their dynamics are particularly elusive as regards the application of territorial benchmarks. Indeed, they look like un-bounded ideo-logical spaces. Take, for example, globalisation; with its cosmopolitan nature, supply chain, gentrifying effects, and *mentalité*, it is indifferent to the very idea of place *also* produced through situated legal practices (see Sect. 5.3.6). Likewise, synthetic jurisdictions such as SEZs or EU functional regions seem to abandon the territorial matrix. Within them, there are neither 'subnational' processes of spatial production, nor 'scalar' belongs of public law (see Sect. 7.2); they are mere instrumentalities of politico-economic powers for specific ends.

Un-bounded spaces share the same features as other ideo-logical spaces. In digital spaces, for instance, the human-physical geographical nexus seems absent. As a part of a broader politico-legal discourse, digital spaces still interact with the real world; their 'environment' is meaningful because the 'third' (i.e. imaginative) geography provides the practical knowledge for manufacturing its unbounded spatiality. Being part of the 'interplay' between digital environments and human agencies, spatial production remains servient to the establishment of legal geographies of power.

There are, then, transnational spaces. The prevalent approach portrays them as abstract logical spaces, where geographical features, once again, are lacking. More probably, they are irrelevant precisely because transnationalism surpasses national jurisdictions. Owing to their cosmopolitan nature, transnational spaces are purely 'based on legal dimensions of globalisation'.[6] Yet, global law still implies (and refers to) concrete contexts. As we shall see, transnational actors code their own interests (and commercial-financial concerns) through the law. Transnational law, evidently, is the generalisation of private legal orderings; ultimately, it is a type of spatial reshuffling, whereby a new spatiality of law (that of markets) is superimposed onto the earth's surface thus creating a new meaningful place.

Even in a time of globalised and de-territorialised spaces, spatial production is still explicable in terms of the matrix. I understand that, far from experiencing a normal phase, we are living in a time of often traumatic changes. In this post-cumulative phase of science, the cross-disciplinary matrix still provides an explanation for spatial production, at least because 'the basic orderings of society require categories and compartments; even in un-bounded spaces, 'borders create order'.[7] In sum, un-bounded spaces remind us of the idea of 'paradigm'. This chapter will ascertain whether spatial production, even in a time of globalisation, is still explicable in terms of 'normal' (and cumulative) phases of science.

---

[6]Husa (2018), p. 49. For a critical account of the legal in globalisation see Amin (2002); Barkan (2011), p. 596.

[7]Newman (2006), p. 143. See also Brighenti (2006), p. 74.

8.2 Geophysical Continuities, Geopolitical Discontinuities

8.2.1 Geophysical Continuities: Atmosphere and Greenhouse Emissions

In the previous section, I stated that un-bounded ideo-logical spaces seem to lack legal geographical attributes. In homogenous spaces, no furniture must be reshuffled; nor is there a territory (or even a community) to be organised.

This alleged lack of striations is patent in the atmosphere. As a part of the geo-ecosphere, it seems unseizable. Although difficult to be physically apprehended, it is not out of our reach. Take, for example, transboundary air pollution such as greenhouse emissions, whose warming effects saturate it. The composition of the atmosphere is influenced by the biosphere. Its chemical elements are mainly 'created by life, particularly plant life', whose deletion would lead to its new composition: '1.9% nitrogen, nil oxygen and 98% CO_2, with a surface pressure of 70 present-day atmosphere'. This would make the earth similar to Venus, which lacks life and surface water.[8]

The atmosphere is involved in a close relation with both the earth and humankind, thus being co-constitutive of our belongs of public law. This is patent in Accursius's *Glossa Ordinaria* (1250 AD ca). When perusing the passage in Digest VIII.2.1 stating that, in case of usufructs, the 'airspace over the highway ought to be free', Accursius made the following annotation: 'Cujus est solum, ejus debet esse usque ad coelum'.[9] Here we find the proprietary traits mentioned above (see Sect. 8.1.1), which are still the tenets of property, international, and outer space law. 'Land hath also, in its legal signification, an indefinite extent, upwards as well as downwards. *Cujus est solum, ejus est usque ad coelum*, is the maxim of the law', Blackstone repeated in his *Commentaries* (1766).[10] As humans, we make the atmosphere a meaningful place by impressing our space-place-environment nexus even into it.

Far from being unseizable, we constantly reshuffle (and impact on) the atmosphere. Concentrations of greenhouse gases are higher than in the last 650,000 years with a 'rate of increase in temperature . . . faster now than previously'. This is due to the 'causal association between greenhouse gases and temperature'; in 1859, indeed, the Irish scientist John Tyndall demonstrated that 'CO_2 absorbs infrared radiation'. The picture is worse than we might expect; CO_2, CH_4, and N_2O (not to mention water vapour) 'absorb some of the infrared radiation that would otherwise stream directly out of space', which causes the greenhouse effect.[11]

[8]Holden (2017), pp. 137–138.

[9]See D.8.2.1: 'quia caelum, quod supra id solum intercedit, liberum esse debet' Accursius's annotation 'may be translated: "Whose is the soil, his it ought to be up to the heaven . . .": Cooper (1952), p. 41.

[10]Blackstone [1766] (2016), II.18.

[11]Holden (2017), p. 177.

Emissions do not occur in a vacuum. They take place in our real world, thus exhibiting geophysical continuity. They occur in the atmosphere, which is essentially a transnational place. Phenomena like convective overturning and atmospheric (or ocean) circulations have planetary impacts and usually interact, as the Walker circulation and El Niño Southern oscillation demonstrate.[12] Evidently, natural geographical attributes do not know 'natural' borders (see Sect. 3.5.1). If the hydrosphere and the atmosphere exchange heat, we affect the whole geo-ecosphere by enhancing (i.e. altering) the 'natural' greenhouse effect, which absorbs solar radiation with the aim of maintaining the earth's average temperature as it currently stands (16 °C; if absorption did not occur, the temperature would be—6 °C). Climate change augments this 'blanketing effect',[13] making the earth more and more similar to Venus.

It is 'Time to Act'.[14] Our legal-ecological commitment should shake off our egotistic carelessness towards the fate of a planet literally 'on fire'.[15] Some actions have already been brought about. According to the 2015 Paris Agreement and the 2021 Glasgow Climate Pact, the increasing in global average temperature should be limited 'well below 2 °C above pre-industrial levels' at least.[16] Yet, we have demised all these arrangements, accepting restraints to emissions be lifted. We were partly influenced by corporate lobbyists and partly worried about our acquisitive lifestyles: indeed, 'markets producing environmental pollution impose varying constraints and crate varying opportunities for governments'.[17] Fascinated by this commodifying state of mind, signatories to the 1997 Kyoto and the 2009 Copenhagen Protocols managed to establish an emission trading system, which commodifies carbon dioxide, i.e. a greenhouse gas. Both arrangements favour 'big emitters' (states with the largest emissions) and international aviation and shipping, which are non-state actors also falling outside the scope of the Paris Agreement.[18] The creation of international legal regimes are indeed 'dominated by businesses operating' at all scales, which 'are generally not in favour of new environmental regulations if their products and services are the sources of pollution'.[19] And this also

[12] Holden (2017), pp. 163–168.

[13] Holden (2017), p. 167.

[14] Christian Climate Action (2020).

[15] Klein (2019). See also Lawrence and Laybourn-Langton (2021).

[16] Art. 2.1(a) of the 2015 Paris Agreement. See also Part IV (Mitigation) No. 15 of the 2021 Glasgow Climate Pact.

[17] Ovodenko (2017), pp. 4 and 7–19 on how (especially global) markets interfere with state environmental policies.

[18] Art. 17 of the 1997 Kyoto Protocol. See Victor (2004) and Popovski (2019). On big emitters see Larkin et al. (2018), p. 692.

[19] Ovodenko (2017), pp. 151–152.

explains why the 2021 draft of the Glasgow Climate Pact 'coals down' our hopes of limiting greenhouse emissions.[20]

8.2.2 Geopolitical Discontinuities

Coded through the law, geophysical continuities impact on our tackling climate change. As the atmosphere is a global belong of public law, it is now time to constrain emitters. Legal arrangements tackling climate change propose a uniform institutional pattern reflecting its transnational character. Yet, this legal and 'methodological architecture . . . in this, irreversibly and irreducibly global, context'[21] still reflects the lack of striations of its own ideo-logical space. In commodifying greenhouse emissions, the Paris Agreement and its Glasgow counterpart still endorse the winning arguments of commerce and trade (see Sect. 5.3.3). Both support the 'construction of a nominally "global" frame' reflecting the 'political and geographical extension of . . . the extant hegemonic bloc' and coincides with 'the capitalist global North . . . its allies and, sometimes, China' and India.[22]

This global legal framework reinforces the dialectic between the geophysical continuity of the atmosphere and the geopolitical discontinuity between signatories of both agreements. The creation of a homogenous framework to tackle climate change reflects the assumption that global environmental policies are basically the same throughout the world. It assumes that the geophysical continuity between the Western world and China entails their geopolitical continuity, at least because the latter possesses a 'model of state capitalism' with the same contradiction of Western capitalism, that is, an incredible ability to produce wealth and commodify everything 'at the expense of the natural environment'.[23]

These arguments fail to recognise that China is a striation within the global fight against climate change. Searching for its support means engaging with a model of state capitalism that is completely different from the Western one. Having replaced the USA as the leading carbon emitter in the world, China could become a leading player in driving forward the struggle against climate change at the international level.[24] Domestically, the Chinese adopt 'authoritarian environmentalism', which 'means a public policy model', i.e. a top-down law-making process concentrating authority and regulatory powers in the hands of the central state. This approach is partly related to (and disguised by) its authoritarian character: it restricts liberties,

---

[20] See Glasgow Climate Pact, Part IV (Mitigation) Nos 17 ('rapid, deep, and sustained reductions in global greenhouse emissions') and 20 ('accelerating efforts towards the phase-out of unabated coal power and inefficient fossil fuel subsidies').

[21] Zumbansen (2020a, b).

[22] Mann and Wainwright (2018), p. 31.

[23] Beeson (2020), pp. 25 and 29.

[24] Beeson (2020), p. 25.

and 'population control policies have been cited as a model for future limits on individual choices related to climate change'.[25] Furthermore, its authoritarian decision-making offers a 'capitalist' model alternative to the Western 'discredited neoliberal variety that is associated with many of the world's [...] environmental problems'.[26] This authoritarian decision-making indeed appeals to those living in the Global South, who are likely to bear the costs of a Western-driven ecological disaster.

8.3 Global Regimes

8.3.1 Places of Acquisitiveness

Transnational phenomena shift the borders between the real world, the knowledge of it, and the way the latter is spatially produced. In ideo-logical terms, producing real-world knowledge poses a 'Problem of Persuasion';[27] the interweaving of economic-financial power, communication, and social media affects spatial processes and impacts on how knowledge is understood by public opinion. Robert F Blomquist discusses this topic in his review of Richard K Sherwin's *When The Law Goes Pop* (2000). Blomquist agrees with Sherwin: the law now 'yields to the media's logic [as] an increasingly popular form of cultural postmodernism'.[28] In this ideo-logical space, spatial reshuffling deliberately forges a legal geography, whose outcome encodes 'legal meanings [flattened] out as they yield to the compelling visual logic of film and TV images and the market forces that fuel their production.'[29]

Our commodifying state of mind actively takes part in this spatial production. We co-produce space because, so to speak, 'we are all postmodernists now'.[30] Blomquist generalises Sherwin's reading of 21st-century American society; being postmodernist means that we 'are quite aware ... that all reality is "constructed" and that there is no such thing as "ideal truth" or no such thing as "objectivist rationality".'[31] Acquisitive ideo-logical space is constructed to foment our never quenched consumeristic thirst. Although aware of its artificiality, we do not fully realise that our contribution to its production is also our weakness. According to Blomquist, the problem of persuasion in a postmodern society is a 'burden we all bear'. He links 'this burden ... to the story of Adam and Eve, who sampled the forbidden fruit of The Tree of Knowledge'. In *Gen* 3:6 the biblical metaphor 'is one *of too much*

[25] Gilley (2012), p. 290.
[26] Beeson (2020), p. 38.
[27] Blomquist (2001), p. 168.
[28] Sherwin (2000), pp. 171 and 31.
[29] Sherwin (2000), p. 4.
[30] Blomquist (2001), p. 174.
[31] Sherwin (2000), pp. 220–221.

sensate, surface knowledge and too little deep understanding'.[32] When, as consumers-citizens, we consciously let the forces of capital influence our behaviours, we certainly accept that all reality is 'constructed'; at the same time, we are unconsciously persuaded that the whole of our life must adapt to this constructed reality.

Persuasion extends over our civic (and climate-change related) commitments. We are induced to think that global warming may be tackled without changing our economically constructed paradigms. Take, for example, the notion of 'ecopiety'. As shorthand for 'contemporary practices of environmental (or "green") virtue',[33] it gives us the illusion that we, as individuals, can actively preserve the environment without altering our ideo-logical space. The argument is deceptive because 'individual action will never be an adequate response to the climate crisis.'[34] What the planet 'will actually notice' is not our individual intervention, but our collective efforts in overturning our economic paradigms supported by 'a heavy dose of government policy'.[35] Armiero terms this choral action 'commoning', that is 'the ensemble of socio-ecological practices which (re) produce commons, transforming it from a "thing" into . . . a relationship'.[36] This also means redirecting the behaviour of those responsible for greenhouse emissions.[37] Convincing us of the adequacy of individual ecopious practices deceives us and conceals their responsibility. At the same time, it puts the blame on us if, as individuals, we fail in protecting the earth notwithstanding our ecopious efforts and creative actions.

Such an illusion is ideologically motivated. What matters most is the maximisation of profits. Even in a time of ecological collapse. We know that capital produces its own desired ideo-logical spaces and legal geographies (see Sect. 5.2.2), within which ecological commitments are influenced by a *commodifying state of mind*. We may also term them *spaces of acquisitiveness*. The forces of capital trade and we, as consumers, commit ourselves to purchase *also* for a greener future. The interweaving of ecological commitments and acquisitiveness is indeed functional to maximising financial returns. These places cannot be 'spaces of doubt'; consumers must commit themselves to acquisitive-individual practices of ecopiety even on the brink of ecological collapse. *Too much sensate, deep knowledge, and conscious understanding* of the catastrophe may cause a drop in commerce, threaten the collection of commodities, and reduce profits. This is particularly apparent as far as climate change is concerned. It is indeed 'a problem that seems to defy not only our control but also our understanding' and is perhaps the most vivid example of what Timothy Morton labels 'hyperobjects', i.e. 'entities of such vast temporal and

---

32 Blomquist (2001), p. 176.

33 McFarland Taylor (2019), p. 2.

34 Klein (2019), p. 81. See also Lawrence and Laybourn-Langton (2021).

35 Wagner (2011), pp. 137–138.

36 Armiero (2021), p. 13.

37 See Romanello et al. (2021).

spatial dimensions that they defeat traditional ideas about what a thing is in the first place'.[38] In order not to alter the function of the market, we tend to reduce the debate about these hyperobjects and their socio-climatic consequences to eco-consumeristic practices. The prospect of ecological catastrophe is marginalised so as not to cause a threat to the comfort zones of commerce.

The prevailing economic interests have legal geographical consequences, inasmuch as they outline the relation between un-bounded spaces and their transnational communities. Like in all legal transnational environments, spaces of acquisitiveness require the creation of a uniform legal framework transcending the striations of domestic jurisdictions. In this sense, this ideo-logical space experiments with its own 'convergence of laws', where the term denotes a vast array of forms whereby its universalist approach is applied within the global community of consumers. The descriptor encompasses several processes, such as harmonisation, homogenisation, transplants, unification of laws, which all, evidently, are inflected after the necessity of smoothing trade and purchase.[39] Yielding to the logic of acquisitiveness, the law becomes the *law of the Buyosphere* (see Sect. 8.5).

8.3.2 The Waste Regime

Counter legal geographies also populate spaces of acquisitiveness, thus reflecting Fish's 'safe oasis', whose demarcation takes place according to Geraóid Ó'Tuathail's act of 'geo-graphing' (see Sect. 5.3.3). The way the oasis is demarcated is thought-provoking. Marco Armiero has recently coined the term 'Wasteocene', which does not point to the 'materiality of waste accumulating everywhere in our environment'. Rather, this geological reference creates an ideo-logical space, where the 'waste regime' establishes a new logical space through new spatial reshufflings. It adopts a new organising principle 'determining what is waste and where it should go'. Out of the metaphor: as 'economic value [is] measured only by price; what [has] no price [has] no value'[40] And 'wasting implies sorting out what has value and what does not'.[41]

Like in any other safe space, the waste regime confines ecological doubts to the background stage. I have said that the Wasteocene is a new organising principle, which cordons off the oasis leaving the dump (and doubts) outside it. In so doing, it draws a divide separating the 'aseptic environments' of cosmopolitanism and the 'griminess and contamination' of places where the dump is expelled. This demarcation process converts wasting into the decision on how spaces and communities must be politically organised. Marco Armiero explains that wasting is 'a social

---

[38] Morton (2013).

[39] See Siems (2022), p. 287.

[40] Standing (2019), p. viii.

[41] Armiero (2021), p. 10.

process through which class, race, and gender injustices become embedded into the socio-ecological metabolism producing both gardens and dumps, healthy and sick bodies, pure and contaminated places'.[42] I understand that the Wasteocene is a by-product of the spirit of the *cité*; it indeed entrenches its winning arguments, so that the elite might live in an oasis surrounded by polluted, marginalised, and segregated territorial communities. Armiero's idea of 'socio-ecological metabolism' is probably fitting when describing the ideo-logical spaces on this side of the wasting regime. In separating gardens and dumps, healthy and sick bodies, and pure and contaminated places, the *mentalité* of the wasting regime echoes the defects Jesus ascribes to 'scribes and Pharisees'. In the Global North we are all 'hypocrites'. 'Even so', Jesus continues, 'ye also outwardly appear righteous unto men, but within ye are full of hypocrisy and iniquity'. Perhaps most striking of all, the places and the *mentalité* of cosmopolitanism 'indeed appear beautiful outward, but are within full . . . of all uncleanness' (*Mt* 23:27–28). The line separating sick and healthy bodies and conscience is also an inwardly sign of how we unconsciously adapt our behaviour to the current constructed reality.

If pollution and climate change are transnational phenomena, our socio-ecological metabolism is not exempt from griminess and contamination. What we usually do is to recollect commodities and food produced out there in the dump. Nanoplastics, for example, intoxicate oceans and are eaten up by fish; served as sushi and nigiri, we swallow up these 'intoxicated fishes', I daresay, with insatiable gluttony, also intoxicating ourselves.[43] Wasteocene is thus related to the 'contaminated nature of capitalism and its endurance within the texture of life'.[44]

Even the Wasteocene reinforces the dialectic between geophysical continuity and geopolitical discontinuity. Take, again, greenhouses emissions. Although a global problem, they also pose a 'situated global issue', stemming 'from local emissions, whose impacts are felt differently in different locations'; it 'spans geographic scales from the rural village or urban neighbourhood to the planet'.[45] Like cross-border economic interests, ecological concerns are shared by transnational communities; unlike economic interests, these are extremely concrete. In an era of Wasteocene, communities and territories are striations upon which globalisation produces its ostensible effects.

Geopolitical discontinuity, by contrast, refers to how transnational phenomena produce localised effects. In generalising them throughout the world, we might think that the territorial effects of global warming are roughly the same, say, in the Global North and in the Global South. But the Waste Regime confirms that its 'cruelest impacts' will predominantly affect less-developed countries: groundwater scarcity and decline in biodiversity for food and agriculture, for example, will mainly impact the Global South; and the 'world's suffering will be distributed as unequally as its

[42] Armiero (2021), p. 10.

[43] Ferreira et al. (2019).

[44] Armiero (2021), p. 10.

[45] Carlarne and Farber (2012), p. 19.

profits'.[46] The discontinuity is also drawn by the 'waste regime', which will locate 'what is waste' outside its logical space and establish 'where it should go', expelling the dump towards the states-as-containers of the Global South.

Although they are 'whitened sepulchres' full of intoxicating food, in a time of Wasteocene, the wealthy will decline in better spirits than the lowly and marginalised.

8.4 Geologic Qualities of Space

8.4.1 Beyond the Wasteocene

Explicitly or implicitly, the term Wasteocene attaches some 'geologic qualities' to spatial production.[47] Armiero uses 'waste' to explain the political organisation separating cosmopolitan 'environments' from places contaminated by our griminess. He is not alone in assuming that humans are a driving force deeply altering the geo-ecosphere and creating a new space-place-environment nexus, willingly or not. The list of geological attributes of our contemporary era is particularly rich. Armiero supplies us with its detailed list; *inter alia*, we find 'Plantationocene', 'Econocene', 'Technocene', Anthrobscene', and 'Manthropocene', which all point to the geological effects of human behaviour.[48]

He then focuses on the geological descriptor 'Capitalocene', which has several connections with Wasteocene. The descriptor was created by Jason W Moore, who situates 'the rise of capitalism, historically and geographically, within the web of life'.[49] 'Capitalism', Moore argues, 'is not an economic system; it is not a social system; it is *a way of organizing nature*'.[50] The physical environment becomes an economically valuable by-product of human agents. Whether we like it or not, we process physical geography into 'nature'; human activity thus blurs the line between culture, nature, and economy. In terms of space, the Capitalocene 'speaks of the origins, [and] the causes, of the socio-ecological crisis' of our contemporary era,[51] whereas the Wasteocene points to its effects, as well as to the 'strata of toxicity … laid within bodies of human and non-human beings' with 'delayed' (but harmful) 'effects … on humans, nonhumans, and ecosystems',[52] which is caused by the 'slow violence' of our acting carelessly.[53] Armiero reappraises the cause-effect

[46] Wallace-Wells (2019), pp. 34 and 121.

[47] The expression 'geologic qualities' is in Yusoff (2018), p. 255.

[48] Armiero (2021), p. 9.

[49] Moore (2017), pp. 608–609. See also Moore (2015).

[50] Moore (2015), p. 2.

[51] Armiero (2021), p. 10. See also Moore (2015).

[52] Armiero (2021), p. 10.

[53] Nixon (2013).

relationship between CO_2 emissions and global warming: the former have not 'created the ecological crisis'; rather, they are the commodities produced by the 'socio-ecological relationships', i.e. the insatiable appetite for accumulation of capital through consumption acting upon the toxicity of their acquisitive organising themes.[54]

8.4.2 New Winning Arguments: **Anthropo*city***

In the midst of such a crisis, doubts resurface—and, with them, the idea of resistance towards places of acquisitiveness. Global warming is altering the organising themes of communities and their ideo-logical spaces. Climate change is deploying several winning arguments that are creeping into the texture of our places of acquisitiveness. Ecological catastrophe has the power to unsettle our beliefs; in insinuating doubts and fears, it challenges our safe places, which become vulnerable to the assaults of new arguments, 'igniting a complex bundle of social kindling'.[55]

To maximise profits, capital must silence the doubts arising from the socio-climatic consequences of our consumeristic ideo-logical space. Doubts must be cordoned off, and resistance to our unsustainable economy marginalised to allow our eco-consumeristic practices to thrive again and again in the unforeseeable future.

The winning arguments of capital are epitomised by the term 'Anthropocene'. The expression has been widely used to denote discrete (and often conflicting) factors, mainly related to how humankind is outcompeting natural processes and changing the earth.[56] Whatever the meaning assigned to it by scholars; and whether it is either a good event of geological magnitude, or an evil, or an expression *bonne à tout faire*, the descriptor has the ability to re-inscribe the figure of 'Man', i.e. the ἄνθρωπος, as the 'universal signifier to claim the geologic flows and processes of the Earth'.[57] Moore is right when he states that 'The Anthropocene is a comforting story with uncomfortable facts'. It is uncomforting, since it 'makes for a familiar story, one of Humanity doing many terrible things to Nature'. But it is also a winning argument, since 'it fits easily within a conventional description – and analytical logic – that separates humanity from the web of life'.[58]

The Anthropocene has the power to ascribe 'to collective human agency the capacity to rewrite the course of planetary history' and a pivotal role in the process whereby the Earth is spatially reshuffled.[59] It generalises our ability to master the

[54] Armiero (2021), p. 10.

[55] Wallace-Wells (2019), p. 154.

[56] See Crutzen and Stoermer (2000) and Crutzen (2002). See also Zalasiewicz et al. (2011); La Viña and Reyes (2022), p. 159.

[57] The expression 'geologic qualities' is in Yusoff (2018), p. 255.

[58] Moore (2017), p. 595. See also Moore (2015).

[59] Matthews (2021), p. 45. See also Dalby (2021).

forces of nature, manipulate the space-place-environment nexus, and produce a place that is fitting to the needs of a global community without 'planetary limits' or 'planetary boundaries', i.e. 'the safe operating space for humanity'; these are 'thresholds beyond which large-scale, comparatively sudden environmental changes will produce major problems'.[60] The 'conceptual framework' which is the Anthropocene thus 'signals simultaneously a deep ecological crisis, and a deep cultural crisis.' Yet, as humans, we make the cultural crisis undergo a process of sublimation, which is servient to both our scientific state of mind and its 'project of human control of the Earth'.[61]

The winning argument lies in our ability to turn the crisis and the earth into an 'area for action'—which means human action.[62] What we sublimate (and therefore generalise) is the performativity of human activities. It is not a matter of climate change-related effects; the Anthropocene focuses on the mighty acts and the 'telluric force' of 'human activity'; as the Italian geologist Antonio Stoppani put it in 1873, it is a force paving the way to the 'Anthropozoic era'.[63] The process of sublimation and generalisation leads to 'The elevation of the Anthropos as a collective actor' encouraging 'several important misrecognitions'; as the sum of the 'human species', the new Leviathan, the ἄνθρωπος 'becomes a mighty, largely homogeneous, acting unit: the 'human enterprise'.[64]

In politico-legal terms, the process of sublimation means activating a process of generalisation, whose roots may be found in the 'problem of sovereignty', which at the global scale means establishing the sovereignty of the ἄνθρωπος as the productor of space.[65] As 'The modern political gaze, for which sovereignty provides the frame, is deeply anthropocentric', its generalisation entails that, if 'political authority is articulated at a transnational or putatively "global" scale, this anthropocentricism persists', and becomes, on its own, the geological force dominating the planetary scene.

8.4.3 *Pride and the Anthropocene*

There are hints of pride in this acting deliberately against-but-through the world, which are traceable back to how, in dismissing chthonic legal traditions (see Sect. 2.1.1), the earth has been Westernised (see Sect. 4.5). Diarmaid MacCulloch offers us an explanation: 'the most plausible doctrine of Western Christianity, depressingly, is

[60] Respectively: Yusoff (2018), p. 266; Rockström et al. (2009), p. 472; Kelman (2022), p. 147.
[61] de Lucia (2020), p. 330.
[62] Yusoff (2021), p. 663.
[63] Stoppani (1873), p. 468. See also Crutzen (2002).
[64] Moore (2017), p. 596, also referring to Steffen et al. (2011).
[65] Matthews (2021), p. 45.

original sin, and at the root of all is pride'.[66] The pseudo-Solomon tries to tone our pride down in the Book of Ecclesiastes. God, he writes, 'hath made every thing beautiful in his time'. Likewise, 'he hath set the world in their heart, so that no man can find out the work that God maketh from the beginning to the end.' (*Eccl* 3:11–12). Within this divine framework, there is no room for the sovereignty of the ἄνθρωπος. It is indeed written that 'The land shall not be sold for ever: for the land is mine, for ye are strangers and sojourners with me' (*Lev* 25:23). Undeniably, 'every man [again, the ἄνθρωπος] should eat and drink, and enjoy the good of all his labour', which 'is the gift of God' (*Eccl* 3:13).

Yet, how is it possible that the ideo-logical space of the Anthropocene falls under the geological telluric force of the ἄνθρωπος? How is it acceptable that tenants overturn the proprietor of the earth, replacing His sovereign rule with their belongs of public law? These belongs, I add, reflect a process of generalisation of the economic interests of the global holders of capital, which are the real ἄνθρωπος in a self-created un-bounded transnational space (see Sect. 8.6). No doubt, 'the Anthropocene is implicated in capitalist power, symbolically and materially'. It is, 'at its core, ... a fundamentally bourgeois concept'; and this 'should surprise no one'.[67] I shall consider how these sectional interests are turned into universal politico-legal principles underpinning the global sovereign rule of the ἄνθρωπος. In this section, I explain why, according to its rule, we must not change our economically constructed paradigms to tackle global warming. Our practices of 'ecopiety' will suffice, and the law of the Buyosphere will entrench them as a fundamental right of the global community of consumers, inflected, evidently, after the necessity of smoothing trade and purchase.[68]

Clive Hamilton detects the pivotal winning argument of the ἄνθρωπος in *The Theodicy of the Good Anthropocene* (2016): 'evil acts, when we take a larger perspective, are necessary to the functioning of the whole'. He explains that this is the generalisation of one of the tenets of Western Christian political thought; 'What may appear to us as monstrous crimes to which God acquiesces must be understood as in the service of his greater, if mysterious, benevolence'. It is not merely a matter of secularising our religious 'response to the existence of evil in a world created by a benevolent God.' The pride lies in our having replaced the '*Theo*dicy', with a sort of '*Antropo*dicy', i.e. a 'good Anthropocene',[69] where the roots of both good and evil revolve around a set of circular justifications. On the one hand, the global sovereignty of the ἄνθρωπος appears to be the sole legal geographical response to universal ecological threats. On the other, this legal geographical response is consistent with the democratic allure of climate change, that transcends borders and strikes without discriminating among places and individuals.

[66] MacCulloch (2017), p. 22.

[67] Matthews (2021), p. 54.

[68] See Siems (2022), p. 287.

[69] Hamilton (2016), p. 234.

8.4.4 Anthropocene as an Ideo-logical Space

There is a further reason accounting for the pride of humankind. The creation of safe places of acquisitiveness gives us the illusion of being at the centre of a bigger space, that of the universe. Yet, at least unconsciously, we realise that life (particularly human life) 'is not the central concern of the cosmos', which is dominated by forces—not to mention Morton's hyperobjects (see Sect. 8.3.1)—that are 'more powerful' than the telluric human force. Such forces 'organize and contextualize [their] appearance, and thus put pressure on the politics of [human] life'—and, in legal geographical terms, on its 'imagined arrangements'.[70] As these forces exceed the might of our human organic life—indeed, they 'do not begin and end in the organism'—,[71] we need to focus of the winning arguments which uphold our ability to recreate (and destroy) the world in a telluric way. We are proud, but at the same time, 'we know'. 'I know', the author of Ecclesiastes states, 'that whatsoever God doeth, it shall be for ever: nothing can be put to it, nor any thing taken from it: and God doeth it, that men should fear before him' (*Eccl* 3:14). Within 'this planetary perspective', 'human life is considered to be but one aspect – and arguably . . . a largely insignificant aspect'. Accordingly, it also 'provincialises human life and the institutions that facilitate "globalism"', as well as 'the earth itself within a broader inter-planetary environment'.[72]

Indeed, we cannot take 'ontological decisions to achieve' the ends of creation.[73] But we do not accept this reality and therefore shield ourselves from it. What we can do is to create a 'geopower as a force that is implicated in the constitution of the planetary' imagination.[74] With the Anthropocene, we give ourselves the illusion of being the actors of the political geology of the earth. And of the cosmos, which we now exploit, making it (and outer space more generally) suitable for an extension of the territorialised paradigm (see Sect. 3.4.2). Yet, 'such a claim is a temporarily insignificant moment in comparison to the earth's own forces, which for much of the earth's geohistory have not supported life, *at all*'.[75] We fear that, when the next mass extinction eliminates humans from the face of the earth, no-thing will be there to remember of our having dwelt 'in the land of the living' (*Ps* 27:13).

It is, probably, the very idea of oblivion that makes us act proudly. We must demonstrate to ourselves that we can arrange the earth according to our logical space. To this extent, we must liberate ourselves from the territorial trap and create connections among territorialised communities. Transcending borders entails producing new spaces without limits to consumption. The Anthropocene is thus an ideo-logical space governed by a global law, which possesses a liberating force

[70] Yussof (2018), p. 205.
[71] Yussof (2018), p. 209.
[72] Matthews (2021), p. 54.
[73] Yussof (2018), p. 208.
[74] Yussof (2018), p. 208.
[75] Yussof (2018), p. 204.

facilitating the harmonisation of laws and stimulating economic development. The sovereign ἄνθρωπος is so mighty that it can exploit the earth even beyond the Earth Overshoot Day.

But: can it also promise a radiant future to humanity? The Book of Ecclesiastes reminds us that God instilled the world in our hearts. The New Revised Standard Version of the Bible (NRSV) provides a slightly different translation of the word 'world': God 'has put a sense of *past and future* into their minds, yet they cannot find out what God has done from the beginning to the end'.[76] Lefebvre teaches us that the past is constantly processed as space: 'The historical and its consequences . . . all of this is inscribed in spaces'.[77] As a telluric force, we establish all the other geological eras, even those preceding the advent of organic life; we measure mineral 'deposits .. that have built over time, and other plants and animal remains', also determining 'which period of time they relate to and how they fit into the global picture'.[78]

In examining the past through stratigraphy and coring, we also project our imagined geographies onto past eras and tracts of lands, thus producing our legal geology with belongs of public law. As an ideo-logical space, the Anthropocene allows us to reshuffle the spatial furniture of the past and eventually get hold of it. Through scientific experiments, the sovereign ἄνθρωπος is thus able to physically apprehend and politically control the earth. This is nothing else than the application of the territorialised paradigm to the un-bounded processes of spatial production, which know neither diatopic nor diachronic limits.

And the future? Is it possible to process it? What we usually do is to extend the scope of the present; that is, to process a space that 'is always, now and formerly, a *present* space . . . complete with its associations and connections in their actuality'.[79] This has prompted the exploitation of the cosmos, making it (and outer space more generally) suitable for an extension of the territorialised paradigm of legal geography. This is the winning argument through which we defeat the forces of the cosmos and the fear of post-human life on earth. Living in a never-ending present, we take our 'ontological decisions' and fatally determine the causes of our foreseeable extinction.[80]

[76] Holy Bible NRSE (1989), p. 563.

[77] Lefebvre [1974] (1991), p. 37.

[78] Holden (2017), p. 80.

[79] Lefebvre [1974] (1991), p. 37.

[80] See also Yusoff (2016), p. 24.

8.5 The Law of the Buyosphere

Within the Anthropocene, the telluric force of the ἄνθρωπος reshuffles the geo-ecosphere and its furniture, establishing new belongs of public law. Rooted, as the latter are, in a new sovereignty,[81] they require a new regulatory legal framework, which reflects the liberating force of globalisation, its unbridled exploitation of the earth's resources, and the smooth flows of goods, capital, and services across frontiers and legal systems.

I have termed this legal framework the Law of the Buyosphere (see Sects. 8.3.1 and 8.4.3). The expression incorporates the lexical item 'Buyosphere' (or sphere of purchase) coined by Anne-Dorothée Herbort in 2012. Sarah McFarland Taylor probes it in *Ecopiety* (2019) with a connotation suitable for my purposes, that of a place whose meaning is compatible with the ideology of the market. The Buyosphere is regulated by a legal framework, which is transnational in character and able to accommodate practices of acquisitiveness and ecopiety. As an acquisitive practice, ecopiety reflects how the capitalist-state nexus encourages the 'popular production of environmental (green) virtue as a commodity', perpetuating 'the logics of global capitalism and market ideology'.[82]

The 'Buyosphere' refers to the final stage of the global supply chain: that of the distribution of commodities. Lefebvre already acknowledged this in his *The Production of Space*: commodities purchased in the Buyosphere 'dissimulate not only the amount of social labour that they embody, but also the social relationships of exploitation and domination on which they are founded.' He then adds: 'Things lie, and when, having become commodities, they lie in order to conceal their origin, namely social labour, they tend to set themselves as absolutes'.[83]

If things lie, money is no exception. Cryptocurrencies like Bitcoin trump the territorial matrix of the real world; as a 'new form of money without the state', they aim 'to create a new world beyond big finance and the corruptibility of state powers'.[84] Their prodigious accumulation dissimulates an incredible number of miners (that is, the producers of Bitcoin), also requiring a huge amount of energy to make the algorithm run its course. It also reflects the social relationships of exploitation and domination populating the flattened and allegedly homogenous geography of digital spaces. 'Not all miners are equal', Katharina Pistor upholds; 'those with high-powered computers and real money to burn for electricity bills are more equal than the others'.[85] Within the Anthropocene, these outputs conceal the way we commercially exploit the earth's resources well beyond planetary boundaries. Commodities 'are fetishized and so

[81] Walker (2020).

[82] McFarland Taylor (2019), pp. 5 and 7.

[83] Lefebvre [1974] (1991), p. 81.

[84] Pistor (2019), p. 197.

[85] Pistor (2019), p. 200.

become more "real" than reality itself – that is, than productive activity itself, which they thus take over'.[86]

With contemporary parlance, global value chains (GVCs) and wages and labour demand[87] are the benchmarks of the generalised, and therefore universally accepted, values of global economic sectionalities. Uniform patterns of globalisation are grounded in processes of generalisation. The law of the Buyosphere reflects this regulatory practice (see Sect. 2.1.2), which makes an idea applicable everywhere irrespective of social, cultural, and legal backgrounds. It is indeed a generalisation of the law that is wanted by the holders of global capital with the aim of getting hold of resources and markets, and is therefore servient to freedom of trade and commerce.[88]

It may seem that the prevalent approach in global law outlines an abstract logical space, where traditional geographical features are absent. Precisely because it surpasses national jurisdictions, it seems purely 'based on legal dimensions of globalisation'.[89] Yet, global law still implies (and applies to) the territorial matrix. Although globalisation disregards the role of territory in organising communities, it reinforces the dialectic between geophysical discontinuity (namely, the presence of borders) and the connections caused by globalisation. Both communities and territories are thus striations upon which globalisation produces ostensible effects.

Globalisation is populated by transnational actors, i.e. sectionalities seeking connections with several legal orders in an attempt to code their own interests through the law. In coding such interests, they preserve the capital-nation state nexus. The Law of the Buyosphere is also applicable to digital spaces (see Sect. 8.6.1), where it generates a "new digital common law"[90] as a variant of the physical global-transnational law, of which the English and the U.S. common law are historically the backbones. Consequently, common-law paradigms are applied globally for processing things as wealth and turning them into capital even in our mixed digital environments.

Transnational actors code their interests in a coordinated way through the constellation of legal orderings which is transnational law. Even at a legal level, although made up of 'decentralised borderless virtual nations', 'cloud communities', and digital polities still have the 'traditional characteristics of political communities'.[91] Furthermore, the process of generalisation is a type of spatial reshuffling, which superimposes its own spatiality of law onto the earth's surface, with the aim of completely replacing concrete meaningful legal geographies with new legal geographies that revolve around different meaningful places, that is, the markets. Under the shadow of linear thinking, they are processing a very concrete global economic geography.

---

[86] Lefebvre [1974] (1991), p. 81. See also Chakrabarty (2018); Matthews (2021), p. 54.

[87] Szymczak and Wolszczak-Derlacz (2022).

[88] Husa (2018), p. 163.

[89] See Husa (2018), p. 49.

[90] De Filippi and Wright (2015).

[91] Orgad (2018), p. 257.

Evidently, the Law of the Buyosphere finds fertile ground in transnational law, a concept which is closely related to the rise of global law and extensively studied by comparative scholars.[92] Apparently, transnational law seems to depart from the universalistic approach of global law, inasmuch as it creates contextual responses to legal issues surpassing state borders (such as environmental policies, mobilisation of goods, services, and capital). Therefore, it has to do with communities (especially commercial communities) whose activities transcend borders and state-related law--making. Yet, the idea of 'transnational legal ordering' resonates with the above-mentioned processes of globalisation. As is evident, these communities code these interests in a coordinated way through the constellation of legal orderings which is transnational law.[93] How the connections work is far from being dysfunctional; if there is conflict between economic and non-economic interests, the former ones prevail by virtue of their coding through transnational law.

As a type of transnational law, the law of the Buyosphere has been forged by the same forces of capital that, as noted above, have traditionally sought a connection with the public sphere, tending to influence and shape its decision-making processes (see Sect. 5.3.1). Concerned, as they are, about their own transnational economic interests, they suggest political community adopt legal devices through which their practical affairs might be organised. To this extent, transnational legal derivatives are outputs of further processes of legal generalisations functional to the protection and enforcement of their own interests.

Through generalisation, the Law of the Buyosphere undergoes a process of 'objectivisation' which denies the relevance of local legal variations. In so doing, it practises a variety of legal universalism, which is aimed at protecting sectional interests and promoting acquisitive practices lurking 'behind the façade of universalist values'.[94] Such a universalist gesture ignores the contexts within which transnational law is applied. Firstly, it replaces former local, i.e. territorial, legislation with new legal devices that possess universal substance and are applicable everywhere through the Buyosphere. Secondly, it denies the relevance of several other interests, such as the ecological concerns of discrete sectionalities, that may be disturb the global supply chain of distribution and purchase.

The law of the Buyosphere is also consistent with the telluric force of the ἄνθρωπος in our time of climate change. Wainwright and Mann term this force *social functionalism*, which is a further generalisation of sectional economic interests. Among the others, there is 'entrepreneurialism', which has 'become an almost universally celebrated quality in capitalist societies'. According to Wainwright and Mann, this type of *social fitness* is the ultimate form of 'individual [i.e. sectional] adaptation to the contemporary moment'.[95]

---

[92] See e.g. Siems (2022), p. 305; Husa (2018), p. 32. See also Maduro et al. (2014) and Zumbansen (2020a, b).

[93] Schaffer and Coye (2020), p. 126.

[94] Khilnani et al. (2013), p. 11.

[95] Mann and Wainwright (2018), p. 70.

Within adaptation, sectional interests compete with other global concerns, such as ecological issues; like striations, the latter are shared by communities that lie across borders. This means that our political communities have become 'transnational'. Yet, they are still territorialised. Like cross-border economic interests, ecological concerns are shared by transnational communities; but, unlike economic interests, these concerns are extremely concrete. Beyond global law, there are usually 'a relatively small number of global law firms and advocates, nationally ranked global law schools and academics, and globally networked judges'; but there are also communities.[96] Climate change is certainly an international, cross-border, and domestic phenomenon. As a transnational phenomenon, it transcends borders; as a concrete concern, it strikes indiscriminately on a territorial basis. To tackle it we do not need the Law of the Buyosphere, but Transnational Climate Law, which ought to be 'multi-scalar, and its responses 'multi-jurisdictional' so as to reflect our troubled reality.[97]

There is a further consequence of how the territorial matrix forges the law of the Buyosphere. Not only do the forces of capital transcend borders and reduce states' ability to control their operativity, but they also influence how territorial communities organise constitutional arrangements. This consequence is a further output of legal generalisations. The paradigm transnational actors propose usually coincides with that of liberal democracy: 'the one thing that has been almost completely absent from the 50 or so cases of attempted democratization since 1974 is experimentation beyond the basic institutions of liberal democracy.'[98]

Liberal democracy perfectly fits into the new universals of the law of the Buyosphere. Indeed, 'Western liberal democracies have acted together to construct the current system that enables the global economy to operate.'[99] More generally, the good governance secured by liberal democracy also matches the universals of global economic and financial dominance. Liberal democracies are thus 'the only viable societal model left, for only they are compatible with economic success and [the] rapidly integrating information-intensive world economy.'[100] The argument is indeed persuasive because it arranges the different sectionalities of our transnational communities around a power-sharing organising principle. It assumes that there is an uneven distribution of wealth; and yet, this does not prevent each member of the community from gaining access to it.

Produced, as it is, by the holders of global capital, the Law of the Buyosphere is thus a transnational phenomenon, which rearranges the territorial matrix and its borders to an imagined global society where, at least on paper, territorial limits to our global economic intervention no longer exist.

---

[96] Walker (2015), p. 42.

[97] Etty et al. (2018), p. 200.

[98] Schmitter (1995), p. 16.

[99] Bohman (2007), p. 28.

[100] Kurki (2010), p. 366.

8.6 Digital Logical Spaces

8.6.1 From Territorial to Virtual

Hints of territoriality are present in another ideo-logical space, i.e. the digital one. Like the global one, it seems 'placeless', 'devoid of constraints in the non-virtual world', and therefore unable to undergo one of the processes of spatial production that make place a meaningful space. [101] The dominant paradigm assumes that digital space and its derivatives (virtual space and cyberspace) have marked the 'end of geography'. [102]

The expression 'end of geography' is particularly fitting as regards spaces of acquisitiveness. Not only have these trumped the territorial trap, but they have also moved to digital environments. Markets are usually 'located where prices are fixed'; in the Buyosphere, transnational conveyance often takes place 'in a computer chip whose geographical coordinates are difficult to identify'.[103] As a telluric force, the ἄνθρωπος has accomplished a new mission, i.e. the creation of the digital space, which multiplies the effects of the Anthropocene well beyond physical spaces. To celebrate this further example of *anthropo*city, the ἄνθρωπος can be addressed with the words of the Psalmist: 'Praise him for his mighty acts: praise him according to his excellent greatness' (*Ps* 150:2).

No doubt, creating digital space is a mighty act. When examining the geographical effects of the Internet, Thomas L Friedman has detected a new feature of the Anthropozoic era, concluding that *The world is flat*.[104] Digital space has changed the very perception of the earth's physical geography. Liberated by the territorial trap, the world resembles a 'digital space ... [with] no actual physical location', thus allowing people to 'easily communicate ... regardless of the limitation of geographical distance'.[105] As a telluric force, we have been able to flatten its surface. Flattening is indeed a type of spatial reshuffling. The Internet is empowering 'people to reach farther, more quickly, more profoundly and [cheaper] as ever before'. Which is topical in a space of acquisitiveness. 'Internet gives to more and more people the tools and the ability to work, to inquire and to relax': the revolution is 'an economic one but also a cultural one'.[106] The digital market is thus a safe ideo-logical place under the law of the Buyosphere; it has introduced 'a uniform, default reality defined by consumption, unhinged from the physical world and its ... environmental disasters'.[107] There, we get whatever we desire, and where

---

[101] Warf (2021), p. 316. See Sect. 3.1.

[102] O'Brien (1990, 1992). See Also Warf (2021), p. 316.

[103] Sanguin (2014).

[104] Friedman (2005).

[105] Miao (2019).

[106] Sanguin (2014).

[107] Crary (2022), p. 9.

commodities are purchased, shipped, and eventually 'recollected in tranquillity' (see Sect. 5.3.3).

The dissolution of physical geographical attributes also extends over methods and instruments of payments. Cryptocurrencies like Bitcoin eclipse the physical setting of financial transactions, but they also disembody their coders. 'Nobody knows exactly who invented Bitcoin', Katharina Pistor warns us. 'Satoshi Nakamoto, the official creator, is an alias for one, or perhaps several digital coders'.[108] Disembodiment is probably caused by cryptocurrencies that, like Bitcoin, use blockchain technology, i.e. 'a tamer-proof ledger that contains a complete history of all state changes in transaction that take place on it,' thus enabling 'the fully transparent storage and continuous updating of information,' as well as 'their unchallengeable concatenation in chain of transactions laid down in blocks'.[109]

Dispersed in sundry digital environments, the 'concatenation of the data and the decentralised nature of the system'[110] impact on our traditional conception of space and law. The quasi-contractual nature of this *lex cryptographia* has emancipated digital legislation from territory, as well as from language and interpretation, now 'replaced by the code of the blockchain'. Essentially, the 'law itself becomes code'.[111] It has also emancipated the ἄνθρωπος from the corporeality of each human being that constitutes it. In regulating legal transactions, it adopts its own rules that reflect 'its globally operative, automated and peer-to-peer legitimized existence'. Being now 'bound . . . by the code alone', it is 'detached from the heteronomous sovereignty of state and law – but eventually from the heteronomy of its own body'.[112] This means that 'we now have turned *posthuman*', thus allowing each of us to make our own 'personal thinking blockchains' or 'mindfiles', thus recording 'every "transaction" in the sense of capturing every thought and emotion of every entity, human and machine, encoding and archiving this activity into life-logging blockchains'.[113]

In digital space, though, hints of the territorialised paradigm resurface here and there. For example, it still possesses the idea of 'a geometrical system of organization which could be measured objectively and scientifically'.[114] Like Lefebvre's space, it is 'absolute' and 'mathematical'; unlike it, however, it is not 'simply . . . an empty idea'.[115] By contrast, it 'actively [shapes] social relations'.[116] No wonder that scientists 'map' it through geographical coordinates or topological systems,[117]

---

108 Pistor (2019), p. 198. See also Manski and Manski (2018).

109 Respectively: Pistor (2019), p. 184; Becker (2022), p. 114.

110 Becker (2022), p. 119.

111 Käll (2018), p. 136.

112 Becker (2022), pp. 120 and 122.

113 Käll (2018), p. 138; Swan (2005), p. 43.

114 Castree et al. (2013), p. 479.

115 Lefebvre (1991), p. 1.

116 Castree et al. (2013), p. 479.

117 Miao (2019).

treating it as a multi-dimensional logical space onto which it is possible to project geographical concepts and attributes as if there were a physical location mimicking the Earth's surface.[118]

This is apparent within a variety of digital space, i.e. 'cyberspace', which is the 'virtual space within Internet technologies wherein interaction occurs'.[119] Despite being made up of 'digital data', it is a hyperregulated environment. Like any other digital space, its 'domains have a rudimentary geography and can be mapped through a process of spatialization'.[120] Activities taking place there become more and more territorialised, 'coming to more closely resemble governance in the non-virtual world, with countries enforcing the same laws online that they enforce offline'.[121] Cyberlaw then completes bordering processes of cyberspace. Our access to it is certainly determined by topological descriptors (such as the IP address), but the physical place from which we actually access it or benefit from public services delivered online is represented by geographical coordinates. These are the coordinates of the non-virtual world geo-referencing where we are in relation to domestically enforced Internet law, cybercrime law, and Internet censorship. All of them outline domestic belongs of public law where the territorial matrix extends over digital space.[122]

To be honest, what is flat is not the digital space, but the market that it hosts and that we inhabit as consumers. In this sense, it looks like an *abstract space*, i.e. a set of 'milieus compatible with the forms of exchange and circulation on which global markets depend'.[123] It also looks flat; the compression of the world onto 'a "plan" existing in a void and endowed with no other qualities' is functional to the creation of a safe trade place that 'simply has homogeneity as its goal' to reflect the smooth flows of goods, capital, and services across frontiers and legal systems.[124]

Digital spaces are strictly connected with the real world. Although we access them from it, they have introduced 'a uniform, default reality defined by consumption, unhinged from the physical world and its . . . environmental disasters'.[125] They are arranged so as to look placeless, leave out of the picture the striations disturbing the flow of capital and goods, conceal the interests of global economic actors, and exalt the illusion of our saving the planet through eco-consumeristic practices. *There*, on the digital side of our mixed places, we get whatever we desire, and commodities are purchased, shipped, and paid for. On this side, money is accumulated at a steady pace. Miners can indeed 'earn additional Bitcoin by offering computing space and providing the electricity required for the algorithm to run its

[118]Dodge and Kitchin (2001).

[119]See e.g., Zhang and Jacob (2012); Castree et al. (2013), p. 90.

[120]Castree et al. (2013), p. 90. See also Kellerman (2016) and Warf (2021).

[121]Warf (2021), p. 318.

[122]Warf (2021), pp. 319–324.

[123]Crary (2022), p. 43.

[124]Lefebvre [1974] (1991), p. 287.

[125]Crary (2022), p. 8.

course'. This, it goes without saying, 'consumes inordinate amounts of energy'.[126] Again, the digital space means surpassing the planetary boundaries to create a safe operating space for commerce. This is an eloquent disturbing striation of the digital space that might obstruct the physical or immaterial flow of capital and must therefore be removed.

These hints of territoriality prompt me to consider digital space a variety of nomic settings. As said, the expression was coined by David Delaney to refer to social constructs like territory, place, and space that become legally meaningful 'by way of inscription or assignment of legal meanings'.[127] Digital space may also be made meaningful through processes of legal and spatial production. Like the Portuguese in the fifteenth century, we are cybernauts following the patterns of *descobrimento*: we locate human achievements in digital space and therefore spatialise it; in so doing, we replicate our matrix in the digital environment (see Sect. 4.2.1). Digital space becomes a 'lived' spatial normativity.[128]

Nothing new under the sun, then. We are getting hold of every possible space. We territorialise it. We assign a meaning to it. 'It is a habitus of action and above all, of reaction', Andrea Brighenti reminds us.[129] Perhaps most striking of all, is what the biologist Jakob von Uexküll pointed out in 1934: 'territory appears as a subjective product which cannot be inferred from mere characteristics of any objective physical environment'. This means, Brighenti continues, that 'Territoriality does not manifest itself as a constantly visible behaviour'. Until 'a member of the same species' displays 'a type of behaviour that is considered intrusive makes its appearance' in the disputed territory, territoriality remains 'latent', which means 'virtual'.[130]

8.6.2 Alternate Reality Spaces

The interactions between humanity and technology have the same features we find when manufacturing legal spaces. Digital space has indeed skirted the line between real and non-real worlds, amplifying the impact of technology also on the legal geographical spectrum. Even within digital technologies, territoriality no longer remains 'virtual'. Quite the opposite: it shifts the borders between the world, the knowledge of it, and the way knowledge is being produced (see Sect. 8.3).

This is apparent when examining how digital spaces blur the vanishing line between the real word and the way popular culture produces knowledge about it. No wonder that the production of space is sensitive to technology, which is thus able to rearrange our interactions with the real world. Processes of spatial production

[126] Pistor (2019), p. 200.
[127] Delaney (2010), p. 59.
[128] Delaney (2010), p. 59.
[129] Brighenti (2006), p. 67.
[130] Brighenti (2006), p. 67.

are also contextualised and therefore governed by social, geographical, and legal variables operating within them. In our digital era, 'space' is produced *also* through technology, which makes it meaningful to human agencies.

The way technology produces new spaces is also related to our living in our spaces of acquisitiveness. When, as consumers-citizens, we consciously let the forces of capital influence our behaviours, we accept that all reality is 'constructed'; at the same time, we are unconsciously persuaded that the whole of our life must adapt to this constructed reality. The spaces we digitally inhabit are mixed: partly online, partly in the real world, they give rise to what I term *Alternate Reality Spaces.*

The expression is a derivative of the so-called alternate reality games (ARG), which accounts for our adaptation to digital reality. Take, for example, the ARG *World Without Oil: Play it – before you live it.* Launched in 2007, the game website defined it as 'a serious game for the public good', which invited everybody to supply 'collective imagination' when confronting the 'risk our unbridled thirst for oil poses to our economy, climate and quality of life'.[131] The game challenged players 'to live their own lives as if they were dealing with the thirty-two days of a massive "oil shock".' Non-renewable resources were outstripped, and players played 'the story of humans' reinhabiting and rebooting earth's systems in a post-carbon future'.[132]

Whereas digital games are completely located in territorialised, albeit non-real, spaces, ARGs are multimedia games, which means that they blend real-world performances with Internet- and media-related content. They are no longer played in virtual spaces, but in the real world. To this extent, ARGs 'are more than merely a call to action; they *are* action'. When it terminated, 'some players continued to play' and live in a sustainable way, creating their networks and continuing "to journal their lives beyond oil'.[133]

World Without Oil is different from other virtual games, which are usually located on the Earth on the brink of an ecological disaster. There were no 'climate horrors', 'sufferings', or 'deprivation'.[134] The Earth was not treated as 'a disposable waste-land, a story of grim embattlement'[135] threatened by extinction, collapse, and 'ecological and political devastation' prompted by climate change.[136]

None of these features were present in the game. Quite the opposite: it placed emphasis on the positive effects of active engagement, which contributed to shaping a radiant pot-carbon future. 'At some point, the game makers portend, we will not simply be playing a shortage'. We will materially experience it. Consequently, we must now learn how to play and join our forces with the most active forces within our

[131] See http://writerguy.com/wwo/metaabout.htm (accessed on 7 March 2022).

[132] McFarland Taylor (2019), p. 252.

[133] McFarland Taylor (2019), pp. 253 and 254.

[134] Wallace-Wells (2019), pp. 198, 200, and 192.

[135] McFarland Taylor (2019), p. 252.

[136] Wallace-Wells (2019), p. 172.

communities. The aim was to 'co-create more ecological cultural sensibilities and ethical action'.[137]

Acting in Alternate Reality Spaces, ARGs sustain the forces of capital in promoting our 'ecopiety'. Like *World Without Oil*, ARGs are saturated with positive 'storied visions of the future' and reflect the idea that 'participatory digital new media and social networking' are the 'prime location for contemporary moral/civic engagement and public debate'.[138] Technologies seem to help tackle climate change and bridge our troubled 'reality to an imagined alternative' where ecological crises no longer take place.[139]

Another example is the digital rendering of 'ecotopian cities', i.e. fictional sustainable cities whose ecosystems are regenerated in an ideal future.[140] The line has been blurred, say, by Marvel Studios; the plot of its two films *Black Panther* (2018) and *Black Panther: Wakanda Forever* (2022) are located in the African Kingdom of *Wakanda*, 'a country whose greatest technological achievement is maintaining its environment'.[141] The city is not a 'master-planned, top-down metropolis'. Far from it: its urban space embeds 'grassroots urbanism where the residents have customized their structures and their communities to fit their needs'.[142] It is, thus, the prototype of 'collective societal public investment in green infrastructure and planning'.[143] In addition, this ecotopian city 'can maintain its ecosystems in part because there are no suburbs'. As Kendra Pierre-Louis argues in reviewing *Black Panther* for *Time* in 2018, this entails less pollution and more eco-practices: 'No suburbs mean fewer roads to fragment ecosystems, making it easier for animals and plants to survive, promoting biodiversity'.[144]

We know that cities like *Wakanda* do not exist in the real world. Despite this, we accept that, 'Before we can build a better world, we need to imagine what one looks like'.[145] And digital media assist us in achieving this goal. This is because technology is so deeply ingrained in our popular culture that we assume that the knowledge of our world can be mediated (and produced) by it. In portraying the green *Wakanda*, digital popular culture adds some flashes of evasion from our distressed world but cannot really distract us from the cause of the distress.

Popular culture bridges real spaces to an imagined alternative created within digital spaces. And makes them meaningful. When they become meaningful, they develop into places, i.e. repositories of knowledge partly produced in the real world and partly in the digital one. In any case, these mixed places become 'profound

[137] McFarland Taylor (2019), p. 252.
[138] McFarland Taylor (2019), p. 16.
[139] Watson (1988), p. 36.
[140] The term 'ecotopia' was first used in Callenbach (1975).
[141] Kendra (2020).
[142] Coleman (2018).
[143] McFarland Taylor (2019), p. 9.
[144] Kendra (2020).
[145] Jerkins (2018).

centres of human experience'.[146] As digital spaces are constitutive of our places, they give us the illusion that we may save our planet through the individual virtuous ecological practices just suggested.

8.7 Spaces of Climate Change

8.7.1 The ἄνθρωπος and the Spirit of the Cité

My examination of unbounded ideo-logical spaces terminates with their production in a time of climate change. These outline counter legal geographies, which are usually juxtaposed to spaces of acquisitiveness in the current ecological crisis. Global nomic settings related to the Anthropocene and digital spaces indeed bear the consequences of the environmental catastrophe. Quoting Richard Mullender again, this makes them 'a normatively appealing, enduring, but not entirely safe politico-legal space' (see Sect. 5.3.3).[147]

I shall consider here why these spaces are 'normatively appealing', revisiting why they are not 'entirely safe' in the next subsection. I have already explained that these places are legally and socially constructed; within them, the global holders of capital practise trade and commerce; we purchase commodities to be enjoyed in the tranquillity of our safe oases. Their force is fascinating, somewhat irresistible; it is a liberating force, which frees us from the communitarian bonds and frees capital from the structures (and strictures) of the territorial trap, as well as from its related legally and socially bounded geographical primitives (see Sect. 6.4). I have also pointed to the telluric force that has created them in our Anthropozoic era (see Sect. 8.4.3): the ἄνθρωπος has flattened (at least ideally) the earth's surface to level the striations that had traditionally opposed the smooth operativity of the global supply chain. Its legal regime, which is the law of the Buyosphere, perfectly matches the generalised, and therefore universally accepted, values of global economic sectionalities.

I perfectly understand that the construction of acquisitive spaces has characterised the story of capitalism. Marxist geographies have highlighted 'the intense relationship between the ownership of power [the elites] and the social production of ideas'.[148] The ruling classes have supported the adoption of such a consumeristic approach. I said that developments in digital computing have placed greater emphasis on how we share spatial data with the creators of digital logical spaces. When sharing our lived geographies, we also share our data with multinational media companies. Their ideological discourse remains apparently unknown, and their commercial purposes are concealed under the technical justification provided by

[146] On places as repositories see Relph (1976).

[147] Mullender (2019), p. 550.

[148] Nayak and Jeffrey (2011), p. 77.

the open-access interface. Yet, our data are meaningful to them; our geodemographic data are used to gain understanding of our behaviours as consumers (see Sect. 3.4.1). Such a fascinating lifestyle fosters individualism, and the use of digital media concurs in altering how we perceive the real world, also contributing to progressively distancing ourselves from societal concerns, ecological issues, and collective action. Regulated by the law of the Buyosphere, whatever lies beyond the lines demarcating spaces of acquisitiveness (such as informal settlements, NIMBY communities, Waste Regime, and so on) is replaced by our mixed Alternate Reality Spaces. There, the consumption of commodities also takes place by consuming electricity (another success of the ἄνθρωπος in replicating natural phenomena), and our individual ecopiety is facilitated by the virtuous practices we experience through them. We prefer to live 'surrounded by so great a cloud' storing digital data rather than to step out of it to engage with the environment and become activists surrounded by another cloud, that of environmental 'witnesses' (*Heb* 12:1).

The spirit of the *Cité* thus resides in the winning arguments of the Buyosphere. Yet, 'only the elite live in' the innermost circle of the *Cité* and breathe its liberating air. We inhabit the space of acquisitiveness. There, we spend our stipends to sustain the economic cycle. The rest of the space is populated by workers living in and 'surrounded by suburbs of what are effectively ... serfs', 'whose job is to be [in the *Cité*] during the day and be away or invisible at other times'.[149] In our social imaginaries, the 'outsiders' and the 'different' are nothing but servants to our ideological spaces of consumption. We watch films like *Wakanda* (see Sect. 8.6.2) because, *there*, there are not suburbs or hints of waste regime because of "the insignificance of whatever life remains unassimilable to [Internet] protocols."[150] Spaces of dump exist neither in Alternate Reality Spaces nor in the *Cité*, from which they are slyly kept out of our sight.

The law of the Buyosphere perfectly fits into a time of climate change. Both are transnational phenomena transcending political borders and physical places. Both are characterised by a strategic use of the law, which makes climate-change policies compatible with profit. The possibility that we tackle the challenges posed by global warming usually reflects a general commitment with universal responses. Paraphrasing Hessel E Yntema, 'the problems of global [in our case, environmental] justice are basically the same in time and space through the world'. His arguments replicate the patterns of universalism: in climate-change-related issues, handling 'extra-legal circumstances' requires a sort of 'convergence-through-pressure' strategy.[151] This attitude does not respect local knowledge or conditions, thus insulating, that is, generalising, ecological concerns from the territorial contexts where they take place.

Furthermore, the global approach revolves around a set of circular justifications, where 'arguments ... assume what they seek to justify'.[152] On the one hand,

---

[149] Welby (2018), p. 145.

[150] Crary (2022), p. 3.

[151] Yntema (1956), p. 903.

[152] Brown (1994), p. 407.

'convergence through pressure' appears to be the sole legal response to universal ecological threats. On the other hand, it is consistent with the allegedly democratic allure of climate change, which strikes without discriminating among places and individuals.

An example is supplied by how we manage the Antarctic cryosphere in a time of climate change (see Sect. 6.3.2). Its fragile ecosystem is now under pressure from a further unsustainable achievement of the mighty ἄνθρωπος, i.e. tourism, which may 'undermine ongoing assertions of environmental authority by offering ways of interacting with the Antarctic environment that are not rooted in science'.[153]

There is something deceitful in how private interests advocate for the role of Antarctic tourism in protecting the Southernmost ecosystems. Economic interests linked to tourism have infiltrated the Antarctic and turned their own environmentally related interests into *the* organising theme of the whole area. The interests of economic sectionalities and limited *élites* that can afford such a journey now count as the interests of the whole. Antarctic tourism, indeed, is nothing more than Edenic and cultivated leisure to observe Antarctic wildlife. Edenic leisure is also correlated to biological decline. In our places of acquisitiveness, biological decline is consistent with our commodifying state of mind, which promises a radiant future to humanity. It is better to safeguard, i.e. exploit, wilderness *for the good of a limited elite* rather than to directly pillage it. It is another mighty act of the ἄνθρωπος: despite the ecological crisis, it assumes that commodification undoubtedly favours the exploitation of these geographical areas which cannot be saved by climate change.

Favouring the production, distribution, and purchase of commodities, the law of the Buyosphere allows us to indulge in extremely refined practices of ecopiety. This influences our situated ecological practices. The purpose of the Law of the Buyosphere is to neutralise the idea of an ecological catastrophe, which is slyly kept in the background so as not to cause a threat to the comfort zones of commerce.

8.7.2 Spaces Beneath and Beyond the Anthropocene

In Antarctica, though, the ice is creaking under the feet of elite tourists. And, through the cracks in the ice, our doubts resurface. Under the veil of the Anthropocene, we are exhausting the earth's resources and surpassing the earth's limits, beneath and beyond which we experience how the space of the Anthropocene is 'vulnerable to the assaults' of the (winning) arguments of climate change impacting on spaces of acquisitiveness.[154] Again, 'Alternate Reality Places' are challenged by variables with 'the power to unsettle existing certainties', opening up a 'space of doubt'.[155]

[153] Howkins (2017), p. 195.

[154] Sennett (2019), p. 268.

[155] Mullender (2019), pp. 551–552.

Beneath and Beyond the Anthropocene, there is the space of climate change. That is why, with Mullender, I label it as 'not entirely safe'. It is not *entirely safe* since, as seen, ideo-logical spaces are always *relative*: economic and financial risk is inherent to every investment and business venture. This space lies beneath the Anthropocene; in countering the tenability of our acquisitive societies, 'Climate change casts a long-term shadow of inevitability' as it projects the telluric force of the ἄνθρωπος onto a logical space which stretches beyond planetary limits.[156] We thought that 'wealth was a shield against [its] ravages;' consequently, we kept on 'mortgaging the ecological future of the planet'.[157] After building 'our way out of nature', we now realise 'how hard, and how indiscriminately,' climate change 'is hitting'.[158]

The impact of climate change also has several consequences on our ideo-logical spaces. Richard Sennett reminds us that, in the course of human history, several *villes* and their ideo-logical spaces (the *Cités*) have been defined by 'waterways' and 'watery edges'. Owing to rising sea levels, 'water has become a destructive as well as a functional and a scenic material', whose 'malign threat' will affect *villes* and their *Cités.* He warns that, 'built at the edge of the seas," coastal cities like New York, Mumbai, or Rio will 'risk episodic flooding'—if not be completely underwater.[159] Far from being *arcifinious*, coasts expose them to submersion and loss.

Human-related consequences are even more distressing. The loss of territories will trigger considerable migrations. Owing to rising sea levels, more and more communities err from their lands, boosting the 'movement of people across territorial borders, the mixing of bodies and places, and the reconfiguration of labour markets'. This has consequences in the field of migration; as it may 'unleash as many as a billion migrants by 2050',[160] borders become blurred. These changes are making us navigate 'an entirely new social and political structure,[161] because 'it will have massive impacts on the way human life on Earth is organized'.[162] Global warming is thus deploying several winning arguments that are creeping into the texture of the *Cité*.

The space of the ecological catastrophe also lies beyond the Anthropocene. No wonder that the insurgent ecological crisis has prompted a renewed interest in what is termed 'apocalyptic thinking'. Already present in Western thought since the 1960s, ecological catastrophe accounts for its revival in the twenty-first century.[163] And the earth's dissolution also calls the Book of Revelation back to the fore. For example, chapter 10 of *Time to Act* written by Christian Extinction Rebellion ('Apocalypse Now – Revelation for a Climate Crisis') visibly turns to it to provide

[156] Sennett (2019), p. 269.

[157] Wallace-Wells (2019), pp. 1, 54.

[158] Wallace-Wells (2019), pp. 71 and 74.

[159] Sennett (2019), pp. 271–272.

[160] Wallace-Wells (2019), p. 133.

[161] Wallace-Wells (2019), p. 167.

[162] Mann and Wainwright (2018), pp. ix–xi.

[163] Di Tommaso (2020).

'the contemporary church [with] a clear call to have hope and courage in times of crisis'.[164] Our troubled times are indeed experiencing a dissolution which is earth dependent. Evidently, 'There is No Planet B'.[165] Parts of our earthly environments will be rendered un-survivable by 'Heat Death,' 'Dying Oceans,' 'Freshwater Drain,' 'Unbreathable Air,' and 'Plagues of Warming'.[166] Politico-legal arrangements presuppose geographical scenarios.

There are also some striking differences, though. 'Apocalyptic imaginaries' are part of the Judaeo-Cristian tradition of the 'beginning' and 'the end of all things and the life everlasting in the Kingdom of Heaven,' that is, the new Jerusalem:[167] 'And I saw a new heaven and a new earth' (*Rev* 21:1). Our 'present day millennialism', by contrast, points to 'an apocalypse without the promise of redemption'.[168] Contemporary apocalyptic visions are 'quasi-theological' because they leave 'behind any hope of rebirth or renewal . . . in favor of an unquenchable fascination with being on the verge of an end that never comes'.[169] As Derrida put it: they are 'without vision, without truth, without revelation'.[170]

Unlike *Revelation*, the passing away of our physical world is *anthropogenic* in nature: the Anthropocene epoch is altering the laws of nature; and the Law of Buyosphere has allowed capitalism to hegemonically confound its own presence and rearranged our communities around its economically driven organising themes.

References

Amin A (2002) Spatialities of globalisation. Environ Plan A: Econ Space 34:385–399

Armiero M (2021) Wasteocene. CUP, Cambridge

Barkan J (2011) Law and the geographic analysis of economic globalization. Prog Hum Geogr 35: 589–607

Becker K (2022) Blockchain Matters–Lex Cryptographia and the displacement of legal symbolics and imaginaries. Law Crit 33:113–130

Beeson M (2020) Environmental populism. The politics of survival in the Anthropocene. Palgrave McMillan, London and New York

Berners-Lee M (2019) There is no planet B: a handbook for the make or break years. CUP, Cambridge

Bird A (2018) Thomas Kuhn. The Stanford Encyclopaedia of Philosophy (Winter 2018 Edition), Edward N. Zalta (ed.), URL = <https://plato.stanford.edu/archives/win2018/entries/thomas-kuhn/>

[164] Christian Climate Action (2020), p. 67.

[165] Berners-Lee (2019).

[166] The chapter headings of *The Uninhabitable Earth* are revealing. See Wallace-Wells (2019).

[167] Embler (1966), p. 65.

[168] Swyngedouw (2013), p. 12.

[169] Jay (1994), p. 33.

[170] Derrida and Leavey Jr (1984), p. 34.

Blackstone W (1766) [2016] Commentaries on the Laws of England, II, of the rights of things. In: Prest W (gen ed) The Oxford edition of Blackstone. OUP, Oxford

Blomquist RF (2001) 'A fascination without scruples': American popular culture and its corrosive impact on the law. Cumberland Law Rev 32:165–186

Bohman J (2007) Democracy across borders: from Dêmos to Dêmoi. The MIT Press, Cambridge, MA

Brighenti A (2006) On territory as relationship and law as territory. Can J Law Soc 21:65–86

Brown HI (1994) Circular justifications. PSA: Proceedings of the Biennial Meeting of the Philosophy of Science Association, pp 406–414

Callenbach E (1975) Ecotopia: the notebooks and reports of William Weston. Bantam Books, New York

Carlarne C, Farber D (2012) Law beyond borders: transnational responses to global environmental issues. Transnatl Environ Law 1:13–21

Castree N et al (2013) Oxford dictionary of human geography. OUP, Oxford

Chakrabarty D (2018) Planetary crises and the difficulty of being modern. Millennium 46:259–282

Christian Climate Action (2020) Time to act. A resource book by the Christians in Extinction Rebellion. SPCK, London

Cohen MR (1927) Property and sovereignty. Cornell Law Rev 13:8–30

Coleman AL (2018) There's a true story behind *Black Panther*'s strong women. Here's why that matters. *Times* (22 February 2018) at https://time.com/5171219/black-panther-women-true-history/. Accessed 11 Nov 2021

Cooper JC (1952) Roman law and the Maxim *Cujus Est Solum* in international air law. McGill Law J 1:23–65

Crary J (2022) Scorched. Beyond the digital age to a post-capitalist world. Verso, London and New York

Crutzen PJ (2002) Geology of mankind. Nature 415:23

Crutzen PJ, Stoermer EF (2000) The 'Anthropocene'. Global Change Newsl 41:17–18

Dalby S (2021) Unsustainable borders: globalization in a climate-disrupted world. Borders Glob Rev 2:26–37

De Filippi P, Wright A (2015) Decentralized blockchain technology and the rise of Lex Cryptographia. SSRN Electron J. https://www.ssrn.com/abstract=2580664

de Lucia V (2020) Rethinking the encounter between law and nature in the anthropocene: from biopolitical sovereignty to wonder. Law Crit 31:329–349

Delaney D (2010) The spatial, the legal and the pragmatics of world-making. Blackwell, Oxford

Derrida J, Leavey JP Jr (1984) Of an apocalyptic tone recently adopted in philosophy. Oxf Lit Rev 6:3–37

Di Tommaso L (2020) Apocalypticism in the contemporary world. In: McAllister C (ed) The Cambridge companion to apocalyptic literature. CUP, Cambridge, pp 316–341

Dodge M, Kitchin R (2001) Mapping cyberspace. Routledge, Abingdon and New York

Embler W (1966) The vocabulary of political theory: "authority". Christ Sch 49:60–76

Etty T et al (2018) Transnational climate law. Transnatl Environ Law 8:191–200

Ferreira I et al (2019) Nanoplastics and marine organisms: what has been studied? Environ Toxicol Pharmacol 67:1–7

Friedman TL (2005) The world is flat: a brief history of the twenty-first century. Farrar, Straus and Giroux (FSG), New York

Gilley B (2012) Authoritarian environmentalism and China's response to climate change. Environ Polit 21:287–307

Godden L (2020) Legal geography – place, time, law and method: the spatial and the archival in 'connection to country'. In: O'Donnell T et al (eds) Legal geography. Perspectives and methods. Routledge, Abingdon and New York, pp 130–148

Hamilton C (2016) The theodicy of the 'good Anthropocene'. Environ Humanit 7:233–238

Holden J (ed) (2017) An introduction to physical geography and the environment, 4th edn. Pearson, London

Holy Bible (1989) New revised standard edition. Anglicized ed. CUP, Cambridge

Howkins A (2017) Frozen empires. An environmental history of the Antarctic Peninsula. CUP, Cambridge

Husa J (2018) Advanced introduction to law and globalisation. Edward Elgar, Cheltenham and Northampton, MA

Jay M (1994) The apocalyptic imagination and the inability to mourn. In: Robinson G, Rundell JF (eds) Rethinking imagination: culture and creativity. Routledge, New York, pp 30–47

Jerkins H (2018) What 'Black Panther' can teach us about the civic imagination. Global EJournal

Käll J (2018) Blockchain control. Law Crit 29:133–140

Kellerman A (2016) Geographic interpretations of the Internet. Springer, Dordrecht

Kelman I (2022) Planetary boundaries. In: Simms K et al (eds) The Routledge handbook of global development. Routledge, Abingdon and New York, pp 147–157

Kendra P-L (2020) 'Wakanda doesn't have suburbs:' how movies like Black Panther could help us save the planet, *Times* (21 September 2020) at https://time.com/5889324/movies-climate-change/. Accessed 11 Nov 2021

Khilnani S et al (2013) Introduction. Reviving South Asian comparative constitutionalism. In: Khilnani S et al (eds) Comparative constitutionalism in South Asia. New Delhi, OUP India, pp 1–22

Klein N (2019) On fire. The burning case for a green new deal. Allen Lane, London

Kuhn T (1970) The structure of scientific revolutions, 2nd edn. University of Chicago Press, Chicago

Kurki M (2010) Democracy and conceptual contestability: reconsidering conceptions of democracy in democracy promotion. Int Stud Rev 12:362–386

La Viña AGM, Reyes JJM (2022) Anthropocene, Capitalocene, and climate change. In: Simms K et al (eds) The Routledge handbook of global development. Routledge, Abingdon and New York, pp 158–169

Larkin A et al (2018) What if negative emission technologies fail at scale? Implications of the Paris Agreement for big emitting nations. Clim Pol 18:690–714

Lawrence M, Laybourn-Langton L (2021) Planet on fire. A manifesto for the age of environmental breakdown. Verso, London

Lefebvre H (1974) La production de l'espace. Éditions Anthropos, Paris. English edn. (1991) The production of space (trans: Nicholson-Smith D). Blackwell, Oxford

MacCulloch D (2017) All things made new. Writings on the reformation. Penguin, London

Maduro M et al (eds) (2014) Transnational law: rethinking European law and legal thinking. CUP, Cambridge

Maitland FW (1964) Township and Borough: The Ford Lectures 1897. CUP, Cambridge

Mann G, Wainwright J (2018) Climate Leviathan: a political theory of our planetary future. Verso, London and New York

Manski S, Manski B (2018) No gods, no masters, no coders? The future of sovereignty in a blockchain world. Law Crit 29:151–162

Matthews D (2021) Reframing sovereignty for the anthropocene. Transnatl Leg Theory 12:44–77

McFarland Taylor S (2019) Ecopiety. Green media and the dilemma of environmental virtue. New York UP, New York

Miao et al (2019) A coordinated view of cyberspace. arXiv:1910.09787 [cs.NI] https://doi.org/10.48550/arXiv.1910.09787

Moore JW (2015) Capitalism in the web of life. Ecology and the accumulation of capital. Verso, London

Moore JW (2017) The Capitalocene, Part I: on the nature and origins of our ecological crisis. J Peasant Stud 44:594–630

Morton T (2013) Hyperobjects. Philosophy and ecology after the end of the world. University of Minnesota Press, Minneapolis, MI

Mullender R (2019) There is no such thing as a safe space. Mod Law Rev 82:549–576

Nayak A, Jeffrey A (2011) Geographical thought. An introduction to ideas in human geography. Routledge, Abingdon and New York
Newman D (2006) The lines that continue to separate us: borders in our 'borderless' world. Prog Hum Geogr 30:143–161
Nixon R (2013) Slow violence and the environmentalism of the poor. Harvard UP, Cambridge, MA
O'Brien R (1990) The end of geography. The impact of technology and capital flows. AMEX Bank Rev 17:2–5
O'Brien R (1992) Global financial integration: the end of geography. Pinter Publishers, London
Orgad L (2018) Cloud communities: the dawn of global citizenship? In: Bauböck R (ed) Debating transformations of national citizenship. Springer International Publishing, Cham, pp 251–260
Ovodenko A (2017) Regulating the polluters: markets and strategies for protecting the global environment. CUP, Cambridge
Pistor K (2019) The code of capital. In: How the law creates wealth and inequality. Princeton University Press, Princeton
Popovski V (ed) (2019) The implementation of the Paris Agreement on climate change. Routledge, Abingdon and New York
Relph E (1976) Place and placeness. Pion, London
Rockström J et al (2009) A safe operating space for humanity. Nature 461:472–475
Romanello M et al (2021) The 2021 report of the Lancet Countdown on health and climate change: code red for a healthy future. Lancet 398(10311):1619–1662
Sanguin A-L (2014) End of geography or revenge of geography? Human societies between a smooth, spiky or flat world. Bollettino della Società Geografica Italiana Roma VII:445–460
Schaffer G, Coye C (2020) From international law to Jessup's *Transnational Law*, from transnational law to *Transnational Legal Orders*. In: Zumbansen P (ed) The many lives of transnational law: critical engagements with Jessup's Bold Proposal. CUP, Cambridge, pp 126–152
Schmitter PC (1995) More liberal, preliberal, or postliberal. J Democr 6:15–22
Sennett R (2019) Building and dwelling. Ethics for the city. Penguin, London
Sherwin RK (2000) When law goes pop: the vanishing line between law and popular culture. The University of Chicago Press, Chicago and London
Siems M (2022) Comparative law, 3rd edn. OUP, Oxford
Standing G (2019) Plunder of the commons. A manifesto for sharing public wealth. Pelican, London
Steffen et al (2011) The Anthropocene: conceptual and historical perspectives. Philos Trans R Soc A 369:842–867
Stoppani A (1873) Corso di geologia del professore Antonio Stoppani: Geologia stratigrafica, vol 2. G. Bernardoni e G, Brigola
Swan M (2005) Blockchain: blueprint for a new economy. O'Reilly Media Inc., Sebastopol, CA
Swyngedouw E (2013) Apocalypse Now! Fear and Doomsday Pleasures. Capital Nat Social 24:9–18
Szymczak S, Wolszczak-Derlacz J (2022) Global value chains and labour markets – simultaneous analysis of wages and employment. Econ Syst Res 34:69–96
Victor DG (2004) The collapse of the Kyoto Protocol and the struggle to slow global warming. Princeton UP, Princeton, NJ
von Uexküll J [1934] (1957) A stroll through the worlds of animals and men. In: Schiller C (ed) Instinctive behavior. The development of a modern concept. International Universities Press, New York, pp 5–80
Wagner G (2011) But will the planet notice? How smart economics can save the world. Hill and Wang, New York
Walker N (2015) Intimations of global law. CUP
Walker N (2020) The sovereignty surplus. Int J Const Law 18:370–428
Wallace-Wells D (2019) The uninhabitable earth. A story of the future. Allen Lane, London

Warf B (2021) Place, space, and cyberlaw. In: Bartel R, Carter J (eds) Handbook on space, place and law. Edward Elgar, Cheltenham and Northampton, MA, pp 316–326

Watson A (1988) Failures of the legal imagination. University of Pennsylvania Press, Philadelphia

Welby J (2018) Reimagining Britain. Foundations for hope. Bloomsbury, London

Yntema HE (1956) Comparative legal research – some remarks on looking out of the cave. Mich Law Rev 54:899–928

Yusoff K (2016) Anthropogenesis: origins and endings in the Anthropocene. Theory Cult Soc 33: 3–28

Yusoff K (2018) Politics of the Anthropocene: formation of the commons as a geologic process. Antipode 50:255–276

Yusoff K (2021) The inhumanities. Ann Am Assoc Geogr 111:663–676

Yussof K (2018) The Anthropocene and geographies of geopower. In: Coleman M, Agnew J (eds) Handbook on the geographies of power. Edward Elgar, Cheltenham and Northampton, MA, pp 203–216

Zalasiewicz J et al (2011) The Anthropocene: a new epoch of geological time? Philos Trans R Soc A 369:835–841

Zhang G, Jacob EK (2012) Reconceptualizing cyberspace: 'real' places in digital space. Int J Sci Soc 3:91–102

Zumbansen P (2020a) Transnational law: theories & applications. TLI Think! Paper 15/2020. Available at https://papers.ssrn.com/sol3/papers.cfm?abstract_id=3601385#. Accessed 11 Mar 2022

Zumbansen P (ed) (2020b) The many lives of transnational law: critical engagements with Jessup's Bold Proposal. CUP, Cambridge

Part V
Conclusions

Chapter 9
Perspectives

9.1 Legal Geography Going Forwards?

In ending the cross-disciplinary journey through law and geography, this chapter outlines some concluding remarks, drawing together the whole of the book. My remarks are both methodologically and substantive. I do not intend to deliver a synopsis of what I have found on my journey into legal geography; rather, it is my purpose to highlight some elements that may favour the transition of what I have termed 'legal geography' from a fruitful encounter among disciplines to a true disciplinary experiment. My methodological remarks also display some prospects for legal geography as an academic hybrid field, which has to necessarily move forwards and assess the scenarios of the production of legal spaces.

In particular, the substantive remarks will end my cross-disciplinary journey by outlining two prospective scenarios of spatial production. Perhaps most striking of all, the first scenario is pessimistic (see Sect. 9.2), whereas the second one veers towards optimism (see Sect. 9.3). In which direction will the production of legal spaces swerve? Owing to the great uncertainties of our contemporary era, this will depend on how we, as humans, respond to strains stemming from the spaces of climate change beneath and beyond the Anthropocene.

My remarks are thus an incisive musing on a broader normative question: 'what is legal geography'? It is, in my opinion, a way of grasping the world intellectually and scientifically. On its own, it is a new ambit of scientific research. It is not a matter of being 'itinerant participants' or 'self-declared' legal geographical scholars (see Sect. 2.5). I feel that my position has become that of a 'Legal Geographer', who has made acquaintance with it via my comparative law toolkit, which is open to cross-disciplinarity. Although the idea of mono-disciplinarity is eulogised within mainstream academic circles, I think that sticking to it runs counter to the very idea of scientific research. We should remain open to the exchange of ideas and opinions, and arguments, whether alluring or disturbing. We cannot assess the world knowledge only by picking up the arguments we prefer and discarding those we do not

M. Nicolini, *Legal Geography*, Ius Gentium: Comparative Perspectives on Law and Justice 105, https://doi.org/10.1007/978-3-031-19410-8_9

like. We must accept that other disciplines provide alternative readings and different insights into our research questions. Like borders, our disciplinary arguments are vulnerable to the assaults of arguments stemming from different areas of knowledge. A methodological change is therefore required. If we intend to deal with legal geography, we must engage in cross-disciplinary investigations.

Throughout this book, indeed, my focus has been on the shift from mono-disciplinary to cross-disciplinary investigations. It is what I have termed is *disciplinary construction* (see Sect. 2.5). As a legal geographer, I have revised the methodologies applied both in law (and, partly at least, in geography) to allow our discipline to acquire the status of an autonomous ambit in scientific research. To this extent, I have interpreted the processes of spatial production trying to sidestep the presence of disciplinary divides. This has meant reading the special relationship between legal geography through the lenses of the territorialised paradigm. Such a paradigm has been particularly productive, inasmuch as it considers how power, space, place, and law are mutually constitutive and describes their co-production. I perfectly understand that reducing the law to a geographical practice may indeed be unpalatable also for legal scholars actively involved in cross-disciplinary research. Yet, we have seen that comparative law is well-equipped for assessing the law as a spatial practice, since it has cross-disciplinary ambitions disengaging the law from formalistic and textual approaches, also showing attentiveness to operational rules beyond the written law. In other words, the practice of legal geography has contributed to challenging the idea of the separateness of the law, assuming that it may be properly understood in relation to processes of spatial production. As seen (see Sect. 2.1), the law is first and foremost a performative practice creating spaces whose meaning is pragmatically enriched by socio-geographical contexts. Space is, on its own, *also* manufactured through legal practices.

As I have upheld throughout the book, disciplinary construction also means that the variables implicated in spatial production cannot be assessed separately as if they were in watertight compartments. Quite the opposite; they mutually interact. The shift from law and geography to legal geography has also entailed challenging how ideo-logical spaces deploy their winning arguments by normalising their own world representation, applying the canons of legal-geographical spatial construction (see Sects. 5.2.2 and 5.2.3). This has also meant that, as a new field of research, legal geography is *subversive*, inasmuch as it discloses the ideological presuppositions of logical space (see Sect. 3.3.2). We have seen that logical space acts as the interface (i.e. the third or imaginative 'geography') between physical and human geographies (Sect. 3.2). Through it, power processes physical geographical data and naturalises its own representation of the world, projecting its own imaginative geography onto the earth's surface—even onto outer space and several ideo-logical spaces, such as that of doubt and terror. Being subversive, then, runs counter to the idea of the acritical normalisation of the processes of spatial production. Being subversive, again, trumps the academic fissure that opposes law and space and makes our 'transdisciplinary project' reclaim a clearer academic status.

Such a subversive approach has entailed reframing our methodological attitude in legal geography. When dealing with spatial production, comparative law and geography must revitalise its pristine empirical, and problem-based, approach to the world knowledge; at the same time, legal geography should become more pragmatic, and therefore develop an even more cross-disciplinary attire. We must remain vigilant and keep the door of the law open to a constant conversation with the world and its environmental concerns (see Sect. 5.2.2). Being cross-disciplinary indeed involves unceasing conversations with the forces within our societies, whose callouts to the public are particularly apparent in several counter legal geographies, as well as in the spaces where climate and ecological emergencies take place.

As an interdisciplinary project, legal geography has twofold consequences. On the one hand, Nicholas Blomley is right when he states that ‘it is disposition, not discipline’. Or, as I put in the *Preface* to this book, legal geographical cross-disciplinary research requires (or, at least, falls under) a good dose of ‘disciplined subjectivity’. This ‘can be found throughout the academy’ as a shared intellectual domain.[1] The world is out there, with its discrete variables. As we saw (see Sects. 1.1 and 1.3), spatial construction is an open texture, which favours the establishment of legal geography as a cross-disciplinary project and opens it to fruitful encounters with discrete disciplines. These join the project because the variety of approaches and methodologies accentuates the cross-disciplinary ambitions shared by the two constitutive parts of such a ‘special relationship’. It is not surprising that the ebbs and flows triggered by cross-disciplinary engagements have led to the proliferation of topics (see Sect. 2.5). On the other hand, the open texture which is legal geography responds to the practical necessities to experiment with a variety of methodologies, perspectives, and topics when examining how spatial construction gives meaning to nomic settings. The way rules, directives, norms of moral conducts, and customs ‘confer significance onto action, events, relationships, and situations’ requires ‘a wide variety of techniques and approaches which makes [legal geography] a rich subject’.[2] In so doing, our spatial normativities ‘are *lived’*.[3]

We have also seen that the change in our methodological mood is explicable in the light of Kuhn’s ‘paradigm shift’. In our case, it points to the changeover to new scientific programmes (the rise of legal geography as a field of research), as well as to the criteria and procedures (the methodology) applied within a given epistemic community. Such a relativist tone perfectly matches the idea of legal geography as a project benefitting from multiple approaches without worrying about which competing methodology is best.

The relativist tone of the paradigm shift makes manifest that spatial production changes over time and place. It also explains why the ‘disciplinary matrix’ of spatial production may change even within the limited epistemic community of legal geographers. The variety of procedures, theories, even of topics does not affect the ‘consensus’ around our disciplinary matrix, which is primarily aimed at explaining how spatial normativities are made meaningful to human agencies operating within

[1] Blomley (2021), p. xvii.

[2] Delaney (2010), p. 59; Holden (2017), p. 13.

[3] Delaney (2010), p. 59.

them. Throughout this book, I have interpreted spatial production through the lenses of the territorialised paradigm, i.e. the matrix transcending time and space. Within the territorialised paradigm, ancestral physical apprehension, colonial domination, modern political appropriation, globalisation, and Anthropocene are processes of spatial production still explicable in terms of 'normal' (and cumulative) phases of science.

This type of scientific mobilisation does not mean breaking the cumulative phases of science, but merely ascertaining whether they are still able to provide new avenues (and paradigms) whereby spaces are produced within our communities. This is the last contribution that a legal geographer with a background in comparative law may give as regards the production of space also through legal practices. The very idea of cross-disciplinary research is, per se, a methodology for exploring this new and intriguing field. Certainly, we will have to revise our existing methodologies to face the challenges the global production of space stages before us. By progressively renewing our toolkit, we may therefore ascertain whether the un-bounded ideological spaces really advocate new legal derivatives, thus giving rise to a new organising principle, which reflects 'the view that, as conceptions of [space and law] change over time, so too should' our legal rules in a time of globalisation, Anthropocene, and climate change.[4] Assisted, as we are, by unceasing exercises of comparative legal geographical imagination, we grasp the world intellectually and explore it with a subversive, empirical, and critical gesture. In this lies our commitment; the law might effectively 'be interpreted and applied in a manner that responds to changing conditions' and challenges that processes of spatial production are likely to pose in the foreseeable future.[5]

9.2 Egalitarian Misery on a Planet 'On Fire'

Which are the production of legal spaces may experiment with in the foreseeable future? In the beginning of these concluding remarks (see Sect. 9.1), I said that I would outline two scenarios: the first one is pessimistic, whereas the second one veers towards optimism.

The pessimistic outlook will be contingent on how digital spaces affect the production of space. I have already said the very thought of being on the verge of the ecological collapse may have severe bearing on profits. If commerce drops because of climate-change-related causes, capital will have to work out further winning arguments. Again. It will have to silence our new doubts arising from the socio-climatic consequences of our consumeristic ideo-logical space. Doubts will be cordoned off, and counter legal geographies (i.e. resistance to our unsustainable

[4] Mullender (1997), p. 27.

[5] Wenta (2019), p. 108.

economy) be marginalised to allow our eco-consumeristic practices to thrive again and again in the unforeseeable future.

In the event of this scenario, we might be experimenting nothing else that the final triumph of the telluric force of the ἄνθρωπος. Like Shelley's *Prometeus Unbound* (1820), it might restore the pristine unity of humankind, thus reshuffling a new logical space:

> The loathsome mask has fallen, the man remains
>
> Sceptreless, free, uncircumscribed, but man
>
> Equal, unclassed, tribeless, and nationless, Exempt from awe, worship, degree, the king
>
> Over himself; just, gentle, wise: but man
>
> Passionless; no, yet free from guilt or pain.[6]

This is nothing else than our individualism stretched beyond planetary limits. Each of us will be the ἄνθρωπος. Lured by the fascinating trend of the Law of the Buyosphere, we will find comfort in our *consumeristic and commodifying state of mind*, which gives us the illusion of breathing the air of the (now greenwashed) ideological spaces of the *Cité*. This space will be characterised by a clear absence of context. Like the *Cité*, it will not have resort to situated legal practices and belongs of public law; rather, it will recreate safe spaces that will be cosmopolitan in nature, therefore exalting the ability of individual logical spaces 'to move easily from place to place, culture to culture',[7] also transcending local customs and traditions.

Spaces of acquisitiveness will become commodities; like them, they will be produced in series to flatter our individualism, appetite, and ambition of being *the only true representative* of the ἄνθρωπος. It does not come as a surprise that such replicability is amplified by Alternate Reality Spaces where technology and reality blend. The smart-city scheme, for example, already outlines places where shifting borders seem capable of solving 'what sociology cannot, sorting and smoothing the relations between people', also prescribing how our places of acquisitiveness might be used.[8] These cities replicate themselves like islands 'within the city'. Richard Sennet mentions Googleplex, which is shorthand for Google complex (of buildings). Situated in New York, it is 'in the city but not of it'.[9]

Googleplex is located in the former building of the Port Authority, which is a 'self-contained' space 'relieving Googlistas [i.e. those working for Google] from exposure to their physical surroundings'. Inside it, what is desirable for every individual is also available to them: laundries, facilities, doctors, gyms, and so on. This space of alluring acquisitiveness is meant to focus 'people on life inside, minimizing non-corporate distractions' and aiming to 'extract labour from otherwise unattached twenty-something'. In sum, spaces where individuals replicate the

6 Shelley (1920), p. 120.

7 Sennett (2019), p. 141.

8 Sennett (2019), p. 144.

9 Sennett (2019), p. 146.

ambitions of the ἄνθρωπος are 'gated communities', i.e. well-bounded orders of the Anthropocene.[10]

Google replicates this pattern around the world, bracketing its workers' lives in unfettered spaces. Insulated from real contexts, these technologically saturated spaces are under the Law of the Buyosphere, which constructs 'a boundary within which interactions' between ideo-logical spaces and the law 'take place more or less independently of their surrounding context'.[11]

Capital also provides us not Googlistas with similar spaces saturated with technology. The pandemic has enhanced our dependence from these Alternate Real Spaces, weakening our resistance within the bracketed contexts of our homes. Our resistance is weakened because the Law if the Buyosphere allows us to practice ecopiety through ecommerce and its 24-hour shipping services. It also takes care of our fitness; workouts video may be used free at home. 'Free', as the geographer Jeremy Crampton acutely argues, does not mean *gratis* (see Sect. 3.4.1). Technical developments in digital computing have placed greater emphasis on the spatial data that we share with open-source applications by taking part in mashups. When sharing our lived geographies, we also share our data with multinational media companies.[12]

Do we really need go out and engage with the rest of the community? Do we really care for the environment? Or, like in *The Matrix* (1999), do we prefer to 'once for all and comfortably sink back into the illusion the Matrix offers'? In his *Visualizing the Law in the Age of Digital Baroque* (2010), Richard K Sherwin sourly asks us: 'does it matter what space I occupy, simulated or real?' Is the mixed space of acquisitiveness 'a paradise or a prison'?[13]

Yet, we are not a 'spiritual machine'.[14] We need doubts and resistance, because the experience of resisting allegedly winning arguments 'is a key aspect of creative work'—to a broader extent, of human creativity.[15] *Sinking into the illusion* of the shifting borders of technology and humanity only entails bracketing our critique of the Law of the Buyosphere and adapting ourselves to its desired spatiality of acquisitive geographies. In Alternate Real Spaces, 'checks and resistance are excluded by design'.[16]

In a time of ecological distress, this pattern reflects a kind of 'egalitarian misery'.[17] We have deceived ourselves, assuming that being bracketed is the winning argument of capitalism and the 'progress' of our time.[18] Yet, we know that the world

[10] Sennett (2019), pp. 147–148.

[11] Blomley (2015), p. 168.

[12] Crampton (2010).

[13] Quoted in Sherwin (2011), p. 36.

[14] Sherwin (2011), p. 36.

[15] Sennett (2019), p. 151.

[16] Sennett (2019), p. 152.

[17] On the notion of 'egalitarian misery' see Mullender (2020), p. 153.

[18] See Felli (2021).

is on fire. To be honest, we do not really care a straw of it. The 'coaldown' of the COP26 clearly demonstrates it (see Sect. 8.2.1). Yet, the decline is still a misery; it sounds so egalitarian because it consumes all our lives in spaces of acquisitiveness. At the same time, democratic misery is the strength of the elites propagating the gospel of consumerism. Once deceived, we adapt to it; as I said before, the wealthy will decline in better spirits than the lowly and marginalised. Disheartened, as we are, we might only argue that 'better counsels . . . will prevail, and a new city rise on the ashes of the old'.[19]

Better counsels are related to the debt that the wealthy owe to our communities, which is the argument when challenging the fascination of the *Cité*. Yet, a debtor can always '*escape payment by migration* . . . at once, *ipso facto*, [shaking] off . . . all his obligations to his municipal creditors'.[20] Such migration, though, is completely different from that of those fleeing from the spaces beneath and beyond the Anthropocene. It is 'the way out' from responsibility and accountability for the damages caused to the world.

This way out consists of establishing a safe normative place. Several newspapers have reported that the world's superrich have been buying estates in New Zealand; its apparent insulation from taxpayers and the ravages of climate change makes it the most fitting place to ride out Apocalypse (see Sect. 8.7.2). Others are also planning to colonise Mars to secure the survival of our species. It is certainly true that, the rich and fugitives will not hear the outcries of their consumers, when their digital devices do not shield them from the sufferings of climate change that will hit even in the fascinating spaces of acquisitiveness on a planet destined to collapse.

In dismissing earthly belongs of public law, logical space will find new spatial furniture (even in outer space) to be reshuffled. And the territorial matrix will soon be applied to get hold of new spaces.

9.3 Spaces of Possibilities

The alternative to our *anthropo*city supplies us (I hope) with an optimistic outlook as regards spatial production. This is related to how we, as legal geographers, may contribute (theoretically at least) to renovating of our disciplinary matrix in a methodological conversation with space of climate change.

We know that, as a cross-disciplinary project, legal geography runs somewhat counter to the 'paradigm of biological decline' which predominates 'among ecologists, environmentalists and conservationists'.[21] I am denying neither the ecological threat nor the loss of biodiversity: 'the extinction crisis is real'.[22]

---

[19] Howard (1898), p. 150.

[20] Howard (1898), p. 142.

[21] Thomas (2017), p. 30 the critique to the loss of biodiversity is discussed in Kelman (2022), p. 149.

[22] Extinction Rebellion (2019), p. 30.

Are there hints of hope? As well as arguments in favour of open futures? Notwithstanding the changes caused by anthropogenic drivers, biologists have demonstrated that, when coping with global warming, ecological biodiversity also 'responds by evolving', proposing a new biological organising theme whereby 'the biological diversity of the Earth' may be increased.[23] In terms of production of space, for example, the concept of climate-change planetary boundaries does not extend to the atmosphere. As said (see Sect. 8.2.1), its chemical elements are mainly produced by interacting with the biosphere, particularly plants. I understand that nature responds by evolving because it knows that, otherwise, it would succumb to the idea of planetary boundary, which would mean a new composition of the atmosphere and the deletion of life on earth. It is the engagement of the biosphere (also *our* engagement) with the atmosphere that prevents the alteration of the latter and the possibility of trespassing planetary boundaries, i.e. the replacement of Oxygen with greenhouse gases.

The same holds true in terms of legal geography. We should shackle off the 'pessimism-laden, loss-only view' of our future.[24] The strategic formalism of the Law of the Buyosphere is a paradigm of the legal and biological 'impoverished habitat' biodiversity of the world.[25] Global logical spaces have homogeneous politico-legal features. Indeed, such features reflect the new economic organising theme, which stimulates business, mobilises mass migration and, at the same time, grant the illusion of taking part in the global distribution of wealth. In a time of climate change, however, it mainly manages (i.e. simplifies) legal complexity.

In a globalised economic world, the Law of the Buyosphere assumes that the legal and biological decline is consistent with the radiant future promised to humanity. What we comparative legal geographers can still do is to 'stir the law', disclosing the ideological underpinnings of our situated legal practices and opening them to mutual interference and interdisciplinarity. In stirring the law, we disclose the domination-resistance dialectic, which is internal to the Anthropocene.

Again, this evokes the idea of an open disagreement as regards the dominant meanings of spatial normativies. The comparison with equity (see Sect. 5.2.3) assists us in redeeming the interests of our communities and in challenging places of acquisitiveness incrementally. This form of pluralising the legal debate may be extremely useful when it comes to addressing spaces of climate change. This might foster 'new regulatory practices', 'forms of governance' such as public participation, or legislative frameworks whereby decision makers stir the law simply by adjusting 'existing legal processes to accommodate changing environmental or social conditions'.[26]

We will be assisted by unceasing exercises of legal geographical imagination that may be done in logical space. *There* we indeed stir the law and normalise a new

---

[23] Thomas (2017), pp. 117 and 9.

[24] Thomas (2017), pp. 117 and 9.

[25] Graziadei (2003), p. 114.

[26] Wenta et al. (2019), pp. 113 and 108.

lexicon through a renovated process of spatial production. The interface between physical and human geography, the space of climate change and 'spaces of possibility' is still represented by the third 'geography' that populates the law of spatiality, i.e. our imaginative geography. With biblical parlance, this leads us to 'a new heaven and a new earth' (*Rev* 21:1), as well as a process of spatial production able to stir the law and open it to the plurality of forces and interests which populate our communities.

9.4 Coda

In a time of climate change, I still label the renovated legal-spatial relationship for a 'Human-Altered World' as 'belongs of public law' (see Sect. 2.6.2). In these new 'belongs', we shall not probably find traditional concepts like ownership, property, lordship, and rulership; nor shall we venture to explore the sovereignty of the ἄνθρωπος as productor of space. Like in Kuhn's 'normal' phases of science, the varieties of actors, methods, and variables coded through the law generate a 'cumulative picture of scientific progress'. Legal geography is still able to account for the unceasing regeneration of the territorialised paradigm with a 'reasonable chance of solving the puzzle'.[27]

The shift is not from territorialised to de-territorialised paradigms, but from monodisciplinary to cross-disciplinary paradigms. As humans, we have the cultural competence to create our own logical spaces and intervene in the world to get hold of it and reshuffle its furniture. As legal geographers, we assess this competence to synthetise the knowledge of the word in terms of legal geographical derivatives. Certainly, we will have to revise our existing methodologies by the time when a global revolutionary phase of legal geography effectively stages a true un-bounded space before us. By then, I assume, legal geography will have already acquired the status of an autonomous ambit in scientific research.

References

Blomley N (2015) The ties that blind: making fee simple in the British Columbia treaty process. Trans Inst Br Geogr 40:168–179

Blomley N (2021) Foreword: what is legal geography? Why, and why now? In: Bartel R, Carter J (eds) Handbook on space, place and law. Edward Elgar, Cheltenham and Northampton, MA, pp xvii–xix

Crampton JW (2010) Mapping: a critical introduction to cartography and GIS. Blackwell, Oxford

Delaney D (2010) The spatial, the legal and the pragmatics of world-making. Blackwell, Oxford

Extinction Rebellion (2019) This is not a drill: an extinction rebellion handbook. Penguin, London

[27] Kuhn [1962] (1970), p. 92.

Felli R (2021) The great adaptation. Climate, capitalism and catastrophe. Verso, London and New York

Graziadei M (2003) The functionalist heritage. In: Legrand P, Munday R (eds) Comparative legal studies traditions and transitions. CUP, Cambridge, pp 100–128

Holden J (ed) (2017) An introduction to physical geography and the environment, 4th edn. Pearson, London

Howard E (1898) To-morrow. A peaceful path to real reform. S. Sonnenschein & Company, London

Kelman I (2022) Planetary boundaries. In: Simms K et al (eds) The Routledge handbook of global development. Routledge, Abingdon and New York, pp 147–157

Kuhn T (1970) The structure of scientific revolutions, 2nd edn. University of Chicago Press, Chicago

Mullender R (1997) Context, contingency and the law of negligence (or from Islands to Islands of Time). Bracton Law J 29:23–33

Mullender R (2020) On the French Revolution and the programmatic imagination: Hilary Mantel on law, politics, and misery. In: Mullender R et al (eds) Law and imagination in troubled times: a legal and literary discourse. Routledge, Abingdon and New York, pp 133–156

Sennett R (2019) Building and dwelling. Ethics for the city. Penguin, London

Shelley PB (1920) Prometheus unbound, a lyrical drama in four acts. C. and J. Ollier Vere street Bond street, London

Sherwin RK (2011) Visualizing the law in the age of digital baroque. Arabesques and entanglements. Routledge, London and New York

Thomas CD (2017) Inheritors of the earth. How nature is thriving in an age of extinction. Allen Lane, London

Wenta J et al (2019) Enhancing resilience and justice in climate adaptation laws. Transnatl Environ Law 8:89–118